I0047047

William Pittenger

The great locomotive chase

A history of the Andrews railroad raid into Georgia in 1862

William Pittenger

The great locomotive chase
A history of the Andrews railroad raid into Georgia in 1862

ISBN/EAN: 9783742833396

Manufactured in Europe, USA, Canada, Australia, Japa

Cover: Foto ©Andreas Hilbeck / pixelio.de

Manufactured and distributed by brebook publishing software
(www.brebook.com)

William Pittenger

The great locomotive chase

Dedication.

To

THE GRAND ARMY OF THE WEST

WHICH, UNDER COMMAND OF GENERAL SHERMAN, IN ONE HUNDRED DAYS OF CON-
TINUOUS BATTLE,

FOLLOWED US

OVER THE LINE OF THE GEORGIA STATE RAILROAD

FROM

CHATTANOOGA TO ATLANTA,

AND

CONQUERED WHERE WE ONLY DARED.

THIS

FRAGMENT OF HISTORY IS MOST RESPECTFULLY AND FRATERNALLY
INSCRIBED BY

THE WRITER AND HIS COMRADES

OF THE

ANDREWS RAID.

PARTICIPANTS IN THE RAID.

Executed June 7th, 1862: JAMES J. ANDREWS, Leader.

Executed June 18th, 1862:

WILLIAM CAMPBELL, from Salineville, O.
GEORGE D. WILSON, Co. B, 2d O.
MARION A. ROSS, Co. A, 2d O.

PERRY G. SHADRACK, Co. K, 2d O.
SAMUEL SLAVENS, 33d O.
SAMUEL ROBERTSON, Co. G, 33d O.

JOHN SCOTT, Co. K, 21st O.

Escaped October 16th, 1862:

WILSON W. BROWN (Engineer), Co. F, 21st O.
WILLIAM KNIGHT (Engineer), Co. E, 21st O.
J. R. PORTER, Co. C, 21st O.
MARTIN J. HAWKINS, Co. A, 33d O.

MARK WOOD, Co. C, 21st O.
J. A. WILSON, Co. C, 21st O.
JOHN WOLLAM, Co. C, 33d O.
D. A. DORSEY, Co. H, 33d O.

Exchanged March 18th, 1863:

JACOB PARROTT, Co. K, 33d O.
ROBERT BUFFUM, Co. H, 21st O.
WILLIAM BENSINGER, Co. G, 21st O.

WILLIAM REDDICK, Co. B, 33d O.
E. H. MASON, Co. K, 21st O.
WILLIAM PITTENGER, Co. G, 2d O.

BUSINESS AND POST-OFFICE ADDRESSES IN 1893 OF THE SURVIVORS OF THE RAID.

1. Wm. Knight, Stationary Engineer and Lecturer, Stryker, Williams Co., O.
2. Capt. Jacob Parrott, Farmer, Kenton, O.
3. Lieut. D. A. Dorsey, Dealer in Real Estate, Kearney, Neb.
4. Capt. Wm. Bensinger, Farmer, McComb, Wood Co., O.
5. Lieut. J. R. Porter, Dry Goods Merchant, Ingall's P. O., Payne Co., Oklahoma
6. Lieut. Wm. H. Reddick, Farmer, Newport, Louisa Co., Iowa.
7. J. A. Wilson, Grocer, Perryville, Wood Co., O.
8. Capt. W. W. Brown, Farmer, Dowling, Wood Co., O.
9. Capt. E. H. Mason, Pemberville, O.
10. Wm. Pittenger, Preacher, of the Southern California Methodist Episcopal Conference, now stationed at Colton, Cal.

---

11. Lieut. Mark Wood, died in Toledo, O., in 1867.
12. Lieut. Robert Buffum, died in Auburn, N. Y., Insane Asylum, July 20th, 1871.
13. Lieut. Martin J. Hawkins died in Quincy, Ill., Feb. 7th, 1886.
14. John Wollam, died at Topeka, Kan., Sept. 25th, 1890.

PREFACE.

THIS book, which is partly historical and partly personal, is written from the standpoint of frank egotism. It is far more easy to tell what the writer thought, felt and did, in the first person, than to resort to tedious circumlocution. As a large part of the interest of such a narrative must consist in describing the sensations experienced in passing through such appalling dangers and tremendous vicissitudes of fortune, it is clear that in a work of plain fact the writer cannot avoid making himself more prominent than his comrades. His own emotions and the incidents in which he participated will be indelibly engraven in his memory, while impressions received at second hand grow dim with the passage of years. It also happened that, in most cases where selection was practicable, the writer was made the spokesman of the whole party, and was thus brought into more frequent contact with both friend and foe. Many instances of this will be noticed all through the story.

Great care has been taken in the citation of authorities. The writer has not hesitated to claim for his own notes and memories the full weight to which they are entitled, and has carefully indicated the sources from which he has drawn all other facts. In no portion of war history of equal magnitude is there more abundant material preserved both on the Federal and the Confederate sides. This is indicated by the accompanying list of authorities; and authority for separate incidents is adduced either in notes or at the

beginning of chapters. With great care in sifting testimony
and constant references to original documents, the writer hopes
not only to give the exact and literal truth, but to carry the conviction of the judicious reader with him on every page.
Whenever conjectures or statements of probabilities are haz-
arded to bridge any chasm in the narrative, they will be offered
for what they are, and always clearly separated from known
facts.

This Fourth Edition contains considerable new matter, such
as the account of the Ohio monument and the history of the
Andrews Raiders to date. New illustrations also have been
added. The greatest improvement, however, has been made
by fusing into the continuous story all the additional material
which had accumulated during the past thirty-one years, thus
doing away with the cumbrous supplement of former editions.
It is believed that no portion of the civil war has been more
fully and faithfully recorded. Abundant references verify
every material statement. The writer hopes that this complete
and permanent edition will aid in maintaining for "The Locomotive Chase" its enviable place in the front rank of all war
stories.

WM. PITTENGER, Colton, Cal., Jan. 1st, 1893.

CONTENTS.

A PARTIAL LIST OF AUTHORITIES REFERRED TO IN "THE GREAT LOCOMOTIVE CHASE."

1. THE OFFICIAL WAR RECORDS.—These are now in course of publication by the Government at Washington. They comprise four series, each containing a number of large volumes. In several of these there are notices of the Railroad Raid, but the principal account is found in Series I., Vol. X., Part 1, extending from page 630 to 639.

2. REPORT OF WOOD AND WILSON IN 1862.—This was made to the Secretary of War, on Nov. 12th, 1862, when the two arrived, as escaping prisoners, at Key West, Fla. A fuller account by the same persons was published in the Key West, Fla., *New Era*, three days later.

3. SWORN TESTIMONY OF SURVIVORS.—The testimony of five of the survivors, Buffum, Bensinger, Parrott, Reddick and Pittenger was taken under oath at Washington, and phonographically reported, by order of Secretary Stanton. It is still preserved in the Archives at Washington.

4. THE FIRST EDITION OF "DARING AND SUFFERING."—This was begun a short time after the return home of the writer, and was founded mainly on personal recollections, aided by a few shorthand notes. Several survivors also contributed personal sketches and incidents to the book. It was published in Philadelphia, by J. W. Daughaday, in October, 1863. In this book the narrative assumed its usual form.

5. UNPUBLISHED LETTERS AND PAPERS OF GEN. O. M. MITCHEL.—These, as well as an unpublished biography by himself, were generously placed at my disposal by his son, F. A. Mitchel, Esq. (A biography of Gen. Mitchel, "Astronomer and General," embracing most of these papers, by his son, is now [1887] passing through the press of Houghton, Mifflin & Co., Boston.)

6. THE ADVENTURES OF ALF. WILSON, ONE OF THE MITCHEL RAIDERS, by J. A. Wilson.—Published in Toledo, Ohio. This is especially valuable as giving the marvelous adventures of the writer in escaping from prisons.

7. CAPTURING A LOCOMOTIVE, by Wm. Pittenger.—Published by J. B. Lippincott, 1881.

8. CONDUCTOR WILLIAM A. FULLER.—Accounts published by this gentleman, in *The Sunny South* and other papers, and still more important personal communications made to the writer, for which due credit is given in the appropriate places.

9. WAR FILES OF SOUTHERN NEWSPAPERS, especially of the Atlanta, Ga., *Southern Confederacy*.

10. "AN EPISODE OF THE WAR."—This is an account of some of the most important events of this history to which he was an eye-witness, by Rev. W. J. Scott, and is published in a volume of essays, with a strong pro-slavery bias, entitled, "From Lincoln to Cleveland," Atlanta, Ga., 1886.

11. THE FOLLOWING MEMBERS OF THE ORIGINAL EXPEDITION have each furnished me with important manuscripts as well as verbal communications: D. A. Dorsey, Kearney, Nebraska; J. R. Porter, McComb, Wood Co., Ohio; William Bensinger, McComb, Wood Co., Ohio; William Knight, Stryker, Williams Co., Ohio; Jacob Parrott, Kenton, Hardin Co., Ohio; W. W. Brown, Dowling, Wood Co., Ohio; William Reddick, Newport, Louisa Co., Iowa.

12. FRANK HAWKINS, Treasury Department, Columbus, Ohio, and Captain Jas. F. Sarratt, of Steubenville, have furnished many written details of the first expedition.

13. MR. ANTHONY MURPHY, Foreman of Repair Shops of the Western and Atlantic Railroad. A very valuable written communication, descriptive of the chase on the railroad was prepared for me by this gentleman.

14. THE ANDREWS RAIDERS, by Frank M. Gregg.—Chattanooga, 1891. A local pamphlet furnishing many narratives of old citizens about Chattanooga.

LIST OF ILLUSTRATIONS.

Tusculum

e.Alj+olcgm

DARING and SUFFERING

CHAPTER I.

A SECRET MILITARY RAID IN THE WEST.

I
T is painful for me to recall the adventures of the year beginning April 7th, 1862. As I compose my mind to the task there rises before me the memory of days of suffering and nights of sleepless apprehension —days and nights that in their black monotony seemed well nigh eter And time has not yet dulled the sorrow of that terrible day, when comra made dear as brothers by common danger and suffering were sudde dragged to a fearful death that I expected soon to share. A man ' has walked for months in the shadow of the scaffold and escaped at almost by miracle will never find the experience a pleasant one to d upon, even in thought. Yet it cannot be forgotten, and the easiest wa: answer the inquiries of friends, and to satisfy the curiosity of the pul is to put the whole matter candidly, faithfully, and minutely on record.

In the spring of 1862 a small secret expedition was sent from the Union lines into the very heart of the Confederate States. In its progress it aroused great excitement, first in the South and afterwards in the North, occasioned the most intense suffering to the soldiers engaged in it, and afterwards gave rise to many eager controversies. Several publications of a more or less ephemeral character have been devoted to it, and no story of the war seems to have fixed itself so firmly in the popular imagination. The present work is the full and complete edition of a small volume by the same writer, with the same title, which was hurriedly written before the freshness of personal impressions had faded, and while the horror and agony came back almost nightly in dreams. The writer is confident that this story, faithfully told, will give a more vivid picture of the spirit, feelings, and awful earnestness of the civil war than any more general war history. To do justice to brave men who perished in a manner ignominious in form but not in reality, to place romantic and almost incredible events, for which there will soon be no living witnesses, upon a basis of unquestionable certainty, and to help a new generation to realize the cost of upholding the Union, is the writer's aim.

The manner in which this volume was produced leaves little room for that unconscious exaggeration to which even candid narrators are liable. The writer made phonographic notes of the principal events very near the time of their occurrence, — mostly on the margins of a small volume of "Paradise Lost." There were enough days of dreary leisure for this or any other kind of work ! On leaving the enemy's territory, he and his comrades were, by order of the Secretary of War, Edwin M. Stanton, brought to Washington, examined under oath, and their answers written down in shorthand, and officially published, together with a lengthened and eloquent report by Judge-Advocate-General Holt. Immediately afterwards the small volume, "*Daring and Suffering*," was written at the solicitation of friends—a crude and hasty sketch prepared before the author had recovered from the sickness that followed his unexampled privations—but preserving the facts in their freshness, and attested in its recital of incidents by all the survivors of the expedition. Now that twenty-five years have passed, and the passions of war and the bitterness of partisanship have declined,—now that the dispatches and letters of Generals and the captured Confederate archives are accessible, and that the author has been enabled to go carefully over the whole ground and explore every source of information, from friend or former enemy, it seems possible to supply all the deficiencies of the earlier edition without diminishing its intense personal interest, and thus to furnish a complete and well-rounded history of the most dramatic adventure of the Great Civil War.

There were two distinct railroad raids into Georgia, which have sometimes been confounded. The first was authorized by Major General Buell in March, 1862, and sent out from Murfreesboro', Tenn. The second starting from Shelbyville, Tenn., in April of the same year under the direction of Gen. O. M. Mitchel, was larger and more important, and had quite a different purpose.

Nearly everything which has been published on the subject refers to the second expedition. I will, however, write the history of the first also, that the relation between them may be clearly seen, and all future misunderstanding be prevented.

The Union cause looked bright in the spring of 1862. It was nearly three years before the Confederates saw again so dark a day. Our eastern army under Gen. McClellan, numbering more than 100,000, was about ready to advance on Richmond. Gen. Grant, after capturing Forts Henry and Donelson, had sent the bulk of his forces to Pittsburg Landing, and was hurrying forward every available man. Buell had occupied Nashville after the enemy—because of the capture of Fort Donelson—had retreated, and was now urged by Halleck to send a large part of his force by water to reinforce Grant. This he declined to do, fearing that the enemy might return to Tennessee and capture Nashville; but he at length divided his force, sending the larger portion by deliberate marches southwest toward Pittsburg Landing, and a single division of about 10,000 under Gen. O. M. Mitchel, supported by 7,000 more in garrison at Nashville and surrounding towns, in a southeast direction. To oppose these powerful forces the enemy had an army in Virginia not more than one half as large as McClellan's, and in the west, at Corinth, the commands of Beauregard and Johnston, which were not yet concentrated, and were much inferior to those of Grant and Buell. Gen. Mitchel had no organized enemy in his front, but was marching into a country of vast importance to the Confederates, which they were certain to strenuously defend. A few comparatively small Union and Confederate armies opposed each other west of the Mississippi, and at various points along the sea-coast and the borders, with the preponderance usually on the Union side; but these may, for present purposes, be left out of account. The main rebel armies, those of Virginia and Mississippi, were united by a chain of railroads running from Memphis, Huntsville, Chattanooga, Knoxville and Lynchburg to Richmond; and this constituted their new and strong line of defense. They had indeed no other railroad communication except a very circuitous and precarious one along the sea coast. At Chattanooga this direct line was intersected almost at right angles by another extending from Nashville to Atlanta, and from there to all parts of the south.

It was the object of the Union generals, while preserving all they had gained, to break this line, and thus isolate the rebel armies and render easy their defeat in detail. The first assault was to be made at Corinth, to which the Tennessee River formed an easy channel of approach. It was defended by a large army, for if captured, Memphis and all the upper Mississippi would also fall into Union hands. Unfortunately, this portion of the West was at that time in two different departments under the command of Generals Halleck and Buell,—both able strategists but slow and timid. For fear of exposing Nashville to attack, the latter declined to reinforce Gen. Grant, who was acting under the orders of Halleck, by steamboat transport over the Cumberland and Tennessee Rivers; but instead marched toward his position through Franklin and Columbia. With the magnified estimate Buell had formed of the enemy's strength in Tennessee, this appeared to be the more prudent course; and had the advance been more vigorous and rapid, the imminent danger of disaster at Shiloh would have been removed. But even with this disposition, Buell feared that an army of the enemy might concentrate by rail somewhere in the direction of Chattanooga or Knoxville, and fall upon his rear. He believed [*] that a large force was gathering at Atlanta and also in Eastern

---

[*] Buell to Halleck, Mar. 23, 1862. War Records. Ser. I. Vol. X., Part 2, Page 60.

Tennessee, from which Nashville might be in great danger. These apprehensions, which delayed and weakened his movement towards Pittsburg Landing and endangered Grant, had at least one good effect. Gen. Mitchel was detached from the main army and ordered to Murfreesboro, with a primary view of guarding against any Confederate advance from the South or East; but he was able afterwards, by using his discretionary power to the utmost, to make the brilliant march upon Huntsville and to accomplish more against the enemy than any Union general with similar forces had been able to do up to this period of the contest.

While at Nashville, Mr. J. J. Andrews, a spy in the service of Buell, proposed to that General a daring plan, which, if successful, would for the time effectually relieve his fears and render a very important service to the Union arms. He offered to take a very small party of fearless men, disguise them as Southern citizens, conduct them to Atlanta, where he would meet a friend of his who ran a locomotive on the Georgia State Railroad from Chattanooga; then to ride with his party as passengers to a favorable point; there to capture the locomotive and to cut the telegraph wires behind him; then to run through Chattanooga and from this point westward, burning the bridges behind him, especially the great one over the Tennessee River at Bridgeport. Whether bridges were to be burned *South* of Chattanooga or not is a point that cannot now be determined; if not, it seems singular that Andrews should have conducted his men so far south as Atlanta; neither is the limit of his proposed operations westward accurately known. That the complete success of such a scheme would have greatly injured the enemy can be seen at a single glance. It would have hindered the concentration of troops and supplies at Corinth a week or ten days before the battle of Shiloh, which probably would not have been fought. It would have relieved Buell from his fears as to a flank or rear attack, or a march upon Nashville, and probably have induced him to reinforce Grant promptly and vigorously. It would have cut the main communication for some days or weeks between the eastern and western armies of the Confederate states at a most critical period. It did not promise the greater positive results of the second expedition, mainly because Gen. Buell was not looking toward rapid and aggressive action in Southern Tennessee. But no candid mind can question the great importance of the results promised, or the daring character of the man who could plan such work 300 miles away in the heart of the enemy's country. We may safely assume that the scheme originated with the intrepid spy and not with his cautious chief.[1] The former probably saw that the commanding General was anxious for the destruction of the enemy's communications, and suggested the means for accomplishing that end. This

---

[1] Gen. Buell confirms this view in a private letter to the writer January 7, 1887.

plan Buell accepted the more readily because it involved so little risk; that is, as Generals count risk,—only the life of a spy and eight [1] men; and General Mitchel, whose division was nearest the scene of the proposed enterprise, was instructed to furnish volunteers for the purpose.

JAMES J. ANDREWS. Engraved from an old Ambrotype.

From this point the expedition comes within the direct knowledge of the writer. As will be explained farther on, I had seen Mr. Andrews previous-

---

[1] Gen. Buell says but six were authorized. In this he is probably mistaken, as eight were engaged.

ly, but did not know his real character. And there were also reasons why the 2nd Ohio Infantry of Sill's brigade, Mitchel's Division, should furnish all the men required, and Co. G. of that regiment, to which I belonged, one half the number. These and many other things will be made clear in the narrative of the second expedition to which this sketch is but introductory.

The securing of volunteers was managed with the utmost secrecy. In addition to Generals Mitchel and Sill, a few of the officers of the 2nd Ohio were informed, in outline, of what was wanted, and they induced men to meet Mr. Andrews; these, after consulting with him, entered upon their strange and hazardous duties without opportunity to give the slightest hint to their comrades. To the common soldiers of the regiment who were not selected, the whole affair was wrapped in mystery. Eight of our best men suddenly disappeared, and we knew not what had become of them. Numberless were the conjectures that floated through the camp and were discussed around the camp-fires. Some asserted that they had been sent northward to arrest deserters; others that they were deserters themselves! But there were few deserters to arrest at this stage of the contest; and the latter idea was contradicted by the character of the men, who were among the boldest and most faithful in the whole regiment, and had been seen in close and seemingly confidential communication with officers just before their disappearance. The most frequent assertion—a pure conjecture, however, suggested by the fact that they were absent without any leave-taking, and that no inquiries were officially made about them,—was that they had turned spies. But this notion I did not seriously entertain, for sending such a number of spies from the private soldiers in the ranks of one company seemed absurd. At the most I supposed that they had gone on some scouting expedition or some attempt to surprise an enemy's post, such as we had been familiar with in Eastern Kentucky the year before. But I was not long left to my own conjectures. Indeed I had two reasons for urgent inquiries, one solely personal and not of a very exalted character, the other less selfish.

My position in Co. G., James F. Sarratt, captain, was then that of first corporal, and I was looking anxiously for promotion to the next grade of non-commissioned officers, that of sergeant. To a civilian these petty grades seem utterly unimportant and almost undistinguishable, but they are not so to a soldier. On many a lonely guard line and dark night or picket, they make all the difference between commanding and being commanded; and authority is sweet anywhere. A sergeant had died at Nashville, and his place would naturally become mine unless some one below me was considered more meritorious, in which case the captain had the authority to carry him, whether a lower corporal or only a private, over my head into the vacant sergeantcy. This would have been bitterl distasteful, not so much because of the increased wages and privileges of a sergeant, as because of the humiliation of being considered less worthy of promotion than a comrade, inferior in rank. One of the missing men happened to be second corporal, a splendid soldier in every respect, competent to fill any position in the company, and a great friend to the captain. I had heard that he might be preferred to me, if for no other reason, because I was near-sighted. Now, some of the members of my army mess said:

"Pittenger, when those men come back with feathers in their caps, the captain will be sure to make Surles sergeant." At the first opportunity I called Capt. Sarratt aside and told him what I had heard, and my own fears. He assured me, somewhat impatiently, that my rights should be cared for, and added, "Pittenger, this is a very little matter of yours. I only wish the men were back in the camp again."

"But where are they?" I asked, "and when will they be back? I would like to know something about it, especially for Mills's sake."

"I am not permitted to tell anything," he responded; "I don't know when they will be back myself, but I know that till they do come I can't sleep much."

The look of weariness on his face smote to my heart, and in view of such anxiety my errand looked utterly contemptible. But my own uneasiness in another direction was greatly increased, and when I left him, with sincere apologies, it was with the resolve to find out where these men were.

Captain David Mitchel of Co. K., 2nd Ohio, was an intimate friend of mine, and a distant relative of our commander. His company had supplied one of the missing adventurers,—my cousin, B. F. Mills, who had been my messmate during the three months service terminating with the battle of Bull Run. It was especially for his sake that I felt such solicitude for the absent men, and this, even more than my own interest, had moved me to speak to Captain Sarratt. I resolved to make an attempt on Mitchel, with stronger hopes of success. The opportunity soon presented itself. I was War Correspondent of the Steubenville *Herald* as well as soldier—the letters being intended principally to inform a wide circle at home of the welfare of sons and brothers. It was time to write again or there would be anxiety by many a fireside among the Ohio hills. Taking pencil and notebook I strayed through the level streets of the white-tented city that had suddenly sprung up in the level fields bordering the clear and beautiful little Stone River. A congenial spot was found on the sloping bank of the stream and I sat down to write. It was in the afternoon of a beautiful day, and the bustle of the camp was all around, but not near.

The warm sunshine and the rest were doubly grateful after the rainstorms and hard marches, knapsack laden, of the two preceding days. Occasionally I would look around to note the occupations of the soldiers, either in the camp, or strolling along the river. But I was especially attracted by the view presented a short distance down the stream, where stood the ruined fragments of a railroad-bridge that had been burned by the enemy a few days before. Now a large working force was engaged in putting the bridge in repair with all possible celerity, and our General was stimulating them by word and example. Large timbers were being hewed, framed, and slowly raised to their places. There was precision and speed such as I had never noticed before in similar work. The general himself seemed nervously eager, encouraging the willing workers, and heartily re-buking those who seemed inclined to shirk. Once I saw him precipitated with a splash into the shallow water, and to those who, like myself, were watching from the shoreward side, it seemed to be done purposely by a soldier whom Mitchel had *hurried* with a rotten piece of wood; but no sign of anger was manifested, and the General urged on the building as before. To a soldier who is "off duty" for a day, nothing is more enjoy-able than to see others work; and the whole afternoon realized to me the ideal life of the soldier. A comrade strolled along and I read to him my letter. It was very boyish, intensely partisan, its attempts at wit not very striking, and its estimate of Fremont as a leader, and of the Union people in the South, not such as would have been made later; but as it may serve to show something of the feeling of the soldiers at that stage of the contest, a portion of it is here inserted; besides, it has a pathetic inter-est, for when it was printed and read by home-friends, the trivial hard-ships mentioned in it were forgotten, and the writer was in a situation desperate as imagination can paint; while those who read believed him to have perished on the scaffold !

LETTER TO THE STEUBENVILLE HERALD.

MURFREESBORO, TENN., March 21, 1862.

"FRIEND ALLISON :—Again we have made a move, and an important one. The divisions of Nelson and McCook started before us, and no one knows where they are gone. But one thing is sure ; the northern frontier of the Gulf States, which has so long enjoyed immunity from the horrors of war, will soon hear the clash of arms, and the Union men of those States (if there are any) will have an opportunity of safely showing their devotion to the good cause. These movements are hopeful and of good omen.

"On Tuesday last (18th), we broke up our camp near Nashville, and with banne waving, and the sun shining brightly on the glittering muskets, we started over a smoo and level turnpike. There was only one drawback—we had so many stores to take wit us that our wagons were filled, and most of the men had to carry their knapsack Transportation is abundant in this country, but our Generals are too *generous* an *chivalric* (*vide* Buckner) to use it. The rugged Fremont was not sufficiently versed i *polite American* warfare to prefer the convenience of citizen rebels to the comfort of hi own soldiers. He would ride up to a house and demand the use of a driver and team fo

CHAPTER II.

GEN. MITCHEL AND J. J. ANDREWS ARRANGE A SECOND RAID.

ANDREWS was undismayed by the failure of his Southern friend, and proposed making another attempt with a larger force, carrying with him engineers and train hands from the Union army. It was never difficult to find men for every kind of work among the Northern volunteers—a fact of which Gen. Mitchel had already made ample proof in repairing and operating railroads.

When Andrews and Mitchel concerted the plans which were afterwards developed in the Chattanooga railroad expedition, no witness was present. Their momentous interview was most probably held in Mitchel's tent at night, where there would be perfect security from interruption and with all accessible maps of the enemy's country spread out before them. Let us, in thought, lift the curtain of that tent, and entering, listen to their discussions, and thus get an idea of the whole military situation. The outlines of such a consultation are not purely ideal, for the communications of Andrews, given to his comrades in many a confidential talk in the two months that followed after secrecy ceased to be necessary, together with the official records, and inferences from what was attempted and accomplished, will guide us with no small degree of probability to the very thoughts and plans that then occupied the minds of the two men. The interview took place either on Sabbath evening, April 6th, or before daylight on Monday morning, April 7th, 1862,—the very dates of the battle of Shiloh or Pittsburg Landing. Gen. Mitchel was then in high spirits, for the obstacles which had so long delayed him at Murfreesboro had been overcome. He had built a bridge across Stone River at that point, from his own resources exclusively, in a period of time unexampled in the history of the army. Gen. Buell, whose excessive caution and delay were far more dreaded by him than all the forces of the enemy, was far on the way to join Grant at Shiloh, where he would be ranked by Gen. Grant. It was reasonable to hope that the impetuosity and daring of the senior officer would more than compensate the chronic slowness of Buell. Mitchel himself was moving rapidly southward, and with the best division in the Western army—made such by his own tireless efforts—there was scarcely anything he dared not attempt. In a letter written home from Murfreesboro a few days before this, he had already expressed the hope that he might at no distant time direct his letters from Chattanooga! On every side the outlook of the Union armies was more hopeful than we, with the memory of the years of bloody struggle that followed, can easily realize. It should never be forgotten—though it often is by our historians—that there were virtually two civil wars; the opening contest fought by volunteers, and the second and final one between soldiers who did not fight purely of their own accord, but were brought into the field by the authority and force of their respective Governments. The first of these wars was now closing, and had been speedily determined in favor of the North. At the beginning there was almost equal enthusiasm; and the greater numbers and natural resources of the loyal volunteers was more than balanced by the better preparation and military organization, as well as by the superior determination, of the Confederates; but the very successes of the latter told against them in the long run. The great victory of Bull Run was no real advantage to the armies of the South; for while it wonderfully inflamed their enthusiasm, it begot such confidence that military efforts slackened; and during the novel (and therefore terrible) hardships of the first winter campaign, for which the Southern soldiers were but ill prepared, their armies gradually dwindled through sickness; volunteering almost ceased; and when the later successes of the Union armies became known, the common soldiers were discouraged and deserted daily. The extent of this feeling is abundantly shown by Confederate reports, even when these try to put the best face on the matter. It is safe to say that on the first of April, 1862, the Confederates had not more than two hundred thousand effective troops in the field. But the leaders were fully determined to succeed or destroy the whole country. They were, indeed, so fully committed to the Southern cause, that they had no hope of mercy in case of Union victory; the wonderful clemency of the conquerors in the day of success being something that no one, loyal or disloyal, dreamed of. But they were not yet ready to abandon the struggle as hopeless. One weapon of tremendous power was within their reach—a weapon which the peculiar organization of Southern society made it easy to wield. The common people were illiterate and accustomed to follow their leaders, submitting often to measures which in the North would have been sternly resisted. So far as the institution of slavery was concerned, there had always been a reign of terror at the South; and whenever that species of property was supposed to be endangered, all sanctions of law were overridden. The natural leaders of the people, the wealthy classes and the large slave-owners, were enthusiastic in the cause of secession. At the beginning of the war, they had no difficulty in persuading the poor to volunteer and fill the ranks, where they made effective

soldiers from the outset, rendering that obedience to their military superiors from life-long habit which Northern soldiers gained by hard and repulsive discipline. They themselves, being wealthier and more intelligent,
made excellent officers, having the instinct of command and all the
personal pride which is nurtured in an aristocratic state of society, and is
a wonderful stimulus to bravery. The Southern troops were as brave
and effective as any the world has known. But now that the spirit of volunteering had declined and the ranks were thinning, it remained to be
seen whether the masses of the Southern people would submit to a whole-
sale system of conscription—something which had no precedent in American history. If not, they were lost. It was the last hope, and it was
faithfully tried on a scale unparalleled in civilized history. A law was
passed putting every effective man between the ages of eighteen and
thirty-five—a limit afterwards greatly extended—into the army! In the
North not much notice was taken of this gigantic measure; it was spoken
of as the last refuge of desperation, a tacit confession of defeat; to escape
it, many citizens, especially in the mountain districts of Virginia, North
Carolina, Georgia, Tennessee and Alabama fled from their homes, and a
few across the lines to the Union armies; but it was generally acquiesced
in as necessary, although the scarcity of arms and proper officers made
the organizing and calling forth of the million of men embraced in this
terrible and far-reaching law a slow and tedious process. But the work
did go on; volunteering became rapid, for it secured some privileges, and
was the only way of escaping the conscription; the drain upon the Southern
armies from desertions and discharges almost ceased. Regiments whose
terms of service expired, were not sent home, but held under the new law;
and an able-bodied man in the South found that he must do one of three
things: submit, and enter the Southern army; hide from relentless pursuit
in the mountains, with the prospect of rebel prisons before him if captured; or flee from home altogether to the Union armies. A feeling of
desperate resolve gradually overspread the whole South—a feeling that
everything was now staked on victory, and that for original Union men,
as well as for secessionists, nothing remained but to fight on to the bitter
end. From this conviction arose the dreadful conflict that followed. But
in April it was only beginning, and did not culminate for some months
after.

The North was affected by the battle of Bull Run and other early reverses in a precisely opposite manner. There was enough of sorrow and
disappointment, but no discouragement. On the contrary there was a
feeling of grim determination. My own experience illustrates that of
thousands. The first three months' term of enlistment expired a few days
before the battle of Bull Run; and though I willingly remained with my
comrades for the battle, yet I refused to re-enlist, as everything connected

with a military life was distasteful; and I felt that there was now no great need of further services. But on the evening of that day, I resolved to re-enter the ranks for the war. All through the fall there was steady enlistment, and the work of recruiting did not flag for the whole winter, though no bounties were then given or promised except that of the old United States law, which few of the soldiers knew or cared about. By spring, at least five hundred thousand Union soldiers, better armed, equipped, and disciplined than the foe—outnumbering him more than two to one—were ranged along the border, and there was no point in the South that did not lie at their mercy, provided only that the advance was made *at once*, before the Confederates could arm and array their million of conscripts. The victory of the Northern over the Southern volunteers was already assured, if no new elements entered into the struggle; and what we may term the first war closed in triumph for the Union arms. But when the remnant of the rebel volunteers was reinforced by the mass of those who were being impressed into the service, and who in three months were as good soldiers as any other, the war entered upon a new phase; and the victory was not won until the North, more than a year afterward, entered with slow and hesitating steps the same path of compulsory military service.

But why should the conflict have been permitted to pass into this second stage? Why not have pressed the great advantages then held so as to prevent the general enforcement of the conscription? *On three men* the responsibilities of thus prolonging the contest will mainly rest. Perhaps from another point of view, we might regard them as agents of Providence in continuing the conflict until all the resources of the South were called out and utterly exhausted, and that section thus made willing to submit to the final overthrow of slavery and re-organization on the basis of liberty.

These three men were McClellan, Halleck, and Buell. They had many traits in common. Each was opposed to interference with slavery, and wished to conduct the war in the kindest and most courteous manner. Each possessed decided talent for organizing and manœuvering armies, talking much about lines of retreat, and bases of supply. Each overestimated the armies opposed to him. There is no reason to suspect their loyalty and good faith, and in subordinate positions they might have rendered valuable service to their country. Their plans were great and plausible, but they moved slowly, requiring vast resources and perfectly drilled troops to execute them. By proper precautions and strategy, with sufficient force, they imagined that the element of risk and the necessity of desperate fighting might be eliminated. They each looked upon the war from the professional soldier's point of view, rather than that of the armed patriotic citizen, who wishes to finish a dreadful but necessary business, and return to his home   But in their strong qualities they were less alike.

McClellan excelled in supplying an army, perfecting its organization, in elaborate reviews, and in securing the enthusiastic devotion of his soldiers. Buell understood well the lines of advance for his own troops and the enemy, and could so well guard every possible approach that he was never surprised. Halleck could clearly see the vital points of a military situation, could embrace in one view the most complicated details, was of tireless industry, compelled his subordinates to clerical exactness, and, when commander-in-chief, could tell at a moment's notice where every regiment in the army was stationed. What magnificent service these men could have rendered to their country if they had only been real warriors!

Each of the two Western generals possessed a subordinate with all the qualities he himself lacked. Halleck sent Grant forward, who achieved brilliant successes, until the arrival of his chief on the field at Corinth, after which he was held in check for many months; Mitchel was fettered by Buell, until fortunately he was so far separated from his chief as to get a little liberty of action at the period our story opens. But it is difficult to repress a feeling of indignation even now at the manner in which these commanders, who looked upon war as a purely professional game to be played regularly and leisurely to the end, caused the auspicious hour, in which success was almost a matter of certainty, to be lost. McClellan held all the Eastern army inert for more than a month before Yorktown during which time the rebels were enforcing the conscription hand in every hamlet in the South, and hurrying the troops ultimately beat him; Buell left Grant unsupported to fight trated forces of the enemy at Fort Donelson, and to come near total defeat in the first day's battle at Shiloh; while Halleck gathered every available man in the West and held them before Corinth in overwhelming array with orders to his corps commanders to retreat at any point rather than to bring on a battle! By the end of this period of inaction, the enemy was able to meet our armies on something like equal terms.

But all this while we have left Andrews and Mitchel alone in the tent of the latter. That they both knew to a greater or less degree of the considerations we have sketched, so far as they relate to events then passed, or the existing situation, is more than probable. Andrews had penetrated often into the South, and while he had a high personal regard for Buell and always spoke of him kindly, the daring nature of his own plans and the consequences he expected to flow from them, shows that he had no sympathy with the policy of delay.

Mitchel was as fully informed of all war topics as any person outside the war department could be. He was indeed a man of whom it is difficult to speak too highly, and if he had possessed the rank and opportunities of one of the great generals we have mentioned, the story of the war would certainly have been different. His fame as a man of original gen-

ius—as astronomer, mechanician, inventor, and civil engineer—was well
established before the war began. He was the better prepared for his
duties because he had graduated at West Point, and served a short time
as professor there, before making so bright a mark in civil life. When
the war broke out he had been appointed first to superintend the fortification of Cincinnati, which was soon finished; and he then obtained permission to raise troops for the deliverance of East Tennessee, always a
favorite object with him; but military etiquette interfered with his march

GEN. O. M. MITCHEL. From a photograph furnished by his son, F. A. Mitchel, Esq.

across Kentucky, which was in another General's department, and the
scheme was abandoned. He was then assigned to the command of a division, and had brought it to the highest state of efficiency. He knew so well
the state of the South that he felt willing to trust himself anywhere in the
enemy's country with that division alone. He lingered with Buell a long
time in front of Louisville, and writes home with the delight of a boy let
out of school when once the order to advance is given; he almost drags
Buell on to Nashville, offering to become responsible " with his head " for
the issue of the events if the permission to go forward is only given; over-
comes obstacles such as rivers, obstructed roads, and burnt bridges with

an ease born of thorough knowledge and boundless energy. At length he is detached, with his own division only, from the army of Buell, and given a very humble task—that of taking such a position as would best protect Nashville from an advance by the enemy (who had just as much thought at that time of going to the moon as returning to Nashville!) But this gave him latitude enough, for as long as he kept the enemy before him Nashville was fully protected, and he at once began that advance which is one of the most brilliant campaigns of the war. He had now reached Shelbyville, and was on the eve of a movement still further into the South.

It should be remembered by those who would fully understand the events that followed, that Mitchel was in no sense a heedless adventurer. On the contrary he was watchful and alert to the highest degree. Every regiment of his division had been taught vigilance by the most energetic effort. I remember an experience of this kind that left an indelible impression on the 2nd Ohio. It occurred soon after Mitchel had assumed command, but was by no means solitary. In camp at Bacon Creek, Ky., on a very dark and stormy winter night, the guard was placed as usual, but along about two or three o'clock had grown careless—more anxious to find shelter than to note everything that was stirring. Suddenly those of us who were "off duty" received a startling surprise. The men on watch had permitted somebody to come up to our post unchallenged, and we knew nothing of it until this person was in our midst, seizing the soldiers in no gentle manner by arm and collar, and shaking them, or tumbling them out of the guard tent, as he exclaimed, "Why don't you turn out the guard?" Some of the soldiers were for resisting, but all were submissive enough when the word passed around, "It's old Mitchel himself!" We were very soon in our places and then we listened to a lecture, as we stood in the rain, *not* on the subject of astronomy! When the General was gone the soldiers grumbled and wished they had an officer "who had not studied the stars so long that he could not sleep at night himself and would not let any body else sleep!" but we resolved not to be caught in the same way again; and we never were! We now knew in our division that the only way to get along in peace with our commander was to faithfully perform every part of military duty. We exercised the soldier's prerogative in grumbling, but we loved and trusted him, for all that, and would have followed him to Mobile or Savannah without hesitation, assured that he would have carried safely through whatever he undertook.

While Andrews and Mitchel considered the advisability of another attempt upon the enemy's communications, there is one point they cannot have failed to understand. Any new enterprise must be on the authority of Mitchel. The original instruction given nearly a month before, to furnish eight men (or six), could have had no binding force now when circumstances were totally different. The purpose Buell had in view was no longer of value. East Tennessee, as Andrews well knew and would not fail to report to Mitchel, was almost completely denuded of troops. Indeed, Mitchel never was haunted by the spectre of a vast rebel army such as continually rose before Buell. No men or arms could be sent from that department either to reinforce Corinth or to attack Nashville. The body of troops gathering at Big Shanty, a few miles north of Atlanta, could just as well be forwarded by way of Mobile. And most conclusive of all, Mitchel intended to move immediately on the line of road from Decatur to Bridgeport, and expected—an expectation fully realized—to have his whole division there within a week. Then he would be able to burn the bridges west of Chattanooga himself, if he did not rather wish to preserve them for his own use. Why did not Mitchel say in view of the situation:

" Mr. Andrews, I give you credit for penetrating to the heart of the enemy's country, even if you did not accomplish what you intended, which, indeed, would be of no value now; and I am thankful for the information you have gained as to the enemy's condition so far within his lines, and especially for bringing back every one of my soldiers in safety; now you may return to Buell and report that there is no further work of the kind he sent you here to do; or, if you prefer, I will gladly have your services as an ordinary spy."

Why did Mitchel not say this, and terminate the conference? Why not indeed, save that he saw a glittering prize before him which Andrews could help him to win — nothing less than the possession of Chattanooga and Knoxville, with all of East Tennessee, thus stabbing the rebellion to the heart, while its right and left hands—the armies of Beauregard and Joe Johnston—were held fast at Corinth and Richmond.[1]

The absolutely conclusive proof that Mitchel wanted Andrews to burn the bridges *south* of Chattanooga is found not merely in the statement of the survivors of the expedition, but in Mitchel's own letters and despatches. There were but two persons present at the conference at which their plans were arranged, and neither of these seems in advance of the performance to have hinted his designs to any persons save so far as was necessary to allow the soldiers who were to accompany Andrews to be intelligently chosen. But on the very day when Andrews captured the train, Mitchel was careful to mislead the enemy so that he might not destroy the great bridge over the Tennessee, and then declares in letters to Buell (and S. P. Chase,[2] Secretary of the Treasury), that he did so in the hope that he might be permitted to march on Chattanooga and Knoxville.

Whether in the case of Andrews completely accomplishing his share of

---

[1] See Chap. XLII. for further discussion with Gen. Buell and Gen. Fry of Mitchel's purpose.

[2] War Records. Series I. Vol. X. Part 2. Page 115.

the work, Mitchel could have carried out this hope, may be judged from what follows. It will be enough to assert here that he thought he could, and that the enemy fully shared his belief.

Let us examine the position of Chattanooga as it lay before these two men, on whose conference hung great issues and so many lives; for Mitchel then revealed to Andrews more of his plans than to any other person, even his own staff officers.

Chattanooga was a mere village in 1862, and was little more than one hundred miles distant in a direct line, but separated from them by almost roadless mountains and formidable streams. By aid of the sketch map facing Chapter I. we may form an idea of the central location of this mountain stronghold and grand strategic position, and thus be enabled to judge whether its capture would have been possible, and whether it would have been of any advantage to the Union arms. If both these questions are answered in the affirmative, and the further admission made that the burning of bridges on the Georgia State railroad would have materially contributed to that end, it will dispose of the question whether Andrews and his men imperiled their lives in a mere aimless raid, or whether as the Confederate newspapers and the United States official authorities declare, it was " planned by genius" and "absolutely sublime."

Andrews could at this time give Mitchel positive information that Chattanooga was denuded of troops, and that the surrounding country for a long distance, more than a hundred miles in every direction, was in the same condition. No recruits could be raised there for the Confederate army at that time. The commander, E. Kirby Smith, under date of March 13th, reports[1] " East Tennessee is an enemy's country; its people beyond the influence of our troops and in open rebellion." The occupation of Chattanooga in force by the Federals would have gone like an electric shock through all this region, and a mountain territory inhabited by more than a million people, comprising East Tennessee, Western North Carolina, and Northern Alabama and Georgia, as loyal as Pennsylvania or Ohio, would have soon given fifty thousand men to the Union cause. Their fear and hatred of the rebellion would have induced them to put forth every possible effort for self-preservation. But could not reinforcements be sent to a great railroad centre like Chattanooga with speed enough to fortify and hold it against every effort of Mitchel's slender force? Just in answering this question was the vast value that Mitchel saw in the new Andrews raid; no longer needed to carry out Buell's design of preventing the enemy's advance toward Nashville, it would be of simply inestimable value if it could, for three or four days even, isolate Chattanooga from its Southern and Eastern connections. Mitchel had already per-

[1] War Records. Series I. Vol. X. Part 2. Page 320.

mission to go, if he could see his way perfectly clear, to Huntsville; this he extended on his own responsibility seventy miles further to Stevenson; and if the way was open by rail, he could as well push on to Chattanooga. No reinforcements could be sent against him from the direction of Knoxville and East Tennessee, for in addition to the fact that any withdrawal of the forces there would have been followed by an uprising of the oppressed mountaineers, and a destruction of the railroad in the mountain defiles, the Union Gen. Morgan was before Cumberland Gap, ready with a superior force to press toward Knoxville if there was the slightest indication of the weakening of his opponents. Besides, a large bridge of the East Tennessee railroad spanned Chickamauga creek near the junction with the Georgia State road, ten miles from Chattanooga, and this would also be destroyed by Andrews. One railroad only was formidable, that leading to Atlanta. Undrilled troops, badly armed but enthusiastic, could be brought from there in a single day, and in two or three days considerable bodies of well-drilled and completely equipped men from Savannah or Charleston. Had the Andrews' party destroyed the bridges on this road and the East Tennessee bridge over the Chickamauga on the day the train was seized, I have no doubt that Chattanooga would have been under the old flag before the next day's sun went down. Then the fall of Knoxville and the opening of communications with Gen. Morgan would have followed as a necessary sequence. Communication was the third day actually opened with Halleck and Buell on the West; so if this *one contingency had occurred*, the whole line of railroad from Corinth to Knoxville, an extent of nearly three hundred miles, would have been occupied by four armies in perfect connection—those of Grant, Buell, Mitchel, and Morgan—a grand total, including their reserves at Nashville and on the march of 150,000 choice troops, opposed to less than 70,000 scattered, isolated, and discouraged rebels! what effect such an achievement would have had on public sentiment at the North, on the promotion of General Mitchel, and the future of the war, are topics not necessary here to dwell upon. But such considerations go far to justify Judge-Advocate-General Holt's opinion, " that the whole aspect of the war in the South-west would have been at once changed." The editor of the Atlanta " Southern Confederacy," well knowing the critical position of Confederate affairs, and the defenseless condition of Chattanooga at this time, wrote the following in his paper of April 15th, 1862:

" The most daring scheme that this revolution has developed has been thwarted ; and the tremendous results which, if successful, can scarcely be imagined, much less described, have been averted. Had they succeeded in burning the bridges, the enemy at Huntsville would have occupied Chattanooga before Sunday night. Yesterday (Monday) they would have been in Knoxville, and thus have had possession of all East Tennessee. Our forces at Knoxville, Greenville and Cumberland Gap would, ere this, have been in the hands of the enemy."

All this may sound wildly extravagant, but the one solid military advantage, not far off or contingent, that determined Mitchel to risk the lives of twenty-four picked men in a desperate enterprise, was that the burning of those bridges would render the capture of Chattanooga perfectly practicable with Mitchel's force, and would have given abundant time for his reinforcement, even if such reinforcement had to be given by orders directly from Washington.

How long Andrews and Mitchel were in reaching such a conclusion—how many maps they turned over together, or if they had more than one interview during the less than twenty-four hours that intervened between Andrews's return to camp and his setting forth at the head of a new and enlarged expedition upon his terrible work—is not certain. The interview we have supposed must have been held at night, for Andrews reached camp after dark on the evening of the 6th, or before the morning light of the 7th, and the same morning all was arranged, and the work of selecting the adventurers begun. In his short time Andrews ceased to be the agent of Buell, and came under the orders of Mitchel, the force at his disposal was trebled, his plan of operations was substantially changed, and he was ready to select his men and set forth. Such promptitude is almost bewildering, especially to one accustomed to the proceedings of such soldiers (with one or two exceptions) as had hitherto controlled the American armies; but the whole of the Huntsville campaign of Gen. Mitchel is of the same character.

CHAPTER III.

SELECTION OF ENGINEERS AND SOLDIERS.

O N Monday night, April 7th, 1862, orders were sent in regular mil tary channels to the Colonels of the three Ohio regiments of Sill': brigade to have a man from each company selected for specia! and hazardous service. Each colonel called a meeting of captains, and getting a general idea of the nature of the work required, each captain, selected a man judged to be best adapted for it, and, returning to his own company, either gave him personal notice or sent him up to his colonel's quarters for instructions. None of the members of the first expedition, who were all of the 2nd Ohio, would volunteer again. The romance connected with such an undertaking had vanished when they found themselves face to face with the terrible risks involved.

The meeting of officers arranged as to the manner of furnishing engineers and firemen and, as these wou in other work, a lil was made. Three ol succeeded in getting through the lines. Two of the men, Brown

WILSON W. BROWN, Engineer. From a war-time photograph.

and Knight, were from the 21st Ohio. That whole regiment was called into line, and the statement publicly made that men who could run engines were wanted for detailed service; and any who had experience in that line were asked to step two paces to the front. Brown thus describes what followed.

"The Colonel ordered me to report to his headquarters for instructions ; when I did, he told me that there was a secret raid being organized to penetrate into the South to burn bridges, and that, to complete the work, a locomotive would have to be captured ; he

added that I had been selected to run this locomotive, though, to guard against all con
tingencies, two other engineers would be along to take my place if anything happened to
me. After giving me some advice he ordered me to report to Gen. Mitchel in person. I
did so, and handed the General a note from the Colonel, which he read, and remained
silent for a moment. Then he said, 'This is a dangerous mission you are going on, and
the utmost caution will be necessary on the part of all of you.' He next questioned me
as to my competency to run an engine. (This work Mitchel well understood.) I answered all his questions, and he then asked me if I had any papers to show. I drew from
my pocket a paper signed by the master mechanic of the Mobile and Ohio railroad, on
which I had run in 1860, a year before the War, and other papers from other roads on
which I had been employed. He carefully read them all and said that they were sufficient
on that point, adding, 'On you rests a great responsibility. You are the first engineer
selected to take charge of the engine, but there will be a reserve of two others from the
other regiments.' I was about to leave him when a sudden impulse led me to say, 'General, I would very much like to ask you two questions about this expedition, if you will
permit me.' He very frankly said, 'You are at perfect liberty to ask any question bearing on this matter.' Then I said, 'What is the object of this raid?' He answered,
'To destroy the bridges over one of the main lines of the enemy's communications. It
will go far to separate their armies, and put them at our mercy.' I said, 'But what do
you think of our chances of success?' 'That depends upon circumstances,' he replied;
'if the enterprise can be carried out as planned by Mr. Andrews, I think the chances are
very good indeed ; but if any delay happens, the difficulty will be increased.' I asked,
'Why so, General?' 'He answered, 'Because as the armies draw nearer, the roads
will be more occupied with troops and stores moving back and forth, and these will be in
your way.' Then again pausing for a minute, he continued, 'Your mission is very hazardous. It is not pleasant for me to send such a number of picked men into the enemy's
power ; but in war great risks must be run, and we are engaged in a war of right and wrong;
armed treason must be met and conquered ; and if you fall, you die in a glorious cause ; I
have great confidence in Mr. Andrews, your leader ; I trust that the great ruler of the
destinies of man will protect you all!' He grasped my hand and terminated the interview. I never saw him again!"

The other engineer describes the manner of his detail very picturesque-
ly. There was always a little emulation between Brown and Knight as to
which was chief and which assistant engineer. One seems to have had
his appointment directly from Mitchel and the other from Andrews; but
they were inseparable companions and always worked in harmony. Knight
had been a long time in subordinate positions on the Pittsburg and Fort
Wayne road and served three years as engineer on the old Pan Handle
route. He says:

"Captain Brewster reported my name to the meeting at the Twenty-first headquarters as the man from his company. We were next called out as a regiment on parade
shortly after dinner, and while on the ground, the Colonel made a call, that if there were any
engineers in the regiment capable of running a locomotive, they should step two paces in
advance. I stepped out thinking that I was going to get some soft snap, such as running
a saw or grist mill ; but it turned out not to be so very soft. He told me to report to
headquarters immediately after being dismissed. The captain offered to go along, and
give me an introduction ; on the way he told me the first I knew of the expedition, and
that my name was already in as one of the men, but said I could use my own judgment

about going. I said I would go, and he introduced me as the man he had for that expedition, and also as having stepped out as an engineer. I was presented to Andrews, who was there, and he asked me if I considered myself competent to take charge of an engine. I said that I was perfectly competent, and showed my papers. He had his maps of the country spread out on the Colonel's table, and showed me on the map where I was at that time, and where he wanted me to go. It looked much better on paper than I afterwards found it on land. Andrews also told me that I would have to take off my uniform and put on citizen's clothes. He stated that if we were caught in the enemy's lines, and they knew us, we would be treated as spies. The Colonel then gave me a pass to go down and procure clothing, saying to take nothing in my pockets that would give me away, if captured; but told me if I was captured to get out by enlisting in the Southern army or any way I could, and it would be considered honorable when I got back. I went to town, bought clothes, got supper at the hotel, and started out the road by myself as he told

WILLIAM KNIGHT, Engineer. From a war-time photograph.

me. Before I got outside the lines I fell in with two others, but as the party were all strangers, I cannot say who they were."

Capt. Sarratt was not asked to select any man. He had long been fretting over the four engaged in the former attempt, and told Andrews, with some satisfaction, that he would not allow any of his boys to go a second time: but his surprise was great when he was simply required to "inform Corporal Pittenger, that he is selected as the man from your company." He did not then learn the reason for what may have seemed to him a strange choice, but probably attributed it to Maj. McCook, under whom I had formerly served. He was not kept long in the dark. While I was busily engaged in my own tent, I was notified that Captain Sarratt was looking for me. Going out into the company street—for the hundred men in a company lodge usually in two lines of tents facing each other at a short interval, which

CAPT. JAS. F. SARRATT.

is the place where the company forms in line for roll-call and other duties; and directly across the end of this street, stands the tent of the officers.

into which common soldiers do not enter uninvited—I met Sarratt and accompanied him to the officers' tent, where he informed me that Col. Harris had just told him that I was to go with Andrews down into Georgia. He said further—what I knew perfectly well—that a soldier is under no obligations to go on such an enterprise, and himself went to the questionable extent of advising me not to go.

This was really a breach of strict discipline; but the depth of his solicitude overbore all other considerations. He urged all the motives that seemed likely to have weight with me, saying that he had enjoyed no peace while the other men were out of camp, and that he was greatly relieved by their return, but if I went it would be as bad as ever. His solicitude affected me greatly, and I somewhat reluctantly informed him that I had already volunteered, and that all was settled.

He still urged me to draw back, only yielding when he found the task hopeless.

After quite a lengthened interview with my kind-hearted captain, who treated me as a brother rather than a subordinate, he wrote me out a pass that I might go to Shelbyville, a mile distant, and purchase supplies needed for the expedition. He also told me that I would probably see Andrews there, and advised me to question him closely, and if he did not perfectly satisfy me as to what he purposed, to return, without hesitation, to camp. This I promised, but I expected to be satisfied; then I took my leave, promising to come and say "Good bye," before leaving the camp.

With a throbbing heart and with suppressed excitement I again walked between the lines of tents looking wistfully upon the old familiar scenes.

It might be that this was the last day of my army life. The preceding day, which was the Sabbath, had brought with it a kind of dim melancholy presentiment as if some great change was impending. There had been nothing in the day itself to make one gloomy, for its quietness, with its warm spring beauty, the greenness of the grass, and the brightness of the sun, are deeply pictured in my mind, and I had devoted the day to writing to friends, with the faint impression that I might not soon again have the opportunity. It is a little singular that the next letter I wrote months after that day went to Jefferson Davis, President of the Southern Confederacy, on a matter of life and death to myself and comrades!

This Monday also was very quiet, for Mitchel had allowed two days of rest and preparation before the exhausting effort of his next dash into the heart of the rebel territory. My comrades, as I walked down the street, were sitting listlessly around the openings of their tents, scouring arms and accoutrements, playing games, telling stories in little groups, or lazily sleeping either within the tents or in the shade just outside. I paced along slowly, observing every object with a kind of languid attention, and almost unnoticed, for no word of our expedition had as yet reached the common soldiers. A few only asked me to stop and play a game with them; but I answered that I could not as I was going to town. Some of them called back, "Get us a pass and we will go with you"—an offer that seldom failed to be made in good faith to any one who had the great luxury of permission to leave camp.

I soon overtook a friend journeying the same way, and had the pleasure of his company to town; but as he was also from Co. G., I knew he could not be enlisted in the new enterprise; so while I enjoyed walking with him around the pleasant little town of Shelbyville, keeping my eyes open for Andrews, I could not talk with him on the subject that filled my mind. I got rid of him as soon as convenient and began to inquire for clothing; others were on the same quest, and as the town only contained two or three stores where such articles could be bought, the proprietors must have wondered to see the Yankees taking such a sudden liking to their goods. I did not find a full suit to my taste, but knew that I could easily supply the deficiency in camp. Among those who were purchasing I noticed with pleasure, Marion Ross, Sergeant Major of the 2nd Ohio. I put a few cautious questions to him and answered as many in return, when we both became convinced that we were on the same errand. This was a pleasure, as I now had an acquaintance in the first stages of the expedition, for, singular as it may seem, I had far more anxiety about being able to find Andrews and get out of the camp in the right direction, than about any of the dangers that lay further on. Our first business was to find him, and learn what steps were next to be taken. Why Andrews had adopted such an indefinite manner of receiving the report of the men who had been assigned to him, I do not know. Probably he wanted to scrutinize carefully those who were searching for him, and notice the manner of their inquiries and approach before becoming committed to them. No such purpose could be accomplished in the case of Ross and myself, as we both had known him before, and his meditative air as well as impressive personal appearance, made him a man impossible to forget. Accordingly we recognized him at once as we saw him coming out of a store, and, gaining his side, we told him that we were to report to him.

Andrews was now in the prime of manhood, being about 33 years of age, six feet in height, a little stooped when not excited, weighing 180 or 190 pounds, with strong and regular features, very clear complexion, an eye dark gray and penetrating, very abundant black hair, and a fine long silken beard slightly waved. In manners and address he was the ideal southern gentleman. He gave to every one the impression of gentleness and strength. His voice was very soft and musical, almost effeminate, never strong, yet with distinctness and firmness of tone which made it well suit the man. His striking personal qualities added very much to

He looked at us sharply for a moment, and asked us what we were to report to him for. We answered that that was the very thing we came to learn. He inquired our names, rank, company, and regiment. A few other questions and answers followed, when being satisfied, he told us that he could not safely say much to us in so public a place, but that he would be a mile or two out of Shelbyville on the Wartrace road shortly after dark, and if we met him then, he would give us full information. His manner was that of one who did not care very much whether we came to the place or not. I have since thought that this plan of appointing two different rendezvous, the one in the village and the other out in the country, and both of so indefi-

nite a character that a person might well have failed to find him in either, was intended to make it easy for any who had misgivings to draw back altogether.

There were some variations in the mode of selection. Porter went to the headquarters of the 21st and there found one or two others with Andrews. The latter entered with some detail upon the plan of his operations—showed a map of the country and pointed out the road over which they were to pass and the bridges that were to be burned. Porter remembers that these bridges were between Atlanta and Chattanooga. They were then shown what they were

MARION ROSS. From a war-time photograph.

in lividually to do, and ordered, if willing to enter on the work, to report at the night rendezvous on the road from Shelbyville to Wartrace.

The sun was shining brightly and the bracing evening air sent the blood coursing cheerily through our veins as Ross and I walked leisurely back to camp. We said nothing to each other of our motives in entering on such an expedition, though I was a good deal surprised to find Ross engaged in it. He was of sentimental character, very fastidious, neat and almost dandyish in dress, fond of parades, and generally of the pomp and glitter of war, and was often teased for these qualities. He could not fail to suffer greatly from the unavoidable discomforts of such a trip as ours, even if we were perfectly successful. But we did talk of the impressions formed of our leader. Ross requested me—with, as I thought, a shade of anxiety in his tone—to give my real opinion of Andrews. I answered with enthusiasm. The strong influence this singular man never failed to exert over those who were brought into contact with him, was already at work. His thoughtful, pensive manner, his soft mild voice, not louder than a woman's, yet with every accent firm and decided, his grace, refinement and dignity, made me at once declare him to be far above the ordinary type of manhood. He did, however, seem more like a dreamer, a poet, or a martyr, than a military leader or dauntless adventurer, yet there was something of each of these in his composition. I would have trusted him to the end of the earth! Ross expressed a similar opinion, and yet along with it a fear that possibly Andrews had now undertaken more than he could carry through. However we did not pursue that thought far, for neither of us knew definitely what he did propose, and felt a strong curiosity for the fuller revelation the night promised.

On nearing our tents we parted to make our separate arrangements. I found that a marked change had taken place in the camp. Listlessness had given place to curiosity. Several soldiers, I was told, had arrayed themselves in citizen's dress and left the regiment. I did not stop to hear conjectures about this, but hurried to the tent of one of the number who had been out with Andrews before, and from him borrowed the articles of clothing needed to complete my outfit. These were never returned! Then I took off the army blue which I had worn so long, and of which I was so proud, carefully folded it in my knapsack, and getting all my arms and equipments in order, left directions with some of my tent-mates to have them turned over to the proper authorities. I might have spared this trouble, as far as I was personally concerned, for all were lost in the terrible battles and marches that followed.

When I was divested of every trace of the army, and clad only in the plain garb of every-day life, I stepped out into the company street. My changed appearance caused a sensation at once. The word was passed from tent to tent, and soon all that were not on guard or otherwise out of reach, came around me and began to shower questions faster than a dozen men could have answered. "Pittenger, got a furlough? Got a discharge? Going home? Going out as a spy "—these are a few specimens only. My appearance confirmed the impression that I heard expressed by a hundred voices: "There's something up!" I did not care to contradict this natural inference, and answering all questions at random, I hurried across the line of tents to the adjoining company, and passing at once to the tent that sheltered my cousin, Mills, who was on the former expedition, I lifted the flap and went quickly in, for this company was also in a state of ferment. Mills, who was uneducated, but very shrewd, asked no questions, as he understood the hubbub outside and my own transformation. He said, with an expression of strong disapproval, not far removed from contempt,

" So ! _you_ are going with Andrews."

I assented, and then told him that my errand to him now was to borrow his revolver—he had a very fine one.

He freely gave the weapon, but added the opinion that if I knew when I was well off, I would stay in camp, closing with the candid words: " Because I was fool enough to go with Andrews, it does not follow that you need be."

When cartridges had been liberally supplied with the revolver, I had everything needed for starting, and it was nearly time to be off. This was to be the final farewell to the camp until the great effort had been made. I had not the slightest notion of trying to recede, even if it had then been possible without disgrace; but a sense of solemnity and awe, a kind of shadow over the inward landscape, was very distinctly felt. I resolved not to try to steal away as the first party had done, though that was the right thing for them to do; but now their report to their comrades, and the excitement caused by seeing the adventurers arrayed in citizen's dress, rendered it impossible to conceal the fact that we were leaving the camp for some kind of perilous enterprise, and there was no harm in the luxury of saying farewell. First I went to headquarters and took Captain Sarratt once more by the hand; he could scarcely say a word, and all his accustomed jests were silenced. I long remembered his troubled, half-reproachful look. Then I wrung the hands of all my old comrades as for the last time I walked down the company street. I had tramped with them over Kentucky and the half of Tennessee; had stood guard on many a dark and wintry night; had slept by their side in the open air when our heads were whitened with frost; had floated with them down a mountain stream on rafts and logs; and it was not easy to leave them, for most of them did not expect to see me again, and I half feared they were right. Some of them did their best to even yet have me give up the journey, but most realized that I had gone too far. Alexander Mills was especially devoted. He was a cousin (and also of B. F. Mills), and from the day of enlistment had been my inseparable companion. So close was our confidence that the only thing in army life I ever kept from him was my desire to go on this expedition. He was determined to keep me back or to go along. Failing in the first he tried the latter, though he had been some time seriously sick with fever. Receiving no encouragement from Captain Sarratt, he hurried up to the Colonel's quarters and pleaded to be permitted to go along so importunately that Col. Harris not only refused but threatened to have him arrested if he tried to leave camp without permission. Then he came back and said "Good bye" with tears on his cheeks. He was refused only because of sickness, for he was efficient in every duty: but in the troubles that speedily came, I would have been sorry to have had him share the hardship and danger, great as the com-

fort of his presence would have been. Alas ! he now lies buried near
top of a beautiful hill in the National Cemetery at Chattanooga, shot w
bravely carrying the 2nd Ohio colors in the storming of Missior
Ridge, eighteen months later—the very battle which assured posses:
of that town toward which our thoughts and efforts were now turned.

above a whisper—the night and solitude and intense curiosity had disposed to silence—Andrews seemed satisfied that no more were coming, and rising said in the low distinct tones that became so familiar to us, "Let us go a little way from the road, boys." He led the way up a slight slope on the right-hand side, through stunted bushes, to a level spot near the summit. We followed, and gathered in a compact cluster around him. Twenty-three were present. Andrews had been authorized to take twenty-four, but this was the number who actually met at this place for consultation. Recent inquiries have led to the belief that at least thirty were detailed, one from each company of the three regiments, but that the others had been lost, as Ross and I nearly were, had refused the service as too hazardous, or had been turned back by Andrews himself. It was reasonable that some latitude should be allowed for selection on both sides.

We now stood in a little thicket of dead and withered trees, with a few smaller bushes around, but the place was sufficiently open to assure us that no listener could be concealed within hearing. Probably the dramatic circumstances of this first meeting were not consciously selected by our leader, but nothing could have been devised to impress more deeply the ardent young soldiers by whom he was now surrounded. A storm was rising, though the afternoon had been so bright, and the wind began to moan at intervals through the naked trees. The mutter of thunder was also heard, faintly at first, but it soon came near and loud, while the flashes of lightning, more vivid in the darkness, enabled us to see each other's faces for a moment, and then left us in total obscurity. Andrews spoke as quietly as before, stopping when the thunder roll was too loud for him to be heard, and resuming the moment it ceased. The storm had little terror for soldiers who were accustomed to stand guard, march, sleep, or fight, by day or night, regardless of the commotion of the elements, or only grumbling a little at the discomfort.

But one noise stands out more vividly amid the sounds of night and darkness than any other—the howling of a dog from the other side of an intervening valley. There is a tinge of superstition in the veins of most people, and the majority of us would have listened with more pleasure to almost any other sound

Two purposes were to be accomplished at this conference. We were to learn enough of the plans of our leader and the risks involved to decide intelligently whether to go on with him or to return to camp. In a sense this was a mere form; for every one of us had already made up his mind to follow Andrews to the death. But to satisfy orders and military custom it was necessary that the offer should be formally made, and as formally accepted or rejected. Acceptance here made every one of us really volunteers although we falsely claimed in the South that we were detailed without our knowledge or consent, and clung to that story without the slightest wavering until the last man was out of the power of the enemy; and it is a touching instance of Andrews's care for the life of his men, that almost in his dying hour he confirmed this claim—although in so doing he seemed to leave his own memory under a cloud. But the truth is that he did state explicitly that if we were detected by the enemy while in disguise beyond our own lines we would in all probability be massacred at once or hung as spies; and declared that we were free to return to our own tents, without other condition than the promise to keep all that he had communicated a secret. No one, however, showed the faintest desire to avail himself of this offer. If any had been detailed who were not willing to incur the hazard and responsibility involved, they had already been sifted out by their failure to report at this point.

The other purpose of the conference was of a more practical character; it was to receive such instructions and information as would enable us to coöperate intelligently with our leader. What was our destination? How were we to get to the scene of action? And how were subsequent orders to reach us? These were a few of the questions that naturally pressed for answer.

The opening words of Andrews to the men who clustered around him were exceedingly informal,—far more like a talk than a set speech, and hardly so loud as an ordinary conversation. Though I listened with burning attention to every word, yet I cannot claim that the language used below is literally exact. There was explanation, repetition, and enlargement of parts not fully understood, with frequent question and answer. Andrews sometimes spoke to all, and sometimes to one or more who wanted information on special points. He said:

"You will break up in small squads of two, three, or four, and travel east into the Cumberland mountains, then south to the Tennessee River. You can cross the river and take passage on the cars at Shell-Mound or some station between that and Chattanooga on the Memphis and Charleston Railroad. You must be at Chattanooga not later than Thursday afternoon, and reach Marietta the same evening, ready to take passage northward on the train the next morning. I will be there with you, or before you, and will then tell you what to do."

"The road," he added, "is long and difficult, and you will have only three days and nights in which to reach Marietta. I will give you plenty of money, and you may hire conveyances whenever safe and convenient. I will ride along the same road that you are to travel, sometimes before, and sometimes behind, and will give you any help in my power. If you should be arrested I may have influence enough to secure your release; but depend on yourselves and be watchful and prudent. Do not recognize me unless sure that we are alone."

Some of our party had travelled enough in the South to know that for inquisitiveness as to the destination of a stranger who comes into their midst, the people of that section are not a whit behind the most curious of Yankees, and therefore inquired, "What account shall we give of ourselves if asked who we are, and why we are coming South?"

Andrews answered;

"The most plausible thing will be to tell them that you are Kentuckians escaping from the rule of the Yankees; and that you expect to join some Southern Regiment. Say just as little as will carry you through, and always have some reason for not joining just then. After you get into the mountains you will be in the track of the Kentuckians who travel South, and will seem to be coming from there rather than from the Union army; so you need not have much trouble. But if you should be closely questioned it will be safe to say that you are from Fleming Co., Ky., for I happen to know that no Southern soldiers hail from that place, and you will not be confronted with any one who knows you are not."

Fleming County was that in which Andrews had resided for several years preceding the war, and it was largely owing to his influence that the State Guard of that county had been preserved loyal to the Union and had furnished no recruits to the rebel army. His directions were listened to with absorbing interest, and the last one turned out long afterward to be a deadly snare. But this contingency could scarcely then have been foreseen.

"But if we are completely cornered and they will not believe our stories, what then?" asked another of the soldiers.

"In such a case, don't hesitate to enlist. It will be far better to serve a little while with the rebels than to run the risk of discovering our plans by holding out. You can probably get away from them some dark night on picket. You are fully authorized to take any course that may seem best, and no one of you will be suspected of desertion even if found among the rebels."

Another inquired:

"Is it likely that a man who can give no satisfactory account of himself will be permitted to join their army?"

"There will not be the least trouble about that," Andrews replied; "the difficulty is to keep out of the Southern army, not to get into it. They are picking up men everywhere, and forcing them to enlist, and are emptying the jails for the same purpose. Stick to whatever story you tell, and as long as they do not get any proof that you are a Union soldier, they will be ready to hurry you into the service even if they don't believe a word you say, as the best way of disposing of you. But I hope that you will not be suspected at all, and will meet me promptly at Marietta. Probably you will not fall in with any Southern troops, and the country people will help rather than hinder you."

Although I listened as if spell-bound to every word, studying the speaker as well as his utterances, I was not disposed to ask any questions as to the first part of the expedition, for I had already learned a good deal about it. But there was another contingency far ahead, which had a deep personal interest to me. I greatly disliked the thought of being left alone under any circumstances—probably because of defective vision—but always felt comparatively easy when I had a trusted comrade whose eyes I could use for distant objects. I was willing to risk the journey Southward with a small squad, for I could see how persons moving in that direction would find it easy to allay suspicion, but was much more solicitous about the return journey, and asked Andrews whether after we had captured the train and used it in burning the bridges, we were to abandon it and try to steal north as we were now stealing toward the south.

He answered me very explicitly, and in so doing revealed still more of the general plan.

"No," he said; "General Mitchel starts south in the morning for a forced march with all his energy, and he will surprise and capture Huntsville on Friday, the very day we are to capture the train; so that when we get back to that point we will find him ready to receive us. If we cannot quite reach him, we will leave the train close to our lines and dash through in a body."

This was glorious. The thought of such a coming into camp after piercing the heart of the Confederacy, set every nerve on fire! But there was another possibility and I wanted to see that also provided for. So I asked again, whether, if we failed to run the captured train through Chattanooga and had to leave it south of that point we would still cling together.

He answered emphatically;

"When we once meet at Marietta, we will stay together and either come through in a body or die together."

This satisfied me perfectly, and there was little but desultory conversation afterwards. Andrews called on the men to form their squads according to their own preferences and then commenced distributing Confederate money among them, giving sufficient to one man of each group for all—though without intending to constitute the man so favored the leader of his comrades.

This formation of these little travelling companies was a somewhat delicate matter, and in the hurry, was not always arranged to perfect satisfaction. I wished the company of Ross, but he asked permission to go along with Andrews as far as he could accompany any of the party, and one or two others making the same request, they were thrown together, and I had to find other companions. I was exceedingly fortunate, as two men of Captain Mitchel's, and one from Company B, the next in regimental line, fell to my portion. We constituted the left wing of the 2nd Ohio! The work of division was now completed; the hour for parting had arrived; and we once more listened to the words of our commander.

"Boys," he began, "we are entering on a very hazardous expedition, but it will be glorious in its results, and will give the enemy the most deadly blow he has yet received. What a grand thing it will be to run through the South leaving the bridges burning and the foe in helpless rage behind! If we burn these bridges, Mitchel will capture Chattanooga the very next day, and all East Tennessee will be open before him. But we must be prompt, for if he gets to Huntsville before us, the road will be so crowded with reinforcements moving against him that our task will be much harder. But if we have the bridges down first they can send no force against him, and he will have everything his own way. The last train leaves Chattanooga for Marietta at five in the afternoon. Be sure to catch it not later than Thursday, and I will either be on it, or an earlier one—good bye!"

About this time the rain began to fall—gently at first—but it soon came down in torrents. One group after another filed off with military precision from the place of meeting, and Andrews shook hands heartily with the members of each as they passed. A considerable interval of time was permitted to elapse between the starting of each squad and the succeeding one, for the separation was to be made from this point, and we did not wish to meet again till Marietta, Ga., was reached, or at least, till on board the last train of cars leading to that point.

As we picked our uncertain way along the railroad, stumbling over the ties which were visible only by the lightning flashes, I looked back and saw Andrews, with none but the three members of the last group near him. He was looking after us, his head bent slightly forward in the pensive attitude habitual with him, and a broad stream of lightning made him at that moment stand out as clearly as the mid-day; the next moment he disappeared in utter darkness, and the crash of thunder overhead drowned every other sound. We hurried on our way and were soon far from the place of meeting.

Seen by the Lightning Flash.

CHAPTER V.

THERE are few more romantic figures among those who rose into prominence during the great civil war than James J. Andrews. The manner in which he inspired confidence among the officers of the Northern and Southern armies alike, and bound to him all the private soldiers with whom he came in contact, is very remarkable. The character of our parting from him at the beginning of the furious thunderstorm

FLEMINGSBURG, KY. From a photograph.

on the evening of April 7, 1862, which was a fitting emblem of the enterprise on which we had entered, shows how strong was the impression he had already made upon us. After that interview I felt no doubt as to the issue of the enterprise. The quiet confidence and matter-of-fact tone of Andrews assured me of success. Similar feelings were expressed by other members of the party.

Who was he, and how had he come to occupy his unique position—neither a soldier nor an officer, yet counseling officers and commanding soldiers? None of our party could have answered this question then, but it is well to give the reader all the information since acquired by diligent search at Flemingsburg and elsewhere.

Flemingsburg is a small rural county seat, situated on the slope of a steep hill, on the border of the "blue grass" region of Kentucky. The population, only 1200, has scarcely changed since the beginning of the war. The branch of a small railroad now leads to it, but then it was almost completely isolated, being seventeen miles distant by turnpike from Maysville on the Ohio River, its nearest point of communication with the outside world.

One day in the spring of 1859, a traveller came either by stage or on foot into this secluded village. He joined a group of idlers near the brick hotel opposite the court-house, and listened for a time in silence to the desultory conversation. At length he made some inquiries, and in response to questions addressed to himself, said that he had just come up from Maysville, and had thought of seeking a position as school-teacher. Noticing the name "Andrews" on a sign across the street, he asked if there were many of those people in town. Being answered in the affirmative, and assured that they were among the leading citizens of the place, he said, "That is my own name, and I think I will go no further, but make my home here for awhile." He further stated that he was a native of Hancock Co., West Virginia, and that he had come down to Maysville on a raft. Though he had little money or apparent means, he was cordially received, and made to feel at home. No opportunity presenting for employment as teacher, he began work as house and ornamental painter, and was very skillful. He had a beautiful voice as a singer, and taught several "singing schools" in the evenings, becoming a general favorite with young and old.

There was always an air of reserve and mystery about him. He seemed, as one of the citizens told me, "like a man who had a story." His wide information and refinement of manner, his manly beauty and easy grace in any company, gave him a marked ascendancy over his companions, even while he tried to be one of them. Nothing definite or positive was known of his former life; and while no one cared to ask him directly on the subject, a story was generally circulated as having originated with himself, to the effect that his father had entrusted him with $5,000, which he invested in a flouring mill with wool-carding attachment, in Ohio, but that the mill had burned without any insurance, and that when he returned to his father's, a sister reproached him so bitterly with this loss that he left home, resolving that his family should not hear from him until he had more than made this amount good. A modification of the same story of a still more romantic type was afterward told by him to some members of our party. I give it as repeated by Jacob Parrot from memory, and, after the lapse of so many years, it is possibly not accurate in all details. It was said that shortly before the burning of the mill, Mr. Andrews had been on very friendly terms with two young ladies, and ultimately became engaged to one of them. When all his property was lost in the flames, this lady wrote him a very chilling letter, asking, in view of changed circumstances, a release from the engagement. He at once took this letter to the other lady, and finding the warmth of her sympathy a grateful contrast, he offered himself and was accepted. But his trials were not yet over. Within a month of the period fixed for marriage, the second lady—he never gave the names of either — suddenly died. The three-fold disappointment— loss of money, the unfaithfulness of his first love, and the death of the second—so wrought upon him that he left his home and resolved to make for himself a new life amid new scenes. His parents had previously removed from West Virginia to Southwestern Missouri.

I have never been able to find the location of this mill or to verify the story in either form. But the absence of all references to his former life in his last letter and bequests,[1] the heartiness with which he accepted the secluded Flemingsburg as his new home, together with a kind of pensiveness and melancholy which were manifest when not actively engaged, all point to some decisive break in his history.

JAMES J. ANDREWS. From a photograph in the possession of Miss Elvira Layton.

Most of the people of Flemingsburg were convinced that Mr. Andrews was very anxious to make money. Yet this conviction seemed to have no other basis than his own words. He was not at all penurious, was strictly honest, and seemed to have done no more than maintain himself respectably.

Some months later, an incident occurred which made Mr. Andrews feel still more at home in Fleming County. An old gentleman, named Lindsey, who lived some seven or eight miles from Flemingsburg, on the Maysville Turnpike, partially rebuilt his house and employed Mr. Andrews to do the necessary painting. While thus engaged, he became intimate with Mr. and Mrs. Lindsey, and they frequently jested with him about

---

[1] See Chapter XXI.

his growing so old without marrying. He answered that he could find no lady willing to accept him, who would not be extravagant and wasteful of the property that he wished to earn. Lindsey rejoined that he knew one who possessed every desirable quality, without being in the least tainted with the defect feared. Andrews, still jesting, promised that if introduced to such a model woman, he would at least try to win her. Mr. Lindsey was a sympathizer with the South in the angry controversies that already began to presage war, while Andrews was as firm on the other side; but

Miss Elizabeth J. Layton. From a photograph belonging to Mrs. Wm. Rawlins.

this did not prevent a great esteem on the part of the older man for the younger, nor make him less in earnest in his match-making proposal; and soon Miss Elizabeth J. Layton, utterly unsuspicious of the serious consequences that were to follow, received an invitation from Mrs. Lindsey to spend a few days with her, to assist in sewing. The lady was tall, graceful and finely formed, of pleasing though not strikingly handsome countenance, and probably a year or two older than Andrews. Her manners were quiet and grave, but she had a very decided character. She was a member of the "Christian church" (Campbellite), and her sym-

pathies were entirely with the loyal side in the rising struggle. How much
similarity of political views had to do with the result, we cannot say; but
soon after becoming acquainted they were warm friends. Delightful
evenings were spent in the large, old-fashioned parlor of the Lindseys;
and Andrews soon confessed to his employer that the latter had judged
well. Before the courtship had ripened into an engagement, the storm of
war broke over the country, and for a time the lovers saw less of each
other. The young man from the North--for Hancock Co., Virginia, and the
whole " Pan-Handle " lying between Ohio and Pennsylvania, fully shared
the sentiments of the loyal States—took no uncertain position, and with
his work in the Kentucky State-guard, and afterwards in more important
enterprises, his time was fully occupied. But soon he found an opportunity of proposing to Miss Layton, and was accepted. At the solicitation of his betrothed, Andrews promised that when he had finished one
more daring enterprise he would retire from military service. The date
of marriage was fixed for June 17, 1862. But a dark tragedy was completed before that date !

When the war broke out in the spring of 1861, Kentucky occupied a
peculiar position. The majority of the people had no wish to secede, but
they wished as little to engage heartily in war for upholding the Union.
The call of the President for troops was received with scorn, and every
county was agitated by the question of secession. For a time the impossible task was essayed of remaining neutral. Troops were enrolled as
a " State-guard," not to be called into service unless the state was invaded,
when they were to be used against the party making the invasion. Such
a position of "Armed Neutrality," was untenable, because refusing to aid
the Government in time of need was disloyalty, and because the position
of Kentucky between the two great sections made it sure that sooner or
later the State would be forced to adhere to one side or the other.

Andrews spoke decidedly in favor of maintaining the Union. He was
never an abolitionist, but wished to see the old flag unsullied, and the
nation undivided. He joined the volunteer organization of Fleming Co.,
and brought all his powers of persuasion to bear in securing its unqualified
adhesion to the Union cause.

The war excitement had brought all business to a standstill. There
was no house building or other improvements during the spring and early
summer months of 1861, and consequently no demand for painting.
Andrews was at this time boarding at the hotel of Mr. J. B. Jackson, who,
finding him out of employment, offered him the temporary position of
clerk. This he held for several months.

At length Andrews made a journey to Louisville, and on his return
announced that he had been appointed deputy U. S. Provost Marshal, and
that his jurisdiction in that capacity would be extended over Fleming and

the adjoining counties. No small dissatisfaction was expressed at this appointment, especially by those whose sympathies were with the Confederacy. But Andrews entered upon the duties of the office with much apparent zeal, purchasing a copy of "Conkling's Practice" to inform himself in regard to modes of procedure in U. S. Courts. But the same evening of his return from Louisville, he informed his friend Jackson that the office was only a blind; that his real business was that of spy for the Union Army. The account he gave of his entrance upon this business was that he met a young lieutenant, incidentally, while on business in Cincinnati, who recommended it to him as an employment in which he could render great service to his country; and that, being at the time out of employment, he went to Louisville, offered his services, and was immediately accepted. Jackson was greatly shocked at this revelation, not because of the nature of the employment, for he was a strong Union man, and in the feeling then existing nothing that could be done against the opposite side was considered too bad; but because of the danger involved. He told Andrews that if he did not give it up, his being detected and hung was only a question of time. Andrews did not deny the probability, but said that he was doing no good now and that he was determined "to make a spoon or spoil a horn." Jackson understood him to mean that he would make a fortune in that business, or lose everything, life included: but it is easy to give the words a nobler interpretation. As the remonstrances of Jackson were unavailing, he told Andrews that he must seek a new boarding-place in order not to compromise his friend. Andrews agreed to the propriety of this, and at once removed to Mrs. Eckles's boarding-house.

Andrews accompanied General Nelson in his expedition into the mountains in Eastern Kentucky, and there rendered considerable service, though not equal to his expectation, for there was not much real work in that place for a spy to do. What he attempted, however, aroused the bitter hostility of the enemy, and his escape from them was, in more than one instance, very narrow. In reference to this he told another friend, Mr. J. H. Cooper, that as the Confederates were so very bitter toward him, he had resolved to do them all the injury in his power, or lose his life in the effort.

It is doubtful whether Andrews up to this time can properly be called a spy, notwithstanding his use of the word to Jackson. He seems rather to have been a scout and agent for secret communication with the Union men of Kentucky. He was known in his own home and generally in the towns through which he passed as a Union man; and while he did not wear uniform, he was not directly in the military service, and seems to have been regarded simply as a Kentucky citizen rendering aid to the Federal army. This exposed him, when beyond the protection of the army, to the hostility of the Confederates, and, at the same time, cut him off from the opportunity of obtaining information regarding the intentions of the enemy. He now resolved on a bolder course, and entered upon it in a characteristic manner. The first step, as told me by Mr. Ashton, present Postmaster at Flemingsburg, who was an eye-witness, is very striking.

Andrews had been absent for several days—at Louisville it is presumed —and on his return was accosted in a friendly manner by Judge Cord, one of his intimate associates, and a leader among the Unionists. He replied coldly and gruffly. As the other approached, and again spoke cordially, Andrews said: "I don't care to talk with you." "What's the matter?" the other responded, still unwilling to accept the rebuff.

Andrews returned:

"I have been behind the scenes the last few days. I saw too much."

"Why, what wonderful things have you seen?" said the other, deeply puzzled by this extraordinary change of front.

"Why, I have seen how this war is carried on, and what it means. It's all a great speculation. Everybody is trying to make what he can out of it, and I will have nothing more to do with it."

Mr. Ashton also was greatly astonished at such language from an ardent Union man, who had been virtually in the military service of the United States; but he saw not far away, and closely observing them, a certain William A. Berry—a bold, reckless man, the ringleader of the rebel element in that entire district. Of course he was greatly interested in this sudden conversion of Andrews.

About an hour afterwards, Ashton observed Berry and Andrews in close and very animated conversation at another part of the street. Andrews was telling of his changed views, and ended by asking Berry to get him admitted to the societies by which the friends of the South were bound together, as it was now his intention to serve the South with all his might. At this proposition, Berry, who had seemed pleased at first, broke out in a volley of oaths, pouring upon Andrews the bitterest abuses and curses, saying, "Andrews, do you take me for a baby or fool? You are nothing more or less than"—with a volley of oaths attached—"an infamous spy! and I will see that your character is made known and will gladly help hang you, as you deserve!"

Andrews quietly said: "Berry, you are excited. You will understand me better after awhile!" and left him foaming with rage and threatening condign vengeance for the treason which he suspected. Andrews continued to avoid the Union men of Flemingsburg in public, from this time, but all his advances were coldly received by the opposite party. He had been too fully committed, and there was a short time when he was in no small degree of personal danger.

At first the leaders of the Union party in Kentucky had held their partizans back from enlisting in the Federal army until the posture of the State had been fully determined, for the sake of their influence and vote in local matters. But soon the ardent young men swarmed into the army in such numbers as in some places to materially affect the preponderance of their cause at home. During this period in Flemingsburg, Andrews was more than ever threatened; and it seemed as if all his great personal popularity had vanished. But he was then, as always, undismayed, and the Union sentiment soon became so firmly established and so aggressive that Berry and his chief friends found it convenient to retire beyond the rebel lines—the former engaging in business in Nashville.

In another conversation with his confidential friend, Jackson, Andrews declared his determination to follow this example, and also go to Nashville. Remonstrances were tried in vain; Jackson saying "Why, Andrews, what will you do about Berry? He wanted to hang you even here; but if he finds you in Nashville where the rebels have it all their own way, he will do it." "I can manage that," Andrews replied, and in due time made his way through the lines.

At Nashville he boldly went to the largest hotel in the evening, and registering as J. J. Andrews of Flemingsburg, Kentucky, went to bed. Before he was up in the morning he heard an impatient rap at his door, and a voice saying, "Let me in. I want to see you!" He opened the door, and his visitor entering, announced himself as Judge Moore of Fleming County, and said:

"What are you doing here? I always heard that you were one of the head Yankees up there!" Andrews had seen Moore, though he had no personal acquaintance with him, and replied:

"So I was at first, but the only difference between you and me is that you saw into this thing a little sooner than I did, and when I turned over to the right side, it got too hot for me at Flemingsburg. The Yankees have it all their own way there now."

Moore grasped his hand, warmly congratulated him on having come over, and welcomed him to Nashville. As soon as Andrews was dressed they breakfasted together, and Moore then took him up to Military Headquarters and introduced him to Beauregard, Hardee, and other prominent officers, as "his friend Andrews from Flemingsburg." The latter told these officers that he proposed to run articles needed in the South through the Union lines—a very profitable, though a very hazardous business, and one that would be of great benefit to the Southern armies, which were in need of medical supplies, especially quinine, and other articles of small bulk but great intrinsic value. The stringent blockade had made all things not manufactured in the South excessively scarce and costly. They encouraged him, giving passes and all necessary facilities.

Shortly after this, probably the next day, Andrews met Berry face to face on the street. The latter stopped as if thunderstruck, and exclaimed with a great oath, " Well ! what are you doing here, anyway, Andrews?" Andrews greeted him cordially, told him that he was intending to hunt him up as soon as he got settled, and adding the same expression that he had used with Judge Moore, " The only difference between us, Berry, is that you saw how things were moving a little sooner than I did." Just then one of the leading rebel officers in passing by, greeted Andrews familiarly, and Berry could hold out no longer. He said, " Andrews, I was a little hard on you when I thought you wanted to play the spy on me, but you must overlook that. I am glad that you are all right, even if I was mistaken in you; and now if I can do you any favor here in Nashville do not hesitate to call on me." Andrews promised to bear that in mind, and as Berry had engaged in the saddler's business, he brought him through a cargo of buckles and other much-needed articles on his first blockade-running trip. Ever after he had no better friend in all the South than Berry.

Andrews now occupied a position where he could do the Union cause the most essential service. His business as blockade-runner gave him free access to all parts of the territory held by rebel armies, and the purchase of new supplies furnished the opportunity and excuse for frequent visits to the Federal lines. The amount of trading that he thus did was duly reported, along with all other matters, to his employers.

Before the general movement of the Union armies began which resulted in breaking the enemy's lines, and forcing him out of Kentucky, Andrews visited Fort Donelson, gaining admission probably as a bearer of medical supplies, and succeeded in getting a complete account of the Confederate forces there, together with a sketch of their works. In order that his information might be in time for the movement which was then imminent, he rode sixty miles in one night.

Andrews formed a business partnership with Mr. Whiteman, a well-known merchant of Nashville, who supplied him with money needed for his purchases and aided him in the work of distribution. No doubt his own pecuniary profits were considerable, and these he took pleasure in representing to his friends within the Confederate lines as the motive which induced him to run such extraordinary hazards. It is probable, though not certain, that the Federal commander was well pleased to have him thus pay himself, and save the army fund the heavy rewards a daring spy might have claimed.

When Buell's advance division under Mitchel reached Green river, opposite Bowling Green, and began throwing shells across, Andrews was in the town. It is even said that he rendered the Confederate commander a slight service by moving out from the station a train of cars which had been abandoned, just as it was ready to start, by engineer and train hands on account of the severity of the Federal fire; and thus won fervent expression of gratitude, and still more absolute trust. He remained in the captured town and was thus able to give precise information of the condition of the flying foe. Then he passed on ahead of the Union columns, and arrived in Nashville, in time to witness the rebel evacuation of that city, and to greet the Union armies on their entrance.[1]

At some time near midwinter (probably about the first of February), Andrews visited Flemingsburg for the last time. The excitement attending his leaving the place had died away—he had returned once or twice previously—and he was very kindly received, the confidence and good will inspired by former acquaintance having overcome the irritation occasioned by his professed change of views. But there were only three persons with whom he held any confidential communications. One of these was his friend Jackson, whom he trusted utterly, and whose warnings and entreaties to leave a business so fraught with danger, he answered by telling him how fully he was trusted by the rebel authorities, and—what Jackson could fully appreciate—what an excellent chance he now had to make money! He assured his friend, however, that he would not continue in the same line of employment much longer.

Another man he trusted was Mr. J. H. Moore, then quite young. He did not talk to him of money, as his motive in his hazardous career, but of the services he was rendering the country, and so enkindled his imagination, that the young man wished to accompany him, and enter upon the same business. As he was intelligent and cool-headed, Andrews did not at first try to hinder him, though he told him that it was a dreadful life, far more dangerous than that of any soldier. Andrews consented very readily to accompany him at least as far as Cincinnati, and the two left Flemingsburg in company. On the trip down the river Andrews was very sociable, and made no objections to Moore's expressed intention of going on with him to Nashville, so that the latter considered himself as fully embarked on the career of a spy. But when they were shown to their room at Cincinnati at night, and were alone, Andrews laid his hand on him and said,

"Young man, you don't know what you have undertaken. I like your spirit, but you are ten years too young. I am going to put the whole matter before you, and then if you go on, it will be on your own responsibility."

Andrews then drew a picture so frightful that young Moore felt all his ardor ebbing away. The sober thought of the morning completed the work that Andrew's evening words had begun, and Moore informed the latter, apparently much to his satisfaction, that his mind was made up to return home and enter the army in the common way. Then Andrews

---

[1] "Andrews was within the Confederate lines when I advanced upon Nashville. He reported to me there."—From a letter by Gen. D. C. Buell to the Author, Jan. 11, 1887.

made a simple request of him, which throws no small light on the character of this strange man.

"Moore," said he, "Mrs. Bright asked me to send her back a dollar's worth of sugar from Maysville and gave me the money to pay for it; but I had so many things on my mind that I forgot all about it, and have her money yet. Will you get the sugar and take home with you for her?" Moore was accustomed to say long after this that Andrews actually seemed to think more of that old lady's disappointment than of all the dangers he was entering upon!

Andrews had not failed to visit Miss Layton during this final stay in Flemingsburg. He tried to make her think that there was no special risk involved in his present employment; but she was not so easily deceived. Intensely patriotic, and having her full share of the war spirit, she was glad that he was serving the Union cause; but there was so much of the hidden and mysterious element of danger in his present employment that she pleaded with him to give it up. He yielded so far as to promise that one more trip should be his last, a trip from which he was to return a considerable time before their wedding-day. The reader may imagine the tenderness and gloom of the parting between the lovers. Even the prospect of marriage four months distant could hardly have much cheer for the waiting woman, who understood but too well the dangers to which her hero was exposed.

At Louisville, Andrews purchased a lady's trunk, large and elegant, and left it in charge of the landlord of the hotel at which he was accustomed to lodge. From his last, and most pathetic reference to this trunk, it is believed that he meant to bear the wedding-presents to Miss Layton in it. He then completed his cargo of articles for sale in the South and returned to Nashville.

The capture of that city had made no change in the lucrative traffic carried on across the hostile lines by Andrews, except that he now took Nashville as a starting-point, and carried goods to Whiteman who had removed to Chattanooga, and to other persons as far south as Atlanta. He was several times seen in the latter city, and still continued to inspire the Confederate officers with unbounded confidence. On the trip preceding the organization of the first railroad raid he was said ' to have brought a son of the rebel Gen. Cheatham through the lines to leave in the care of some friends at Nashville for education. At the very time of which I am now writing, he rode a horse borrowed from Mr. Whiteman, and had also received 10,000 dollars for the purchase of quinine and other articles that he was to bring through to Chattanooga. This employment did not, however, make him the less watchful of the interest of his real employers—the Federal Generals.

CHAPTER VI.

FIRST LESSONS IN DISGUISE AND DUPLICITY.

NO start on a long journey could be conceived more discouraging than ours. The night was pitchy dark and the rain poured down. The Tennessee mud, which we had pretty fully tested on our army marches, was now almost unfathomable. While we clung to the railroad this difficulty was avoided, but the danger of falling was still greater, and the blinding flashes of lightning gave a very uncertain illumination for avoiding cattle guards, and other hindrances. We hoped to pass Wartrace, our first stage, which was beyond the Federal pickets, before daylight. A walk of a dozen miles on the first night, would put us well on our way, and was possible by clinging to the railroad, though exceedingly difficult.

But an hour or two of toilsome trudging added to the fatigues of the day and the intense nervous strain we had endured ever since our detail, rendered the thought of rest almost irresistibly attractive. We unanimously resolved to find a house and make up for the delay by an early start and a hard day's work on the morrow.

But we did not wish to lie down in the rain, or under the shelter of a tree. There were none of the rude conveniences at hand which were never quite wanting in camps—no tents, water-proof blankets, or means of making a fire. We preferred to find a house, or at least a barn. This seemed very simple, but in practice was not so easy. The country was thinly peopled; and a mist which began to creep along the ground prevented us from seeing a rod before us. We continued on the railroad until we felt almost sure that there was no house that way before reaching Wartrace at the junction; and we did not wish to find lodging in the village. So we turned out on the first road that crossed the railway, and watched more narrowly than ever for the shelter so much desired.

For a time we had no better success than before. But as we plodded along in the mud, we suddenly heard the barking of a dog, which ceased as suddenly as it began. We huddled together and disputed as to which side of the road it was on; then halloed to try to provoke more barking, but in vain. No question could be more puzzling than the simple one of finding the direction of a house indicated by the ear alone. But we heard no further sound, and taking one side of the road at random, we formed a line with long intervals, and passing frequent signals to keep from being separated, we moved forward and swept around in circles through the darkness and rain.

A barn was our first discovery. A little earlier we would have been willing to accept this; but now we were so wet and chilly, and were so anxious to be sure of an early breakfast, that we resolved to look further for the house. Leaving one at the barn to prevent losing it—for even the lightning could scarcely penetrate the fog—we swept around in ever-widening circles till the house was reached, and the dog then was roused and made noise enough. Even with this aid it took some time to waken the old farmer, who on seeing us was greatly alarmed, and looked as if he would refuse us entrance if he dared. But the request of four strong men, in such a storm and at such an hour, had a good deal of persuasive force.

The house, as we saw in the morning, was a rude log structure, consisting of two pens built twelve or fifteen feet apart, and connected by a roof, thus making an open porch between them—a style of house very common in the South at that time. When admitted, the smouldering fire in the large open fire-place was raked into activity, and fresh wood was thrown on, so that as we surrounded the hearth, there was soon abundant light without the use of candle or lamp, and a most grateful heat. Then our host, who insisted on providing us something to eat, though we told him that we needed nothing before morning—that all we now cared for was to get dried and go to bed,—questioned us as to who we were, and our motive in travelling so late in such a dreadful night.

It was a very good opportunity to commence our drill in deception. If detected no special harm could follow, for we were stronger than our questioner, and being still near the Union pickets, he could not get help against us; while if we succeeded, we would be more confident when our life was at stake. We said we were Kentuckians from Fleming Co., and that we were travelling in the night to evade the Union pickets, and would not have stopped till we were in safety, if it had not been for the rain. He inquired our reason for leaving home in these troubled times and coming South. We replied that we were disgusted with the tyranny of the Lincoln Government, and meant to fight in the Southern army.

His reply gave us great pleasure, which, however, we did not express:

"You might as well save your trouble, for the whole South will soon be as much under Lincoln as Kentucky is."

This avowal of Union sentiment was the more grateful, as we, like most of the Northern soldiers, doubted the sincerity of Union people in the South. He advised us to go back home and try to content ourselves there, for our errand was bootless, and that the rebellion would soon be put down as it deserved to be.

5

We assured him that we would never submit; that we would die first. He laughed and said it was easy to talk, but that time would show.

While we partook of a plain but good meal, and still more, when we again gathered for a short time around the fire preparatory to retiring, we continued to argue the great questions of the day. It was a novel experience, trying to maintain that the rebels were right in seceding and our own people all wrong; but we did the best we could. I noticed that George D. Wilson, who spoke with great ease and force, seemed dissatisfied whenever Campbell or Shadrack took part in the conversation, as they did several times, expressing very radical Southern views with great emphasis of language; and that he would repeatedly interrupt them in the midst of their statements, and called on me for something in another direction. When our host and his wife, who also had been aroused to minister to us, had retired, Shadrack asked him the reason of this, which had been quite marked. Wilson replied that they were overdoing their parts,—as he expressed it, " making fools of themselves by being better rebels than the rebels themselves;" then he complimented me for being always moderate in statement and telling the necessary stories in such a way that they were sure to be believed. The truth was that the whole business of such false representations was distasteful, and I did just as little of it as would suffice to make the stories plausible. I supposed the very frank words of Wilson would be resented, and tried to think how best the disagreement could be smoothed down; but to my surprise, both Campbell and Shadrack admitted the justice of the criticism, and declared that they would hereafter, in the presence of strangers, limit themselves to endorsing all that Wilson and I should say. This they faithfully did through the whole of their journey, always waiting for one of us to take the lead in conversation.

As we sit around the fire slowly undressing for bed, and enjoying the languor caused by the supper and the grateful heat, it may be a good time to sketch these companions with whom I was so long and intimately associated. There is a deep sorrow in the task, for not one of them emerged from the gloom and darkness into which we were entering.

George D. Wilson was the most remarkable man of all who enlisted with Andrews. He was not highly educated, and had spent many years as an itinerant journeyman shoemaker. He was 32—nearly ten years older than the others, which increased his ascendancy over us. He had travelled and observed much and forgotten nothing. In vigor and force of language I never knew a man who surpassed him. He delighted in argument on any topic—social, political, or religious—and was an adversary not to be despised. In the use of scathing and bitter language, in hard, positive, unyielding dogmatism, in the power to bury an opponent under a flood of exhaustless abuse, he excelled. In coolness and bravery, in natural shrewdness and quickness of intellect, he was fully equal to Andrews; no danger could frighten him. His resources always rose with the demand, and on one memorable occasion he was carried to the very summit of moral heroism, and in the whole war no death was more sublime than his. Our friendship, which began on this first night, increased to the end, though we often engaged in heated discussion.

Wilson was tall and spare, with high cheek-bones, overhanging brows, sharp gray eyes, thin brownish hair, and long thin whiskers. The accompanying photograph was taken ten years earlier.

Perry G. Shadrack was about twenty-two, and came from Pennsylvania to Knoxville, Ohio, when the 2nd Ohio Regiment was being reorganized for the three years' service. He was not large, but plump and solidly built, merry and reckless, with an inexhaustible store of good nature. His temper was quick, but he was very forgiving and ready to sacrifice anything for a friend. His wit was frequently the life of the whole party, and his merry blue eye sparkled with mischief on the slightest provocation.

As far as Wilson excelled all the other members of the party in intellectual strength and acuteness, so did William Campbell in the more tangible quality of physical strength. His muscular feats were often marvellous. He weighed two hundred and twenty pounds, was of fine build, and with his great weight was as agile as a circus actor. Danger seemed to

GEORGE D. WILSON. From a photograph ten years before the war.

WILLIAM CAMPBELL. From a war-time photograph

have an innate attraction to him, and the thought of death but little terror. He was not in the least disposed to be quarrelsome, and often reproved wranglers. But it was said that the use of stimulants, which he did not touch so far as I know on this expedition, very considerably changed his nature. He was a native of Salineville, Ohio, and had led an irregular life, being in Louisville when the Union army passed through. He had not formally enlisted as a soldier, but was on a visit to Shadrack when the latter was selected for the Andrews expedition, and had requested permission to go along. Being fully trusted, and well qualified for dangerous work, his request was readily granted. As was natural, he and Shadrack were inseparable friends.

At length our luxurious chat was over and we lay down for rest, two in a bed. I think Wilson slept at once, but I lay awake for some time, watching the fitful light of the declining fire as it flickered over the bare rafters and rough side logs of the room. I thought of many things which this eventful day had brought forth; and there was one line of thought which it may be as well to record with some care—more than I gave it then,—for it is sure to arise in the mind of many readers.

How could we reconcile to our consciences the falsehood involved in the very nature of our expedition? This question becomes more urgent as the years go by, and the stormy passions of the war are quieted; and especially as good feeling between the sections, and charity for former opponents, is restored. No such question was asked during the war; but it has been a thousand times since. It does not justify our leader or ourselves to say that we were placed in such a position that the preservation of life required us to deceive the enemy; for the real question is as to the moral right of putting ourselves in that position. The historian is not bound to justify what he records even of his own actions. But it is required that these actions should be fairly weighed.

Let it never be forgotten that we did not look on ourselves as upon the same plane with the enemy. We were not fighting against a nation armed with all the rights of independence. In our view our opponents were nothing but rebels, and we regarded rebellion itself as a crime which forfeited all rights and was justly punishable with death. We did not think that men who had associated together against our government had acquired any more rights by that association than a band of pirates or murderers. To kill them was a public duty—the very purpose for which we had left our homes. To defeat them in their criminal design by falsehood, seemed just like throwing a murderer off the track of his intended victim by strategy. In other words, we looked upon the rebels as out of the pale of all law by their own act. Men who would have shrunk with horror from the thought of deceiving any one in private life, found nothing but pleasure in outwitting the destroyers of the nation. In

Flemingsburg, among the old neighbors of Andrews, no expression was more common than, " He was a *true* man. You could depend upon *every word he said.*" But to deceive the enemy was accounted a virtue, not a crime.

The manner in which the enemy carried on the war intensified this feeling. Their soldiers were imperfectly uniformed at best. They had encouraged guerilla warfare, and on more than one occasion Federal uniform had been used to get within our own lines and work injury, as in the case of the capture of Gallatin by Morgan but a few weeks before this date.[1] Their citizens were often found living in apparent innocence within our lines, and yet sending complete information of all Union movements southward, and taking the oath of allegiance only to break it—often seizing their rifles and acting as soldiers against us without any organization. It would have been as easy to carry on the war, while keeping the law of perfect truth, as to observe the law of love, which forbids all injury ! Our party, in their disguise, went forth to play at the enemy's own game, knowing full well its hazardous character, but feeling that the enemy would have no reason to complain.

But there is a broader aspect of the question. Every war between the most civilized of powers must involve an element of deceit and fraud almost as prominent as that of violence and destruction. The moral law and the golden rule are set aside in both instances; and therefore some Christian denominations consistently take the ground that war can never be lawful. With them we have no controversy, for the subject is too large. Writing false dispatches, making movements which have no purpose but to deceive, using flags of truce for pretense or delay, sending out trusted soldiers as deserters to carry false intelligence, and employing every other possible *ruse* to mislead an adversary—all these things will be done by the most honorable general, and if they succeed are considered worthy of especial commendation ! A general will urge a soldier by large rewards to become a spy, and put lies in his mouth. On the well-accepted principle, that "the receiver is as bad as the thief," the general cannot be better than the spy ! If the laws of war remove all odium and danger from the general to fix it upon the spy, it should be remembered that the generals have made the laws of war, and naturally favor themselves; this, however, cannot change the abstract right in the case.

The further consideration should be kept in view that, with the exception of two or three, we were very young, not members of any church, and that we held ourselves amenable only to the common laws of army morality, which, so far as the enemy was concerned, were not very stringent—the common sentiment being that a rebel had no right to anything —not even to the truth? Our commanders, from General Buell down,

[1] War Records, Series I., Vol X., Part I., page 31.

had sanctioned our expedition, knowing all that it implied, and the convenient army morality which gives the blame as well as the praise to the
officers was as applicable on this as on any other occasion. We had been
directed to put off our uniform, and to make representations corresponding to our changed costume. We were also authorized to enlist in the
rebel army—a measure which would have required the taking of an oath
of allegiance to the Southern Confederacy—a most distasteful form of
deceit, and one that would fall short of perjury only because no rebel
could legally administer an oath to United States soldiers or citizens. We
felt that we were serving our country in the way that the country itself had
pointed out, and that if there was any wrong, the country, which was making the war, was responsible. We were absolutely true to each other, and
to our rightful allegiance in the most trying emergencies, but we drew a
decisive line in our dealings with the enemy—telling them the truth when
there was no motive to do otherwise ! Yet I never liked to deceive them,
though constant use made it less grating, after a time; and there even
came to be a certain pride of dexterity in doing it well, which Wilson stimulated to the utmost by putting me forward on all possible occasions as
spokesman, first of our own little squad, and afterward, when we were
united, of the whole party. On such occasions his praise for being able
to say plausibly and exactly what we had agreed upon beforehand was not
altogether unrelished.

We passed the night in quiet and safety. Our host had been pledged
by us not to inform the Federal pickets of our presence—though it would
have been an enjoyable practical joke under other circumstances to have
let him bring in a squad and have us arrested—and after breakfast we
went on our way. For a time the sky was clear but it soon clouded, and
we were compelled to suffer the inevitable drenching which befell us every
day on this weary journey. We reached Wartrace in the midst of a pelting storm. At first we intended to go around the town, as it was the last
station on the Union picket line. But it was raining so hard that we
thought we might manage to slip along the street unobserved. On making the attempt, however, we found that Mitchel's soldiers were too vigilant for that, and we were promptly halted. For a time we tried to per-
sonate the innocent Southern citizen, but were compelled to wait under a
sheltering porch until a messenger had ridden to brigade headquarters and
brought an order for our release.

Then we travelled onward, wading swollen creeks and plodding through
the mud as fast as we could We were crossing what might be called the
neutral zone between our lines and those of the enemy. The next stage
ahead was the town of Manchester, at which point we were really to enter
the enemy's country. Early in the afternoon we reached Duck River
opposite the town, and as the river was at flood height, we crossed by a

novel kind of ferry—being taken up one at a time by a horseman, and thus carried through the torrent.

We found the population of the town in a wild ferment. Some of the citizens had reported that an approaching band of Yankee cavalry were even now visible from the public square. We repaired thither with all speed to witness the novel spectacle of the entrance of National troops into a hostile town from the Southern point of view. Mingled were the emotions expressed, fear and hatred being the most prominent, but some people looked not unpleased. Soon the terrible band loomed up over the hill which bounded the view, when lo! the dreaded enemies were seen to be only a party of negroes who had been working in the coal-mines a little further up the mountains. Some of Mitchel's cavalry had made a raid eastward to divert attention from their real movement southward, and having destroyed the works, these contrabands were run off here for safe keeping. The feeling of the town's people may be better imagined than described as they dispersed, with curses on the whole African race.

Here we saw several others of our own party. There was no personal acquaintance between us previous to being detailed—I could not have given the name of a single member of the whole band aside from our cluster, and the one occasion when we had been together the previous evening had not been favorable for intimacy;—but there was something in the manner of each by which it was easy to recognize comrades. Reddick and Wollam had lodged for the night with a Confederate family who were afraid at first to harbor them, for fear of Yankee vengeance should they be discovered; but being persuaded to take the risk, they not only gave supper and breakfast, but furnished a guide who piloted them by a lonely path over fields and woods, around both Union and Rebel pickets, so that they escaped all molestation, crossed swollen creeks on fallen logs, and Duck river in a wagon, and finally reached Manchester in advance of all the others. We greeted several of our squads in Manchester, but did not long remain in company. Andrews also passed on horseback, but as all was well, we did not speak to him.

As we pressed on in the dreary afternoon we realized fully that we were in the enemy's country. We obtained the names of the most prominent Secessionists along the route, and from this time had nothing more to do with Unionists. It is difficult for the reader of to-day, when our country is truly united, to form any idea of the mysterious horror which clung about the South at the period of the opening of the war. Slavery had been separating the two sections more and more for a generation preceding that event. Terrible tales of outrage upon suspected abolitionists were freely told. If a Northern man travelled in the South, he was obliged to carefully conceal his sentiments on the subject of slavery or run the risk of being dragged from his bed at night and whipped, tarred and feathered, or even hung. These tales may have been exaggerated, but they were believed, and had some foundation. Three years before I had been turned back from an intended trip as a teacher to Kentucky, by the fear of violence. Now the smothered flame had broken out openly; the sword had been unsheathed; and we were in the midst of deadly foes with no protection but the flimsy veil of falsehood. It is not easy to describe the half fascinating, half terrible sense of danger with which we passed from house to house and from village to village where we would have been torn to pieces in a moment if our true character had been suspected. But being entertained by those whose Confederate loyalty was above suspicion, and always inquiring for others of a similar character, it was taken for granted that we were like them, and few questions asked. We paid our way wherever money would be accepted; more frequently. it was refused; the people declaring, " It is a privilege to do something for the gallant Kentuckians on their way to fight for the liberty of the South."

CHAPTER VII.

THE HEART OF THE CONFEDERACY REACHED.

THERE are but few incidents of this downward journey upon which it is worth while to linger. That night we were still some miles from Hillsboro, having been greatly impeded by the muddy roads and swollen streams.

The gentleman with whom we lodged this (Tuesday) evening was a slave-hunter, and hunted negroes with bloodhounds for money, as we heard for the first time from the lips of one who practised it. Our host said he had seen some one dodging around the back of his plantation just as it was getting dark, and that very early in the morning he would take his hounds and hunt him up; if it proved to be a negro he would get the reward always allowed for a fugitive slave. He said that he had caught a great number in that way, and regarded them as perfectly fair game, and the business as highly profitable. The idea that there was anything cruel or dishonorable about it had not occurred to him.

We had to agree with all he said; but I well remember that the idea of hunting human beings with bloodhounds for profit caused a thrill of horror and detestation. Not long after we found that these hounds were equally serviceable for other human game !

The next morning we continued our journey, and after walking about an hour, found a man who agreed for an exorbitant price, and for the good of the Confederacy, to give us a ride in his wagon for several miles. We were anxious to avail ourselves of every help, for it was now Wednesday, and the very next day we ought to reach Marietta. It was clear that we had no time to lose. But this conveyance was a great aid, and we trotted briskly along, becoming very lively. Several others whom we overtook were invited to share our good fortune, and jokes and laughter rang out in merry wise. A listener could scarcely have believed that we were leaving our homes as exiles to engage in war, as we claimed; still less that we had actually entered upon an enterprise as desperate as any forlorn hope that ever mounted a breach.

Soon we came in sight of the Cumberland mountains, and to me no scenery ever appeared more beautiful. For a short time the rain had ceased to fall and the air was clear. The mountains rose before us as a mighty rampart of freshest green, and around their tops, just high enough to veil their loftiest summits, clung a soft shadowy mist, which, gradually descending lower, shrouded one after another of the spurs and high mountain valleys from view. The beautiful scene did not long continue: soon the mist thickened into a cloud, and again the interminable rain began to fall. And as if to add to our discomfort, the driver of our wagon, about the same time, declared that he could go no further, and we were obliged again to plod along on foot.

At noon we stopped for dinner at a miserable hut close to the road. There was nothing inviting about the place, but no other house was in sight. The owner belonged to the class of "poor whites," whose condition in the old slavery days was little if any better than that of the slaves. They owned no property of their own, seemed to be devoid of any ambition to better their condition in life, and eked out a scanty subsistence by hunting or fishing, only working for a day or two occasionally when driven to it by hunger. The terms "sand-hiller," "clay-eater," or "poor white trash," conveyed a terrible reproach, for even the negroes looked down upon them. Of course our entertainment was of the plainest, but we ate our half-ground and half-baked corn bread and strong pork cheerfully, paid a round price, and passed on our way.

Soon after dinner we fell in with Mark Wood, who had possessed himself of a bottle of apple-brandy and imbibed too freely. He was talkative, and in no fit state for meeting strangers. But we walked him along rapidly and gave no one a chance to say a word to him until the fumes had passed away, when he was so much frightened by his imprudence and its dangerous possibilities as to have no inclination to repeat the offense. This was the only instance of that kind during the whole trip; and there was but little even of moderate drinking in the company.

We had now reached the foot of the Cumberland mountains and addressed ourselves to the task of climbing the steep slope. While going up the first long hill we overtook a Confederate soldier of the Eastern army who was at home on a furlough. He was quite a veteran, having been in a number of battles, and among them the first Bull Run, which he described very minutely. Little did he think that I too had been there, as we laughed at the wild panic of the Yankees. He expressed great delight to see so many Kentuckians coming out on "the right side," and contrasted our noble conduct with that of some people in his own neighborhood who still sympathized with "the Abolitionists." When we parted he grasped my hands with tears in his eyes, and said that he hoped the time would soon come when we would be comrades, fighting side by side in one glorious cause. My heart revolted from the hypocrisy I was compelled to use, but having begun there was no possibility of turning back.

On we clambered up the mountain till the top was reached; then across the summit, which was a tolerably level table-land about six miles in breadth; then down again over steep rocks, yawning chasms, and great gullies. This rough jaunt led us into Battle Creek Valley, which is delightful and picturesque, being hemmed in by projecting ridges of lofty mountains.

While here they told me how this valley obtained its name, which is certainly a very romantic legend, and probably true.

In early times there was a war among the Indians. One tribe made a plundering expedition into the country of another, and after securing their booty retreated. Of course they were pursued, and in their flight, were traced to this valley. There the pursuers believed them to be concealed, and, to make their capture sure, divided their own force into two bands, each one taking an opposite side of the valley. It was early in the morning, and as they wended their way cautiously along, the mountain mist came down (just as I had seen it do that morning), and enveloped each party in its folds. Determined not to be foiled, the pursuing bands marched on, and meeting at the head of the valley, each supposed the other to be the enemy, and at once attacked with great vigor. Not till nearly all their number had fallen did the survivors discover their mistake, and then they slowly and sorrowfully returned to their wigwams. The plunderers who had listened to the conflict in safety, being further up the mountains, were thus left at liberty to carry home their booty in triumph.

But we had no leisure for legendary tales.

The sun set behind the heavy masses of the mountain, and we again sought a resting-place for the night. We soon found the house of a rabid secessionist whom our soldier friend on the mountain had recommended to us. He received us with open arms, and shared with us the best his house afforded, giving us his only bedroom and sleeping with his family in the living-room or kitchen. This reception of open-handed hospitality, which we knew was given by those whose dearest hopes we were laboring to overthrow, was even more painful than the plentiful falsehoods to which we were compelled. We spent the evening in denouncing the Abolitionists, which term was used indiscriminately, like that of "Yankees," to designate all who did not advocate the acknowledgement of the Southern Confederacy. Practice had rendered it nearly as easy for us to talk on this side of the question as on the other; and a little observation had shown us just how the Confederates liked to hear us talk.

There was one truth we told, however, that made more impression than all our falsehoods. In the character of Kentuckians we informed our host that we were, so far as the Yankee sympathisers in our state could accomplish it, forever exiled from our homes by the expatriation law recently passed by the Kentucky legislature. This act made the reasonable provision that any person going south to fight in the army of the rebels should lose all rights of citizenship in the state which he thus forsook. The old man thought this was unparalleled oppression; and in the morning before we were out of bed, came into the room and desired that I should write that law down that he might show his Union neighbors what the Yankees would do when they had the sway. As soon as I was up, I wrote it, and we all afterward signed our names to it. No doubt that document was the theme of many angry discussions in the houses among the spurs of the Cumberland during the months of rebel domination that followed. So thoroughly did we deceive the old man, that when two days after, the railroad capture fell upon the astonished Confederates like a clap of thunder out of a clear sky, he would not believe that his guests were part of the men engaged in it. One of his Union neighbors afterward told us that to the last our host maintained that we were true and loyal Southerners. I would have greatly enjoyed visiting him in my true character, but never have had the opportunity.

It was now Thursday morning. That evening we should have been in Marietta, and here we still were in the spurs of the Cumberland, more than a hard day's journey from Chattanooga. But a momentous decision, full of disaster, had been arrived at the afternoon before. In our company we had discussed the slowness of our progress and had resolved to get supper, and then making the best excuse we could, to set out for a night journey, getting so near Chattanooga, or a station west of it, that we could reach there in time for the down train for Marietta. The prospect of such a night journey was not pleasant—far less than that of sitting by the fireside, discussing politics, and sleeping on a soft bed; but it was not an extraordinary hardship, and we would have endured it without complaint. Yet it was a considerable relief when we learned, from a squad that came along just after supper, that Andrews had postponed the enterprise one day on account of the unexpected hindrances of the weather. The result of this in increasing the difficulty of the final achievement fourfold, could not have been certainly foreseen; but delays are always dangerous. Why did Andrews take this risk? The men could have easily been urged on to their work; and as will be seen hereafter, those who did not get the word of postponement were on time at the Marietta rendezvous.

The answer is easy. Andrews knew Gen. Buell better than he knew Mitchel. With probably any other officer in the war except Stonewall Jackson on one side, and Mitchel on the other, it would have been perfectly safe to calculate that such a series of down-pours as we had experienced ever since we had left camp, would have caused more than one day's delay in three days' march of an army; and if Mitchel was delayed, it would be possibly better for us not to be too early on the ground. But Mitchel, with a fertility of resources almost incredible, and with inflexible determination, had pressed on; and neither drenching rain, fathomless mud, nor bridges swept away, delayed him an hour! The calculation of Andrews, which would have been right ninety-nine times, failed in the hundredth !

About noon on Thursday we came to the town of Jasper, and walked quietly up the street to the principal grocery of the place, where we rested a while and talked with the idlers gathered around on the state of the country. We told them that Kentucky was just ready to rise and shake off her Yankee chains. They gave us ready credence, and in turn communicated some wonderful items of news.

Having been now three days outside of our own lines we were extremely anxious for any kind of intelligence. Nothing could be heard in regard to Mitchel, which was a little disappointing, as we thought that his movement southward would by this time have caused some excitement; but it was so silent, and all communication with his columns was so completely cut off, that until the blow fell the next day, scarcely anything was known of him.

But we heard the first indistinct rumor of the battle of Shiloh or Pittsburg Landing. Of course it was believed to be a great rebel victory, in which thousands of Yankees had been killed and innumerable prisoners taken, as well as scores of cannon. It was the impression that the armies of Grant and Buell were totally destroyed. This did not cause us any great degree of uneasiness, for we placed a low estimate on the accuracy of Southern news—being in this almost as extreme as some of the negroes we afterward knew in Atlanta, who made it their rule to believe the exact opposite of whatever their masters asserted. One countryman gravely assured me that five hundred gunboats had been sunk ! I told him that I did not think the Yankees had so many as that, but was not able to shake his faith.

From Jasper we journeyed directly to the banks of the Tennessee River with the intention of crossing in the morning and taking passage on the Memphis and Charleston Railroad. We were recomended for the night to a kind of rude country hotel well known in the neighborhood as "Widow Hall's." The entertainment here was excellent; and as we believed the harder part of our travel to be over, we were in high spirits. Andrews and several others joined us, and for the first time, as we spent a social evening together, we had a chance to become in some degree acquainted. The large guest-chamber with its great roaring fire, and two beds in the corners opposite, was exceedingly comfortable; and after a smoking hot supper, we gathered around the open fire and began to talk. The family were with us, and we personated strangers who had met for the first time. Many stories were told of our home life in Kentucky, and the different parts of Fleming county from which we came, and in these, imagination naturally played a larger part than memory. Andrews was, according to his wont, rather silent and reflective, but appeared greatly to enjoy the conversation of others. Especially did he show that he appreciated the wit of Shadrack (which seemed to pour forth in an unending stream) remarking: "That man never opens his mouth but he says something!" Wilson as usual gave us copious information on every subject that was broached. He let it be known that he had travelled widely, and dominated the conversation in all serious and political matters—for we did not shun the war politics of the day—as much as Shadrack did in sport and humor. Dorsey, who like myself had been a school-teacher, formed the idea that it might be better for him to appear to know but little, and carried out this notion during all the expedition as well as this evening. He was amused and felt complimented when told by a member of the squad who had followed him, that some of the citizens of Jasper had referred to his party as a "lot of country fellows, who scarcely knew enough to come in when it rained." Songs were sung and the nine or ten of us who were together began to be acquainted—not the less so that each knew all the others to be acting fictitious parts. There was no drinking; for that would have been an element of danger, destroying all hilarity. Just such another evening I never spent. The absence of all restraint in speaking of our former life; the equal latitude with which we discussed our plans and hopes; the presence of some admiring auditors who either believed all we said, or, thinking that we were only using a traveller's privilege, were too courteous to contradict us; and another portion of the audience who knew we were acting, and could appreciate and imitate from that standpoint;—all combined to make the whole like a stimulating and highly interesting game. Our gaieties were prolonged to a very late hour, but no opportunity came to repeat them; before another day closed we were in very different scenes.

Several of our number were in advance, and four spent the same evening south of the Tennessee—two who had not been notified of the change in our plans even reaching Marietta. These latter were Porter and Hawkins: and as they found that the party were not on hand to start Friday morning, they remained in their hotel waiting for us.

Reddick and Wollam lodged on Wednesday night with an exceedingly hospitable and influential rebel who lived in a large double house. They won his entire confidence even too completely, for he was very anxious that they should enlist in a cavalry company his son was commissioned to raise for Confederate service. Fortunately they had told him a slight modification of the common Kentucky story—claiming that they were already enlisted in the regiment of a Kentucky Colonel Williams, and were now on their way to join him. Their host offered to send at once to Colonel Williams and get them excused, entertaining them in the meanwhile at his house, if they would only consent to join his son; adding that some other Kentuckians had just passed by who had left their homes in the same way, and he had sent on to Jasper to persuade them also to return. They still refused to yield. He then told them that it was two days' hard walk from where they were to Chattanooga, but that he could put them on a plan by which they could reach there that same evening.

"It is against the law," he added, "for any man in this country to ferry, except the regular ferryman, but to help you along I will take the responsibility upon myself. My sons are ferrying bacon across the river for the army; you go down and tell them that I said for them to put you over. It will then be only half a mile to Shell-Mound station, and you will have time enough to catch the twelve o'clock train for Chattanooga."

He then gave them a guide to the ferry, and his boys obediently set them over and directed them to the station. They had to wait long for the arrival of the train, which was several hours late. When it came they found a whole regiment of rebel soldiers on board, who had been sent on to Corinth to reinforce Beauregard, but for some reason

WILLIAM REDDICK. From a war-time photograph.

that they did not understand had been turned back. Reddick and Wollam did not fail to get on board and were kindly received, but managed to say little till they reached Chattanooga. They were just in time for the down train for Marietta, but as they had learned that the attempt was to be laid over for another day, they preferred waiting till morning.

At the Crutchfield House to which they repaired, there was so great a crowd that the only bed they could get was in the same room with two very sick Confederate soldiers. These had some kind of fever and kept calling for water continually. The bell-knob had a card on it, saying that twenty-five cents would be charged for every time that bell was rung during the night. Our comrades, however, had money, and not only remained awake most of the night ministering to their foes, but invested a good many quarters in their behalf. The sick soldiers' gratitude and promises to return the favor if ever in their power, were, under the circumstances, very touching.

The next day the raiders strolled for a time about the town, being in both the commissary and ordinance departments. This was scarcely prudent, but they seem to have acquired perfect confidence. They also witnessed the burial of some officers with the honors of war, who had been killed at Pittsburg Landing. Happening to go to a photographic gallery they were seized with the desire to have their pictures taken. The artist asked them to wait awhile, as he was engaged in whittling out a frame with his penknife from a cigar box, explaining that since he had been "cut off from communication with Yankeedom," he had been compelled to make everything for himself. The frames when completed were very handsome, and the pictures that were then taken for them were prized the more because of this home-made setting; but they were afterwards presented to the wife of Swims, the Chattanooga jailer.

In the evening they were at the depot when the remainder of the party came in. Andrews gave them a warm grasp of the hand, and assured them that all was right.

There were three other men who reached Chattanooga in advance of the larger party,—an important contingent, for without them the whole expedition would have been helpless. Our engineers, Brown and Knight, were with this squad. They met at the night consultation, and from that time throughout the expedition were inseparable. They had the same experience with the rain that has already been narrated, sleeping that night in a barn. At day-break they got up and walked five or six miles before breakfast, and then stopped with a strong "Southern rights man," which was just the kind of men they wanted to find. They told him that they were from Fleming Co., Kentucky, and were going to Chattanooga to see if they could find any regiment from their own State in which to enlist. Knight continues:

"We asked him for the names of some good men we could trust along the line of our journey, so that we could keep away from all Yankee sympathizers. These he gladly gave, and guided us past the last Federal outpost. Soon after he left us, we came to a rebel picket in the bend of the road, and were upon them before we saw them. They were armed with double-barrelled shot guns and were not slow in bringing them to bear on us, and demanding that we should give an account of ourselves or be blown through. The usual story proving satisfactory, they dropped the guns and presented a quart bottle, which, being less formidable, we did not refuse.

"That day we crossed the Cumberland Mountains and took dinner just beyond. An old lady and two daughters were the only persons in the house, and there was a small Union flag over the mantel piece. We told them that they were displaying the wrong kind of a banner, but they stood up for the old 'Stars and Stripes' royally, and it went sorely against the grain for us to disagree with them. That was the last Union flag we saw for eight months.

"That night we lodged with a Colonel who was a violent rebel, and gave us a terrible downsetting on the sly. It would not have been safe for him to say what he did against Union soldiers if we had been sailing under our own colors ; but now he had the advantage, and we took it meekly. He sent us for the next night's lodging to a Major, who was more quiet and seemed to be reading our thoughts in secret. He could not have succeeded very well, for he gave us a letter to a squire, (all the people in the South had some kind of title!) and directed us to reach his house by the trail that led over the mountain spurs, through a most desolate part of the country. It was rough travelling, and we did not see a man or a house all day ; but it brought us out all right in the evening, and only five or six miles from Chattanooga. The squire proved to be a good entertainer, which after our dinnerless jaunt was well appreciated ; he seemed also well posted in army movements, telling us how the Yankees were moving on Huntsville, which was no news to us, though we had not before heard it intimated by any of the citizens. He said that the Southern army was moving back only to get the Yanks in a trap, which they had already set. He told us that we might see the last one of them in irons in Chattanooga before many days. We did not believe him then, but we certainly did see some of them in that condition. The squire was kind enough to give us a letter in the morning, which he said would pass us over the river, and introduce us to a Colonel in Chattanooga. We did not need to use it for the first, as there were no guards set that day ; and we had no inclination to use it for an introduction.

"During the day we strolled around the streets of the village—for at that time Chattanooga was nothing more—and saw whatever we thought worth looking at. Toward evening we saw many of our party coming in. I did not know them personally, as I had not yet had an opportunity of becoming acquainted, but I could have picked them out from the whole Southern army. We were at the depot in time to take the evening train, with the others, for Marietta."

Thus there were on Friday morning at least seven of our party south of the Tennessee, and it was necessary that we should overtake them before they arrived at Marietta, or join them this evening at that point. Notwithstanding our protracted gaiety at "the widow Hall's," we were up by daybreak, and Andrews mounted and rode back into the country without waiting for breakfast. It was but a few hundred yards to a place where an old flat-boat was kept, by which we intended to cross the Tennessee. So we waited for our breakfast and then went leisurely down to the river bank, for we had ample time to reach the station on the other side before the noon train; not to be in a hurry was a great luxury. But while the owner of the flat-boat was bailing it out, a man rode up with a stringent order to permit no person to cross, on any pretense, for three days. The only explanation given was the rumor that the Yankees were coming. We had no fear of the Yankees, and wanted to get over very badly; but our urgency could make no impression. We might have forced a passage, but we wished to abstain from violence. To build a raft and get over was possible, but if seen would have been reported at Chattanooga, and might have led to a searching investigation. The best thing we could think of was to go up the river to Chattanooga, and try to get over directly at the town. This involved a laborious and hurried half day's journey mostly over mountain paths and rugged valleys. The river is very crooked and as we journeyed we were sometimes on its banks, and again,

6

miles distant.   At length we turned into a road that was more travelled, and which led down a valley directly to the river bank nearly opposite Chattanooga.   Travellers were now more frequent and from them we learned many items of news.   The accounts of the battle of Pittsburg Landing were not quite so rose-colored as the day before, but still they told of a wonderful victory, though not won without considerable loss.

One item of news from the East was still more interesting: it was that the Merrimac had steamed out, and after engaging the Monitor for some time without decisive result, had thrown her grappling irons on the latter, and towed her ashore, where of course she fell an easy prey.   Our infor-

Preparing to Cross the Tennessee River.

mant claimed that now, as the Confederates had the two best gunboats in the world, they would be able to raise the blockade without difficulty, and burn the Northern cities.   I need not say that the histories of the war have all neglected to record this wonderful capture.   From this time forward we heard of almost continuous Confederate victories, although it was not infrequently stated, after a battle, that their own forces had fallen back for strategic reasons.

On reaching the river shortly after noon we saw a large number of persons, several of them belonging to our own party.   The ferryman was also here with his craft, a little frail boat driven by horses, but such a fearful wind storm was raging that he feared to attempt the passage.   We waited as patiently as possible.   Others of our number came up and the danger of detection correspondingly increased. We urged the boatman to try the passage, and not to be afraid of a little breeze, but he was unwilling to do this until the wind moderated, as he said it was sure to do with the setting of the sun. This was well enough, had it not been that our train on the other side—the last which would be of any service to us —was to start at five o'clock in the afternoon, and it was drawing uncomfortably near that time. At length, when we could not persuade him, we tried another plan—that of making him angry. We talked *at* him, speaking of his cowardice, which we contrasted with the bravery of Kentuckians or even of Ohio Yankees ! Soon he grew very indignant, and told us that if we would only show some of our skill by helping him to get out in the stream, he would put us across or drown us in the attempt; and for his part he did not care much which. We promptly accepted the not very gracious invitation, and when all who were willing to risk the passage were on board, we took hold with pushing poles, and also pulling on overhanging limbs,—for the river was flooded, and the current very swift, against which the beating of the wind raised sharp and ugly waves,—we succeeded in getting a good start up the river and after some tossing landed safely on the other side.

We cared very little for the danger by water, because we feared another more formidable on the land. We had every reason to anticipate a strict guard on the Chattanooga shore, and supposed that at the rebel headquarters they would be less easy to satisfy than the citizens in the mountains. No pass had been asked for by the ferryman; but this was natural enough if the guard was on the other side.

We looked with keenest interest across the turbid water toward the town. We saw no sign of fortifications, and at this date there were none. What then was there in its situation which made this place of so great importance, and caused rivers of blood afterwards to be shed for its possession ?

Chattanooga is not on a high mountain—indeed much of it lies so low as to be easily flooded by the Tennessee—but there are sufficiently steep hills of moderate height immediately around it to render fortification easy. At a distance it is surrounded by mighty mountain ramparts, from which it commands the egress in almost every direction. Many valleys converge at or near Chattanooga—the upper and lower Tennessee, North and South Lookout, Chickamauga, and Sequatchie. The sides of these are all very steep, and some absolutely impassable to an army. This is the reason that all common roads and railroads through a large section of the country converge at Chattanooga. Now it is a military maxim that an army is never so easily attacked and destroyed, or, on the other side, is never so weak in aggression, as when emerging from a mountain defile. Chattanooga, if adequately fortified and held by a moderate force, was well-nigh impregnable, and dominated all the surrounding country; it furnished the best starting-point for a hostile movement in any direction. When to this is added the fact that it was the natural centre of a great mountain district passionately loyal to the old flag, its importance to the Federals early in the war is seen to be immeasurable. Had Mitchel reached Chattanooga a day or two after we stood gazing at the town, *and with the bridges south of it burned*, two or three things would have followed with

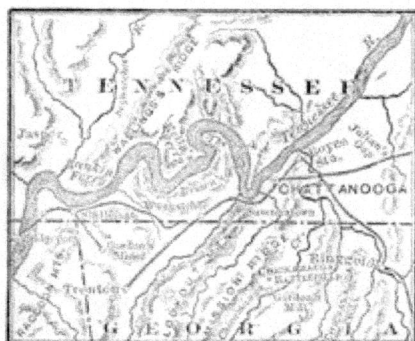
Surroundings of Chattanooga.

almost mathematical certainty. He would have skillfully, and quickly fortified the place; down every mountain defile would have streamed loyal recruits to his banner; and East Tennessee, already in insurrection, would have driven out Gen. E. Kirby Smith with his small army. No doubt the enemy would have put forth tremendous efforts to regain a place of so much importance. But Mitchel would have had railroad communications open with Nashville and the North, while the enemy could not have come nearer than fifty miles by rail. Mitchel's army would have been at once increased; and he, not the languid Halleck, would probably have been the great man of the day, *and the first Commander-in-Chief !*

But the great and immediate question with us was, " How shall we pass the guard on the further side?" Judge then of our delight when, on crossing, we saw no guard whatever and were permitted to pass unquestioned, and without entering the town, directly to the railroad station, where we had less than an hour to wait for the train. I have never learned certainly the reason of this sudden relaxation of vigilance. The simplest explanation would be to suppose that the sentries were withdrawn because no one was expected to cross in such a storm; but it is not common for soldiers on guard to be permitted so much indulgence. Probably all attention was called westward to meet the alarming advance of Mitchel in the direction of Bridgeport. The panic produced by his occupation of Huntsville and his headlong rush eastward on the railroads with train-loads of soldiers, was intense. and the attention of the enemy was about equally divided between preparing to resist, and preparing to evacuate—either of which called for the employment of every disposable soldier. The occupation of this road also cut Chattanooga off from direct communication with Beauregard at Corinth, leaving only the circuitous route through Atlanta; and when this also was destroyed for a time the next day, the excitement knew no bounds.

At the station we found several of our party who had come earlier,

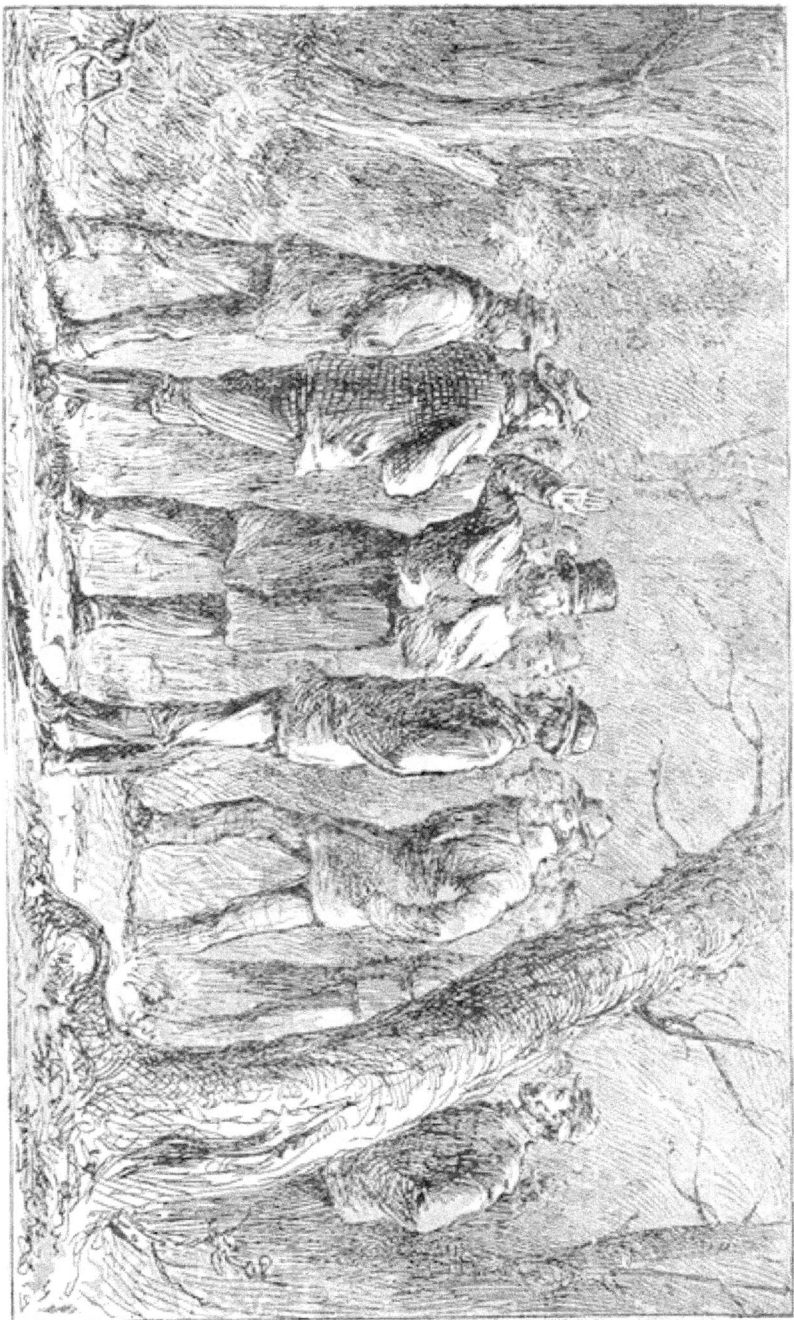

Consulting in Darkness and Storm.

and, like us, were waiting for the train. We also found a large number of passengers, many of them soldiers. Of the town itself we saw almost nothing.

When we had purchased our tickets, and had ceased to walk about the station for the purpose of looking for others of our party, we got on board. Many of the passengers were furloughed soldiers, who were going back by the southern route to join Beauregard. The conversation still turned to the mighty battle of Pittsburg Landing, and the spirit of the soldiers seemed to be wonderfully stimulated by what they regarded as a great triumph. We took part in the talk, and expressing as much interest as any, our true character was not suspected. There was no system of passports then in use on that line. or, indeed, in most others in the South, and travel was entirely unrestricted. Our raid, however, wrought a complete change in this particular.

The sun was about an hour high as we glided out of the depot, and it soon sank to rest behind the hills of Georgia. On the northern end of the road which frequently crosses the crooked Chickamauga creek, there are many bridges, and one additional over the same stream on the East Tennessee Road—eleven large ones within thirty miles ; and as we ran southward we could not help picturing our proposed return on the morrow, and the probabilities of the destruction we intended to wreak upon them. Darkness gradually closed in, and on we went amid the laughter and oaths of the Confederates, many of whom were very much intoxicated. I had been standing, but now procured a seat on the coal-box, and gave myself up to reflections, naturally suggested by the near culmination of our enterprise. Visions of former days and friends—dear friends both around the camp-fire and the hearth at home, whom I might never see again, floated before me. But before these had deepened into sleep, I was aroused by the call of "Dalton," the supper station. It was after dark, for the train had been making very slow time--whether it was behind or not I do not know, but the running on all the Southern roads was but moderate. There was a great rush for places at supper, and as I was near the door and excessively hungry, I managed to be among the foremost, and secured a good meal. I remember it the more vividly as it was the last regular meal to which I sat down for more than eleven months! Not all of the passengers were so fortunate. There was not even room for all at the second table, though the conductor was very patient. Buffum raised quite a laugh by a dexterous manœuvre. He was small and agile, and when a large rebel officer rose from the table, Buffum stooped down and, rising under his arm, dropped into his place, just as a half dozen famished persons rushed for it. "That's a Yankee trick !" was called out with some indignation, and a good deal more truth than the speaker imagined, for Buffum was a native of Massachusetts !

After supper I felt too comfortable to trouble myself much further either with speculations or memories equally vain; so, as the night wore on, I sank into a refreshing slumber, having obtained a regular seat in place of the one on the coal-box.

Near midnight we were awakened by the conductor calling " Marietta." The goal was reached.   We were now almost directly in the centre of the Confederacy, with our deadly enemies all around us.   Before we could return many miles toward our own lines we were to strike a blow that would either make all rebeldom vibrate to its centre, or be ourselves at the mercy of the merciless.   It was a time for very serious thought; but most of us were too weary to indulge in it.   In the Tremont House, the greater part of us registered names—either our own or others—and were soon sleeping soundly —the last time we slept in bed for many weary months !

Andrews was with the larger party in the hotel near the railroad station, while four others, among whom were our three engineers, were in the other hotel at some distance.   Two of these were not awakened in time for the next morning's train, and the other two were barely able to get over.   Had they been a few moments later the great railroad adventure, with all its excitement and tragedy, would not have been!

In order that the desperate chase of the morrow, and all the causes that affected its issue, may be fully understood, it is now necessary to narrate what had taken place since the adventurers had left their camp four days earlier.   We do this the more willingly because no other campaign in the whole civil war displayed more of genius in conception, or of energy in execution.

CHAPTER VIII.

THE BLOODLESS VICTORIES OF MITCHEL.[1]

WHEN Gen. Mitchel broke camp at Shelbyville on Tuesday morning, April 8, 1862, his destination was a profound secret. Some of his soldiers may have conjectured that he meant to march to Huntsville and seize the Memphis and Charleston Railroad, but he communicated his intention to no one. Even Gen. Buell was not positively informed, so far as appears; for that officer, after having written the comprehensive letter of instruction dated March 27, 1862,[2] that summed up the views on which the two commanders agreed, had left him with virtually an independent command. He was now about to undertake a bold but not unduly hazardous advance, which, if successful, would regain more territory for the flag than any other movement hitherto made by the Union armies, except that on Fort Donelson. At the worst he would but be attacking the foe, who would otherwise attack him.

(The accompanying map will give a good idea of the campaign.) Two railroads leading south diverged at Nashville, and after running about a hundred and twenty miles struck the Memphis and Charleston Railroad, which ran east and west to Decatur and Stevenson, points about ninety-five miles apart. There is a great bridge over the Tennessee River at Decatur and another long double bridge, with an island in the centre, at Bridgeport, likewise on the line of the Memphis and Charleston railroad. Between these bridges the Tennessee River arches south like a great bow, of which the railroad forms the string. Huntsville was situated on this railroad, nearer to Decatur than to Bridgeport. The first part of Mitchel's plan was to move quickly between the east and west railroads to Huntsville, and then, dividing his force, to occupy all the Memphis and Charleston Road north of the Tennessee; the second part was still more extended.

Mitchel's army consisted of the three brigades of his own division with infantry and cavalry, amounting in all to about 10,000, in the highest state of efficiency; also Gen. Negley's brigade at Columbia with additional scattered regiments designed to guard Nashville and protect his

---

[1] Special obligation is acknowledged to F. A. Mitchel, Esqr., for use of papers relating to his father's campaign. The substance of some of his descriptions is closely followed.

[2] War Records, Series I., Vol. X., Part 2, page 71.

communications, numbering 7000 or 8000 more. These, however, were
to be under his direct command only in case of emergency; but they did
add materially to his strength.

On Monday night[1] there was an almost tropical rainfall, which continued with slight intermission the following day. This made the process
of striking tents and preparing for marching Tuesday forenoon far from
pleasant. Preparations were made very leisurely and no indication of
hurry given. A civilian can form but little idea of the discomfort and

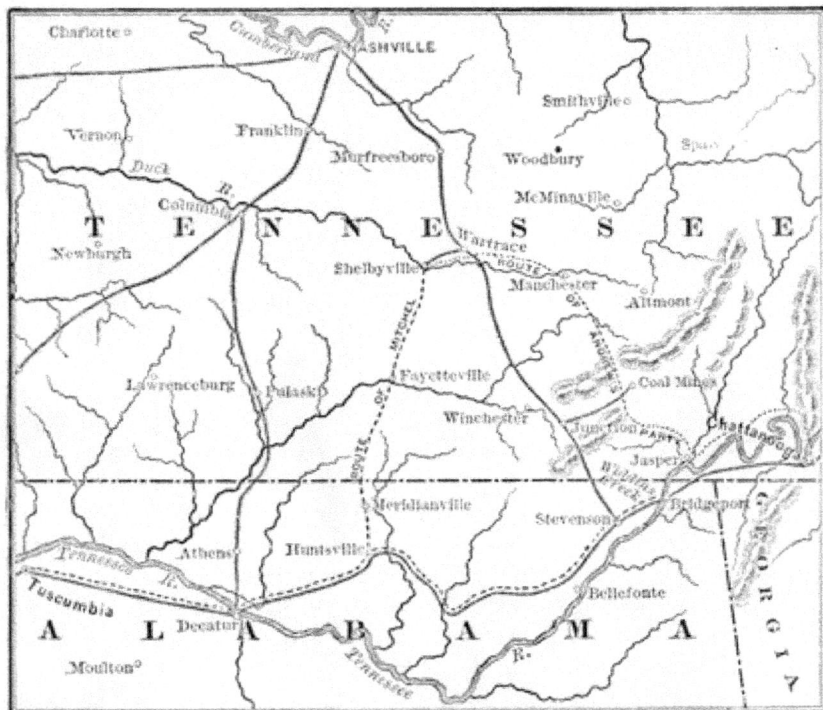

The Campaign of Gen. Mitchel.

dreariness of moving camp in the midst of a rain-storm. The tents are
wet through as soon as struck, the clothing of the soldiers is in the same
condition; the knapsacks are wet outside and too often inside also; the
three days' rations carried in the haversacks must be watched very closely,
or sugar, salt, and crackers will melt into a paste-like mass. The forty
rounds of ammunition and the loaded musket must be kept dry at any cost.
Then the slow ploughing of the wagons through the mud, the sticking fast

---

[1] The Andrews expedition left camp the same night.

of cannon, the roads soon trodden into thick and almost bottomless jelly —these are but ordinary discomforts, of which a soldier has no right to make special complaint. A forced march under such conditions is, however, a terrible hardship, which only a General of iron will and great ascendancy over his men can exact. But Mitchel had taught his troops to endure "hardness as good soldiers," and they plodded on. If the great weight of water-soaked garments and blankets, added to arms and equipments—a total of perhaps sixty pounds per man—with the mud under foot, compelled them to go slower, they only marched the more hours; if the wagons with the camp equipage could not get up to the camp at nightfall, the men laid down unsheltered on the flooded ground; and when the heavy wagons became immovable, the soldiers took rails and pried them out. So the great column moved wearily on. The cavalry was thrown far out on the flanks to bring in all travellers, and guard against any sudden dash of an enemy, while chosen companies of infantry, also in advance of the solid marching columns, were scattered as skirmishers over the hills and through the woods, keeping even pace with their comrades in the road, and watching to see that no foe was lurking in ambush. Thus the great army with its many hundreds of wagons and cavalrymen, and thousands of marching men, wound its way steadily over the gentle hills and valleys and flooded rivulets of middle Tennessee.

Fayetteville, a distance of twenty-seven miles, was reached in two days. Here one brigade was left behind as a guard for all the baggage and wagons not immediately needed in the still more active service beyond. The other two brigades rested till noon of Thursday, April 10th, and then, being in light marching order, moved more easily and swiftly forward. The rain did not fall on this day and a strong wind soon made the roads more passable. One brigade continued longer on the march than the other, pressing on and on, till at dusk they turned into the fields for rest, pitching no tents and kindling no fires, but sleeping on their arms. Now they were within ten miles of Huntsville. Pickets were thrown out in all directions, and every person stirring was arrested, both to gain information and to prevent any being carried to the enemy. The Huntsville mail coach was captured, and brought into the circle of peopled fields.

From twilight until the moon went down about two o'clock all was quiet, and the soldiers, who were not on the terribly severe outpost duty, rested as best they could. Had the advance of the army been watched, the report would have doubtless gone forward that they were safely encamped for the night. But this interval was one of intense anxiety for the commander, who was alert and at work all that night. He visited pickets and personally questioned citizens. At midnight came startling intelligence. A negro just captured made the assertion, on the authority of his master, that 5000 troops had arrived at Huntsville, and what was still worse, that they had been warned of the coming of the Yankees.[1] There was nothing improbable in such a body of the enemy being there, for Huntsville was in direct communication with the principal armies of the South, and Mitchel afterwards said that he fully credited the presence of the troops, but did not believe, in view of his precautions, that they could know of his own coming. The outlook was far from cheering, for he had not more than three thousand men in his advance-guard, with which the town must be won, if he was to realize the advantage of a surprise. But he did not for a moment hesitate. The cavalry was divided into three bands. The first and second, accompanied by two or three light guns, were to diverge in opposite directions from the main body and strike the railroad some distance on each side of town, for the purpose of preventing the escape of any engines or cars, and also by cutting the telegraph wires to keep the capture a secret until Mitchel was ready to report it to the enemy by a further advance. The third detachment was to gallop directly to the telegraph office and depot, stopping for nothing else, in order to capture all dispatches and prevent destruction of property.

Before two o'clock the men were called to their feet without roll of drum or note of bugle, and Mitchel, who was a ready and inspiring speaker, briefly addressed each regiment as it filed past into the road, telling what a glorious morning's work was before them, and asking them to preserve perfect silence till the moment of action. Then they glided on their ghostly way like an army of shadows—even the horses seeming almost conscious that they were stealing on a sleeping foe ! No conversation was permitted; the wagons were left behind; even the few cannons were moved so steadily and carefully over the muddy roads that, except on striking an occasional stone, or rumbling over a bridge, they gave no sound.

As the morning hours wore on, the soldiers, in spite of the thrilling nature of their advance, grew tired and sleepy. Half way the little village of Meridianville was reached, and soon silently left behind, seeming like a city of the dead, for no voice or sound was heard, and the people could scarcely credit it when told that an army had gone by while they slept ! The coming of the dawn was watched for with the greatest eagerness by all the soldiers. At the first tinge of day, a party of cavalry galloped off to the West and was soon out of sight. The danger of the enemy's escape was greatest in that direction, as on the eastern side the railroad bent northward toward them. The success of this first party was complete and eventless.

Now Huntsville is but four miles away, and the railroad leading east-

---

[1] The negro's story was in part true as to the troops. Over 3,000 had passed through Huntsville the day before. War Records, Series I., Vol. X., Part 1, page 643.

ward not half that distance. The dawn broadens into clear twilight and the advance-guard hasten their steps; the infantry take the side of the road while the few pieces of artillery roll rapidly down the centre. Another party of cavalry at full speed rushes eastward over the level fields. All restraint as to noise is removed, for speed, and not silence, is now the aim. Then followed an exciting scene. The scream of a locomotive —the regular eastward train to Chattanooga—rises shrill on the morning air. The engineer sees the army, and stops for a moment only, then presses on under full steam. One cannon is thrown into position—the mark is large but the distance great, and the time short! The cannon booms, and the locomotive answers with a defiant scream—unhurt! and goes roaring and rattling away among the hills. One train has escaped! Another locomotive follows, but the second gunner is ready with better aim. Another long range shot and the engine is disabled and the engineer killed. Now the cavalry have reached the track and there can be no more escape for the valuable engines shut up in the enemy's town.

But the third party, accompanied by Gen. Mitchel himself, have not checked their horses to see the result of this duel between cannon and locomotives. They gallop at breakneck speed into Huntsville, followed fast by the infantry, who have now forgotten all weariness, as soldiers will when the sound of guns is heard. If any enemy is in the town he must be given no time to rally; if not, all property must be saved, and at the telegraph office secrets of the enemy may be found of greatest value. Citizens are roused from their morning slumbers, and doors and windows are thrown open, while the cry goes up from a thousand voices, "The Yankees are coming." Men rush into the streets almost naked, women faint, and children scream, while the negroes laugh, because it is only what they have long been hoping and praying for!

Fifteen locomotives and eighty cars,[1] two southern mails, all the telegraphic apparatus, and one cypher message of priceless value—if it could be read—were captured. There was no key to the cypher, but Mitchel's wide knowledge came to his aid. He understood the principles on which cyphers were made; and laying it on the table and calling some of the young men of his staff to his aid, they soon made it give up its secret— when lo! an appeal from Beauregard at Corinth to the rebel Secretary of War for help! giving the exact number of his effective men, and stating that he was utterly unable to resist an advance in force by the Union army! This was dated April 9[2]—only two days before—and as Huntsville was a repeating station, it had been copied, and the paper, which looked like a jumble of nonsense, was left lying on the desk. It was sent at once to Buell and Halleck, and if they had been energetic soldiers, it

---

[1] War Records, Series I., Vol. X., Part 1, page 641.

[2] War Records, Series I., Vol. X., Part 2, page 618.

would have sealed the doom of the rebel army at Corinth! But against stupidity, it is said that even the gods fight in vain! The despatch was only published in the Northern papers, and gave the Confederate War Department great trouble and perplexity, in trying to make out how it could possibly have been obtained and deciphered.[1]

But Mitchel's work was not finished. The capture of Huntsville would be almost valueless if he did nothing more. At once he reorganized the railroad management, and spent the day in getting ready two other expeditions—this time by rail, where danger might be greater but hardships to the soldiers would be far less. Whether the first suggestion of running a train of cars through an enemy's country while they were surprised and deceived as to its character, came first from Andrews in his proposed railroad expedition, we do not know.

The first expedition now planned consisted of a single regiment. This was placed on a train at six o'clock in the evening, and steamed slowly westward to save the Decatur bridge, and if possible to open communication with Buell, by means of which a part of that army, which was now clearly not needed at Shiloh[2] (unless for an immediate attack), might be used for great enterprises eastward. This expedition was perfectly successful.

But to the east there were greater obstacles. The distance, even if Mitchel only advanced to Stevenson, was greater, being some seventy miles, while to Decatur was only twenty-five; and the train which had escaped would naturally warn the enemy that some kind of a demonstration was being made at Huntsville, though they could not know whether it was only a dashing raid, or an advance in force. But another matter in which Mitchel was still more interested might either be of the greatest help or a serious hindrance. The Andrews party he had sent into Georgia, would early this day, if on time, start toward Huntsville; and Mitchel could not on Friday have forgotten the men who had been picked from a whole brigade for desperate service on Monday. They could not reach Huntsville before the middle of the afternoon, but if they came then and had succeeded in burning the Chickamauga bridges, all his work and the grandest opportunity of the war would lie clear before him. They would bear exact information of the number and condition of the enemy's troops all along the line, and the state of the road over which they had passed. Then he could load his trains, using perhaps one of Andrews's engineers, and run, not simply to Stevenson or Bridgeport, but to Chattanooga at once; then with railroad communications opened with Halleck and Buell, and also with Nashville and the North;—with Beauregard despairing and demoralized at Corinth; with Lee outnumbered in the East; with the sea-

---

[1] War Records, Series I., Vol. X., Part 2, page 439.
[2] In the captured despatch Beauregard said he had but 35,000 men.

coast towns from which the rebels alone could draw any immediate supply of men, cut off by the burning of the Western and Atlantic bridges;—what might not be hoped? The right use of the Western army would then end the war in a few weeks! It was such considerations that prompted Judge Holt to say,[1] "The expedition itself, in the daring of its conception, had the wildness of a romance, while in the gigantic and overwhelming results which it sought and was likely to accomplish, it was absolutely sublime. * * * * * * The whole aspect of the War in the South and Southwest would have been at once changed;" and also led the Atlanta "Southern Confederacy" of April 15, 1862, to exclaim,[2] "It is not by any means certain that the annihilation of Beauregard's whole army at Corinth would be so fatal a blow to us as would have been the burning of these bridges at that time by these men."

But alas! on that Friday, when the whole road from Chattanooga westward would have been open for us, and while Mitchel and his heroic army would make all the hills of Huntsville ring with cheers over our arrival, we were only going toward our destination as ordinary passengers, with all our work still to do! However, it was not yet too late for all the consequences described, if the work was still done; but Mitchel, in running eastward, would need to go very slowly, with a sharp lookout lest he should encounter our belated train. He could go to meet us, and at the same time hold the railroad according to his original plan—though without news from us, he would not wish to venture beyond Bridgeport. It was hardly an accident that Gen. Sill's Brigade—that from which the Andrews party had been selected—was chosen for this service. The troops, perfectly equipped, were loaded on long flat cars with low sides, that there might be no obstacle to their firing immediately or debarking to meet a foe. Two cannon mounted on a flat car pointed diagonally ahead on each side of the locomotive and were at least formidable in appearance.

When every preparation had been completed, long after dark,[3] the train, with 2000 men on board, and Mitchel himself stationed with the engineer, moved silently from Huntsville on its perilous way. No other such advance over an enemy's railroad directly into an enemy's country was ever made during this or any other war. It was a perilous novelty. The progress was necessarily very slow. With a clear track the seventy miles to Stevenson might easily be made in two hours; but nearly five times that long was required. At every bend an ambush might be found, or an armed train, sent out by the rebels to learn why no

---

[1] War Records, Series I., Vol. X., Part 1, page 631.

[2] Ibid.

[3] There is a great apparent discrepancy in the date of this expedition, some accounts placing it on the 11th, others on the 12th. Probably it started near midnight, Friday evening or Saturday morning.

Gen. Mitchel's Armed Train.

trains or telegrams came from Huntsville, might bear down upon them, and cause a frightful accident. The Andrews party might yet come, followed by a rebel train—in short the unknown road before them was full of alarming possibilities. A timid General would never have undertaken such a run. He would have preferred to advance on foot, or at least to keep cavalry guards ahead of the train, and in consequence would have encountered far more real danger.

On the morning of April 12, the little division of Mitchel presented,—had an eye been able to look over all the military field—an imposing spectacle, such as no equal body of men afforded during the war. They were spread over more than two hundred miles of railroad. The Andrews party had captured their train and were running northward toward Chattanooga; another portion were far to the west, pressing on toward Decatur and Tuscumbia, two-thirds of the way from Huntsville to Buell's army; a stronger detachment was running eastward toward Chattanooga, certain to reach that point if the north-bound train did its work; while the remainder of the division which had been left behind on the forced march, was closing up, overland, to Huntsville! Such unparalleled activity showed that one commander at least wanted to finish the war.

All obstacles were surmounted by the train which bore Sill's brigade, and early in the afternoon it drew near to Stevenson. Still no word of the Andrews train, which had been having a day of wild adventure and terrible vicissitudes of fortune far beyond that which came to any of the other railroad parties! At Stevenson there were 2000 of the enemy, but these fled in all haste on Mitchel's coming, without firing a gun.

Mitchel remained a short time securing the six additional locomotives found at Stevenson, and then, getting aboard the train, steamed seven miles further to Widden's Creek.[1] This was the extreme point of his advance. Here he waited and pondered the situation. If Andrews had finished his work on Friday he would have reached Huntsville the same day. Even if he had burned his bridges on Saturday, it was now past time for him to have reported at this advanced post. It would not be prudent for Mitchel, in the absence of any intelligence, to advance further, at least until his western expedition had fully accomplished its work of joining communication with Buell. He could easily go on four miles further and reach the Tennessee River; but there was great danger that if he went that far, and then did not at once rush on and over the bridge, that the enemy would be alarmed, and burn it, which he was very anxious to prevent, as he says (letter to Chase, April 21, 1862), "I spared the Tennessee bridges in the hope that I might be permitted to march on Chattanooga and Knoxville."[2] He also says in his original report to Gen.

[1] E. Kirby Smith's Report, Series I., War Records, Vol. X., Part 1, page 643.
[2] War Records, Series I., Vol. X., Part 2, page 115.

Buell, April 12 (by mistake dated April 11), " I also ordered the destruction of a small bridge between Stevenson and Bridgeport, which we can replace, if necessary, in a single day; " [1] and adds on April 29th his strong desire to strike a blow at Chattanooga, for which purpose he had spared the Tennessee bridge. [2]

From these dispatches we can almost watch the operations of Mitchel's mind as he paused at this bridge. By burning it, the enemy would think that he meant to go no further, and would spare the great bridges over the Tennessee, which would thus be ready for his use as soon as he should need them; if Andrews still came, he with his party could cross those bridges, and their engine be either abandoned, destroyed or left there until this bridge, " in a single day," had been rebuilt.

At length, feeling that further delay was useless, Mitchel ordered combustibles placed under the bridge at Widden's Creek, and the fire to be kindled. He watched the rising flames for a time, and then turned away. It was the culmination of his own star, which from that time began to pale. He rendered excellent service to the country afterward, but from the hour in which he turned back on the road to Chattanooga, his opportunity to become the great general of the war ended. The fault was not his own, but the result was none the less sure.

---

[1] War Records, Series I., Vol. X., Part 1, page 642.
[2] War Records, Series I., Vol. X., Part 2, page 619.

CHAPTER IX.

CAPTURE OF THE TRAIN.

THERE was a great panic in Chattanooga on Friday, April 11th. The day before, Gen. Maxey, who had commanded at Chattanooga, was sent to Corinth with three regiments and a battalion. These just succeeded in getting by before the capture of Huntsville. Only four regiments were now within reach—two at Bridgeport and two, one of which had not yet been armed, at Chattanooga. This great diminution of force, with the changes it required, was the probable reason for our party finding no guard at the ferry when we crossed at Chattanooga on the 11th. There was a regiment at Dalton, some three or four at Big Shanty, and six more somewhere on the way from Charleston. But all these except those at Bridgeport and Chattanooga could be brought up only by the line of railroad connecting with Atlanta. The slender and partly unarmed garrison of perfectly raw troops was under the command of Gen. Leadbetter, a man whose only celebrity was in the terrible cruelties he had inflicted on the Unionists of East Tennessee, many of whom had been hanged by him with scarcely the form of a trial, and multitudes of others subjected to the most barbarous imprisonment. He soon came to be as much despised by the Confederates for his cowardice as he was hated for his cruelty by the Union people.

The train which had escaped Mitchel at Huntsville, Friday morning, arrived at Chattanooga about noon. They could only say that they had been fired on by artillery, and had seen charging cavalry. From their story it did not appear whether a lodgment had been effected on the line of the road or not; and the opinion was general that nothing more than a dashing raid for the destruction of property was intended. But telegraphic communications were interrupted; no more trains came in, and none were suffered to go out until definite intelligence could be received. Toward evening the excitement increased, though the officials tried as much as possible to prevent knowledge of the trouble from getting among the people. Soon additional offices were cut off from the telegraph line; and the military authorities promptly made arrangements for removing valuable stores from the city, lest the enemy should suddenly sweep upon it as he evidently had done upon Huntsville.

It is interesting to consider what would have happened if Mitchel *without waiting for his raiders* had pushed on to Chattanooga the next day. Gen. Smith has expressed the opinion that he might as well have done so as not, "If he had not been too timid!" The full published reports enable us at this distance of time to understand the situation as neither Mitchel nor Smith did then.

Mitchel was advancing with two thousand choice troops, thoroughly disciplined and organized. For the present, this was about all that could have been spared from other parts of his line. He would have met at Chattanooga, or on the way, four regiments commanded by a coward, all raw, and one of them armed only with squirrel rifles and shot guns. The result of the first onset would not have been for a moment doubtful, and on Sunday Mitchel would have possessed Chattanooga. But all of the enemy who escaped would then have been reinforced by the six' regiments that Beauregard had telegraphed as being on the way from Charleston to Chattanooga; four or five more scattered along the road from Atlanta; and four which Smith says could have been spared from East Tennessee for offensive operations. Some sixteen or eighteen regiments could have gathered around the little band of Ohioans within two days, and have selected their point of attack—either at Chattanooga, or anywhere on the line of communications toward Bridgeport, or beyond. The highest genius could scarcely have saved Mitchel, or have secured time for the slow help of Gen. Buell.

But with the Oostenaula and Chickamauga bridges of the Western and Atlantic, and the Chickamauga bridge of the East Tennessee Railroad burned, all this is changed. All the regiments from the sea-coast and Big Shanty would then be stopped fifty miles away. Smith and his four regiments could not come nearer by rail than ten or twelve miles of Chattanooga. They could form no junction with the other regiments, and could attack in one direction only. Thus Mitchel would have time to bring up his own reserve force of five or six thousand, join hands with Buell at Corinth, and with all the forces of the North at Nashville; and if the South chose to further reinforce, and make the final issue of the war to be determined around Chattanooga, the Federals would have no reason to complain. But enough of military speculation as to what might have been!

The Andrews party were greatly crowded in the large hotel at Marietta on Friday night, having to sleep three or four to a bed, but soldiers are not fastidious, and the greater number slept soundly. We had unbounded confidence in our leader, whose part it was to provide for all contingencies.

Andrews scarcely slept at all that night. He first went to the hotel and saw that those who lodged there had made arrangements for being

---

[1] Official Report of E. Kirby Smith, April 13th, 1862. — War Records, Series I., Vol. X., Part 1, page 643.

called on time in the morning. Porter and Hawkins, who had come down the evening before, and had gone to bed much earlier, were not seen, and as they had not paid the waiter any fee for rousing them early, they were left behind; a diminution of our force much regretted, as they were both brave men and Hawkins was an experienced engineer. This left us but nineteen men out of the thirty that, I judge, had been originally selected.

We were all roused promptly at the railroad hotel, a little before daybreak. Andrews, who came back to us, now went from room to room while we were dressing, seeing every man, giving him exact orders as to his part in the work of the morning. There was suppressed fire in his low, almost whispered words, a calm confidence in his tones that was con-

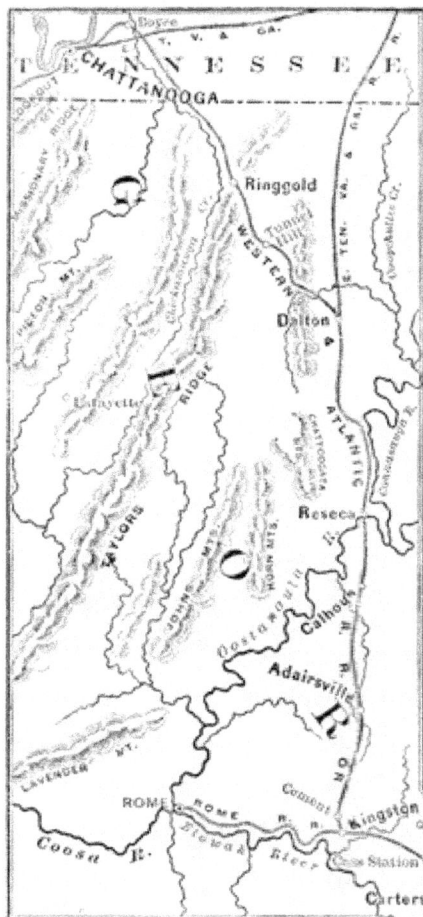

The Western and Atlantic, or Georgia State Railroad.

tagious. There seemed to be no doubt, hesitation, or shrinking on his part, but, on the contrary, an eagerness and joy that the time was so near at hand.

When we were ready, as it still lacked a little of train time, we gathered in Andrews's room for an informal council of war. Some were seated on the edge of the bed, one or two on chairs, and the remainder stood around as best they could. We did not speak very loud as we wished no sharers in our plans. Andrews gave no exhortations—

the time for that had passed—but rather cautious to prevent too precipitate action. He said:

"When the train stops at Big Shanty for breakfast, keep your places
till I tell you to go. Get seats near each other in the same car, and
say nothing about the matter on the way up. If anything unexpected
occurs, look to me for the word. You and you,"—designating the men,
—"will go with me on the engine; all the rest will go on the left of the
train forward of where it is uncoupled, and climb on the cars in the best

Bed-room Consultation at Marietta.

places you can, when the order is given. If anybody interferes, shoot
him, but don't fire until it is necessary."

Sergeant Major Ross, the ranking man of the party, and as brave as
any, offered a respectful protest against going further. He said that
circumstances had changed since we set out; that it was a day later than
planned; that many more troops were at Big Shanty than formerly; that
we had noticed the crowded state of the road as we came down, and that
Mitchel's movements would make the matter worse. For all these reasons he thought it better to put off the attempt, or give it up altogether.

Our heads were very close together as we talked, and the words softly spoken; the door was locked, and the windows overlooked the railroad, so that we were sure to see the train coming. Andrews very quietly answered the objections of Ross, admitting all the facts he stated, but claiming that they only showed our way the clearer. The military excitement and commotion, and the number of trains on the road would make our train the less likely to be suspected; and as to the troops at Big Shanty, if we did our work promptly, they would have no chance to interfere. Capturing the train in the camp would be easier than anywhere else, because no one would believe it possible, and there would therefore be no guard.

Andrews could always find a reason for everything; but these plausible arguments were not perfectly convincing. Several others, among whom was J. A. Wilson, joined in a respectful protest against proceeding. Then Andrews, speaking even lower, as was his wont when strongly moved, said:

"Boys, I tried this once before and failed; now, I will succeed or leave my bones in Dixie."

The words and manner thrilled every hearer, and we assured him that we would stand by him, and, if need be, die with him. He grasped our hands and we hurried to the platform, for the train was now almost due. I had said nothing in the discussion, for I felt that we were under the leadership of Andrews, and should simply obey, leaving the responsibility to rest on him. I am not sure that, on a later critical occasion, we did not carry this principle a little too far.

Although we only needed tickets to Big Shanty, we purchased them to various points along the line that attention might not be attracted by such a number bound to one place. As the train came up, we noticed three closed box-cars attached. Every passenger train, as I have since been informed by Conductor Fuller, was at this time required to carry empty cars northward which were brought back filled with bacon and other provisions, vast quantities of which were then being gleaned out of Tennessee and stored in Atlanta. We all took our places close together in one car, that we might be ready to help each other in case of need. Knight sat near the front door, and says that on looking back he saw that most of our men were pale yet resolute. The passengers had that listless and weary air always seen in the early morning on board a train.

The conductor, whose name we afterwards learned was William A. Fuller, entered and began to take the tickets. He looked narrowly at us, for it was an uncommon thing for so many persons to enter in a body as did at Marietta; besides, he had been warned very recently to watch that no conscripts used his train for the purpose of escaping, and ordered, in case of suspicion, to telegraph for help at once. No doubt we looked soldierly enough, but he afterwards told me that he did not suspect us of being conscripts. We also scrutinized him carefully, for it was possible that he might, if his suspicions were in the least aroused, endeavor to prevent us from taking his train.

He was quite young for a conductor, being, as we afterward learned, only twenty-six, though he had been for seven years in that position. He had a frank, genial, but resolute face, was of medium size, and looked active and strong.

We had little leisure for looking at the grand form of Kenesaw mountain, which rose on our left, and around the base of which the road describes almost a half circle, and then turns away before it reaches Big Shanty. Here was fought one of the severest battles of the war between Sherman and Johnston; but this, with their prolonged struggle over the whole line of this railway, did not come until two years later. The question of deepest interest to us, and one which would be quickly solved was, "How much of a fight will we have at Big Shanty? If the train is left guarded during breakfast time we will have to overcome the guards; if anybody sees us going on the engine, and a rush is made to prevent, we will have to fight sharply and at close quarters—the most deadly kind of fighting." Every revolver had been carefully examined at Marietta before we slept, and every preparation made, so there was nothing to do but to wait as patiently as we could.

CAPTAIN WM. A. FULLER. From a war-time photograph.

It was a thrilling moment when the conductor called out, "Big Shanty! twenty minutes for breakfast!" and we could see the white tents of the rebel troops and even the guards slowly pacing their beats. Big Shanty (now called Kenesaw) had been selected for the seizure because it was a breakfast station, and because it had no telegraph office. When Andrews had been here on the previous expedition, few troops were seen, but the number was now greatly increased. It is difficult to tell just how many were actually here, for they were constantly coming and going; but there seems to have been three or four regiments, numbering not far from a thousand men each. They were encamped almost entirely on the west side of the road, but their camp guard included the railroad depot. As soon as the train stopped, the conductor, engineer, fireman, and most of the passengers hurried for breakfast into the long low shed on the east side of the road, which gave the place its name. No guard whatever was left—a fortunate circumstance for us, but not at all unusual on Southern roads even when not so well guarded by soldiers as this train was. Now was our opportunity! yet for a moment we were compelled to keep our seats and wait the appointed signal by our leader. It required a strong effort of will to keep from rushing forward. We had no desire for eating as we saw the passengers leaving their seats around us and pouring in to breakfast. The moments seemed hours; for we knew that when the signal was given, we must do our work in less than half a minute or be slaughtered on the spot; we also knew that any one of us who failed to get on board with the rest would be lost; but we did not know how long during the twenty minutes Andrews would wait. If anything could be gained by waiting five or ten minutes we were sure that he, with his marvellous coolness, would wait and expect us to do the same. It seemed already a considerable interval, for the last passenger who wanted breakfast had left the train and disappeared within the room.

But Andrews did not mean delay. He had been absent from the car for a time as we came up the road and had only just returned, and taken his seat close to the door. Now he quietly rose, and without turning his head toward us, stepped to the door with the crowd that was pouring out. Engineer Knight, whether from natural impulsiveness, or at a signal from Andrews, rose also and went out with him. These two got off on the side next to the camp, and opposite the depot. They walked forward at an ordinary pace until abreast of the locomotive, which they saw at a glance to be vacant—engineer and fireman had gone to breakfast. That was very good! Andrews walked a few steps further forward with Knight still at his side, until he could see ahead of the engine that the track was clear as far as a curve a little way up the road which closed the view. Then they turned and walked back until just in advance of the first baggage car and behind the three empty freight cars, when Andrews said with a nod, "Uncouple here and wait for me." Knight drew out the pin and carefully laid it on the draw bar. Andrews came back to the door of our car and opening the door, said in his ordinary tone, not a shade louder or more hurried than usual, "Come on, boys; it is time to go now." Our hearts gave a great bound at the word, but we rose quietly and followed him. Nothing in this was likely to attract the attention of the few passengers who still remained in the car; but it mattered little, for the time of concealment was now past. Andrews glided forward very swiftly, and Knight, seeing him coming, hurried on before and jumped on the engine, where he at once cut the bell-rope and, seizing the throttle bar, stood leaning forward with tense muscles, and eye fixed on the face of his leader.

Andrews did not follow, but stood a step back from the locomotive with one hand on the rail, looking at his men as they ran forward. Brown

and Wilson (the other engineer and fireman) darted forward at the top of their speed and took their post beside Knight on the engine. As soon as the rest of us reached the hindmost box-car we saw that its door was wide open. Whether this was a mere happy accident, or whether, as is more likely, Andrews had gone forward before we reached the station and opened it, with his usual audacity, I do not know. But he motioned with his hand to us saying, "Get in! Get in!" We needed no urging. The floor was breast high, but the hindmost shoved and lifted the foremost and were themselves pulled up in turn. I helped to throw Shadrack up and had my arm almost pulled off as I was dragged in by him a second after. All this time a sentry was standing not a dozen feet from the engine quietly watching, as if this was the most ordinary proceeding, and a number of other soldiers were idling but a short distance away. They had not made up their minds what to say or do, and we were hidden by the train itself from the view of persons at the depot. The first report of the Atlanta papers speaks of four men only as taking the train—no one at Big Shanty seeming to notice any but the four who boarded the engine. All this work was of seconds only, and as the last man was being pulled in, Andrews stepped on board, and nodded to Knight, who had never taken his eyes from his face. Quick as a flash the valve was thrown open and the steam giant unchained!—but for an instant which seemed terribly long the locomotive seemed to stand still; Knight had thrown the full power on too suddenly, and the wheels slipped on the track, whirling with swift revolutions and the hiss of escaping steam, before the inertia of the ponderous machine could be overcome. But this was an instant only; none of the soldiers had time to raise their muskets, give an alarm, or indeed to recover from their stupor before the wheels "bit," and the train shot away as if fired from a cannon!

We were now flying on our perilous journey. The door of the box-car was pulled shut to guard against any shot that might be fired, and while partially opened afterwards to give us some view of what was passing, it was always closed again whenever we neared a station.

This capture was a wonderful triumph. To seize a train of cars in an enemy's camp, surrounded by thousands of soldiers, and carry it off without a shot fired or an angry gesture, was a marvellous achievement. There are times when whole years of intense enjoyment seem condensed into a single moment. It was so with us then. I could comprehend the emotion of Columbus when he first beheld, through the dim dawn, the long-dreamed-of shores of America, or the less innocent but not less vivid joy of Cortez when he lifted the cross of Spain over the halls of the Montezumas. My heart beat fast with emotions of joy and gladness that words labor in vain to express. It was a moment of rapture such as will never return. Not a dream of failure cast a shadow over us. We had been told that to reach and take possession of the train would be difficult, but that all the rest of the enterprise would be easy. It would have been on the day originally fixed !

Various manifestations of triumph were made as soon as we were off. Dorsey sprang to his feet crying, " Boys ! we are done playing reb. now ! We are out-and-out Yankees from this time on." But Geo. D. Wilson, who was older than the rest of us, cautioned him, saying, " Don't be too fast, Dorsey; we're not out of the woods yet."

And indeed it soon seemed as if we were to have serious trouble at the outset. The engine ran slower and slower, until it finally came to a full stop. We were not yet far from camp. There had been just one burst of speed, and then this sickening and alarming failure of power. We asked eagerly of those forward what it meant, and the answer was far from reassuring— " The steam has gone down." In a few moments we learned the reason. The dampers were closed on the engine fires when the stop for Big Shanty was made, and they were not opened by our boys in the hurry of the start; consequently, the fire was almost out. A little oil and some fresh wood promptly mended matters. No time was lost while stopping here in this enforced manner, for we had started ahead of time, and had leisure to obstruct the track.

JOHN W. SCOTT. From a war-time photograph.

The telegraph wire was also cut. This was necessary, for though there was no office at Big Shanty, a portable battery might be found, or a swift messenger be sent back to Marietta, and a single lightning flash ahead would blight our fondest hopes. Breaking a wire is not as easy as it seems; but we adopted a plan which worked all day, and took up no time that was not also utilized for other purposes. John Scott, who was agile as a cat, ran up the pole, and knocking off the insulating box at the top, swung down on it. A small saw found on the engine easily cut the tightly stretched wire close to the box. This did not take more than one or two minutes.

At this first stop, Andrews, who had not shared our uneasiness about the fire in the engine, came back and clasped our hands in ecstasy, manifesting more excitement than I ever saw in him before; exclaiming that ve had the enemy now at such disadvantage that he could not harm us or save himself. " When we have passed one more train," he declared," we'll have no hindrance, and then we'll put the engine at full speed, burn the bridges after us, dash through Chattanooga, and on to Mitchel at Huntsville. We have the upper hand of the rebels for once ! "

By saying that we had only one more train to pass *before doing this*, Andrews did not mean that there was but one train coming toward us. There were three, which had already left Chattanooga; but only the first of these, a local freight, which might be met at any point between this and Kingston, was a real obstacle. Andrews knew the time schedule of the other two, and could plan to meet them at any given station, even if we were far ahead of our own time. Had there been none but these three trains, his triumphing would have been well warranted.

The following is the basis upon which Andrews made his calculations: he believed that no engine could be had for pursuit nearer to Big Shanty than Kingston on the north, or Atlanta on the South, each about thirty miles distant. If the rebels pursued toward Kingston the best they could do was to follow us on fleet horses, and the time, allowing for delay in starting, and the state of the roads, could not well be short of three or four hours, by which time we ought to be out of reach, with all our work done. If they rode or sent back to Marietta (where we had lodged for the night), which would seem to be their best plan, that would take at least an hour; then a telegram to Atlanta could very soon start a train after us, but it would be forty or fifty miles behind; and long before it could come up, bridges would be burned, the track and telegraph cut, and the road completely destroyed. We expected to run on our regular time to Kingston, which would thus take about two hours, but to obstruct the track at several places on the way; then with the local freight safely passed, hurry on to the Oostenaula (or Resaca) bridge twenty-four miles further in half an hour more, burn that, and sweep on over the eleven bridges of the Chickamauga, and leaving them in flames (also the one of the East Tennessee road, over the same stream), pass by Chattanooga on the " Y" running over to the Memphis and Charleston road, and press as rapidly as possible westward to Bridgeport, and on to Mitchel wherever he might be. It will be seen that leaving out of account any accident to our train, and any difficulty in passing the trains we were to meet, our calculations were almost dead certainties. With two experienced engineers, and caution in running, accidents were not likely to occur; and Andrews trusted to his own marvelous address (and not vainly, as the result showed) to disarm suspicion from any trains met. On Friday, so far as human vision can now penetrate, these 'calculations would have worked out with the precision of a machine, and all the results indicated have followed. To-day there were new elements which were to task our powers to the utmost, but of these we as yet knew nothing.

All careful and prudent preparations were now made for a long run. A red flag placed on the last car showed that another train was behind, and served as a kind of silent excuse for being on the time of the morning mail. The engine was also carefully inspected by Knight, whose mechanical knowledge was most useful, and found to be in excellent working condition. It was thoroughly oiled. Then we moved leisurely onward, until we came near Moon's station, where some workmen were engaged on the track, and the opportunity of getting necessary tools was too good to be lost. Brown sprang down and asked a man for a wedge-pointed iron bar with which he was prying. The man gave it at once and Brown stepped back with his booty, but a little disappointed, for one of the bent, claw-footed bars, for pulling out spikes, would have been worth much more, but they had none. The bar taken was the only one of their tools that seemed likely to be of value, or more would have been borrowed—by force, if not otherwise.

As we went on, Andrews cautioned his engineers not to run too fast, which they inclined to do; all of us would have relished more speed in this first part of the journey. But running on all Southern roads in war-times was slow; our train was not scheduled at over sixteen miles an hour. The road itself was exceedingly crooked, with abundance of short curves, and, having but light iron rails, was unfitted for high speed. We were anxious to get past the local freight that we might test the road's capabilities. Those on the engine were very much amused, as we ran by station after station, to see the passengers come up with their satchels in their hands, and then shrink back in dismay as we sped past without a sign of halting. But when by, we would stop and cut the telegraph wire so that no suspicions or inquiries could be sent ahead.

Thus we passed through Ackworth, and Allatoona, and then stopping again to cut the wire, also endeavored to lift a rail. While we were sure that no train from Big Shanty could follow us, we wished also to make it difficult for one from Atlanta, if any should be sent from there, to run rapidly ; and what was of equal importance, we did not wish the local freight to proceed Southward after we met it, to be turned back by any pursuers. A lifted rail is almost sure to throw an unsuspecting train from the track; and we put such an obstruction before each train that we met on this journey. Yet the process of taking up a rail, though we made much of it, was far from easy with the imperfect tools we possessed. A single tool —a bar constructed expressly for drawing out spikes—would have enabled us to baffle all possible pursuit. But this we did not have, and more than five minutes were consumed for each rail taken up, in battering out some spikes with our iron bar, and afterwards prying the remainder loose with handspikes, and with the rail itself. This delay was of no great importance now, for we had a superfluity of time; but in the quick and terrible struggle further up the road, when seconds were decisive, it was far otherwise. The rails when lifted were carried away with us, and the break thus left was for a time a barrier (to a train not supplied with track-laying tools) as absolute as a burnt bridge. The feeling of security after such obstruc-

Tearing up Track.

tion was very delightful and not unwarranted. In no case did a pursuing train pass a place where we had torn up a rail in time to do us any damage.

There was an exultant sense of superiority while running along in the midst of our enemies in this manner, such as a man in a balloon might feel while drifting over hostile camps, or over the raging waves of the ocean. As long as all is well with his balloon the man need not care what takes place in the world below; and as long as our engine retained its power and the track was clear before us, we were in a similar state of security. But a knife blade thrust in the silk globe overhead, or the slightest tear in the delicate fabric, will in a second take away the security of the man in the clouds. So the loosening of a bolt, or the breaking of a wheel would leave us powerless in the midst of our deadly enemies. It was such possibilities, always so near, that imparted thrilling interest to our passage through towns and fields and woods in the heart of the enemy's country.

At length we reached the Etowah River and safely passed over the great bridge at that point. No stop was made, though the first serious cause for anxiety was here visible. Hitherto everything had worked exactly as we had calculated, and our confidence in our leader and in final success was correspondingly increased; but on a side track which connected with a little branch road that ran up the river about five miles to the Etowah iron-works and rolling-mills, there stood a locomotive! It was but a short distance from us, and the smoke from the funnel showed too plainly that it was ready for work, thus constituting an element of the most dangerous character which had not been embraced in our calculations. It was named the "Yonah,"—a private engine used by the owners of the works for their own purposes. Thoroughly as Andrews had explored the road, he had no knowledge of its existence until the moment when he saw it standing on the side track not a dozen yards away, and looking as if it was ready to enter upon a race with our "General" on equal terms. It was still thirteen miles to Kingston, and the enemy, if there was any direct pursuit, would be able to get an engine that much sooner than we had supposed possible. Several men were gathered about it, but not enough to make an assault seem very formidable to our party. At the first sight, Knight said to Andrews, "We had better destroy that, and the big bridge," but Andrews refused with the remark, " It won't make any difference."

Nearly all critics of the expedition who knew of the presence of this locomotive,—for a long time I did not, as I was shut up in the box-car—are disposed to think that here Andrews made a most unaccountable mistake. But this is far from certain. Probably he had an aversion to the shedding of blood when not clearly necessary; but this he would not carry so far as to prevent anything he thought best for ultimate success. There is no reasonable doubt that we could have overcome the men about their engine, and then have caused it to jump off the track, or precipitate itself from the bridge; and a very few minutes more would have sufficed for setting the bridge in flames. The morning was damp and it had already began to rain slightly, which would have delayed the burning; yet all might have been done in twenty or thirty minutes, and when accomplished we would have certainly felt safer,

But, on the other hand, it must be remembered that the burning of this bridge formed no part of Andrews's original plan, and could have accomplished nothing more toward the furthering of Mitchel's plans than the burning of the Oostenaula bridge. The local freight train was now due, and if it came in sight while we were engaged in destroying the "Yonah" or the bridge, and getting the alarm as would be almost inevitable under the circumstances, should get away from us and run back to Kingston, or should run on us and cause a wreck, our situation would be far worse than with this engine left behind us. But even if we could be assured that the local would not come, but remain for us at Kingston, still the attack here would alarm the enemy, and we would be followed from this point as readily as from Big Shanty, but eighteen miles further up the road. It would be but little more than an hour's gallop to Kingston, where a train for pursuit would surely be found. The capture at Big Shanty assured us of a longer start under any circumstances than seemed possible if we stopped to strike a blow here.

To understand how clearly the case may be presented on the negative side, suppose the most favorable circumstances as actually taking place: We burn the bridge and go on to Kingston, and safely pass the local freight, which goes on southward and, coming in sight of the smoke, is warned in time, learns who has done the mischief and comes back after us ! For throwing off or disabling trains an injured track was far better than a burnt bridge, with its column of smoke visible for miles; though the latter was a far more serious and lasting injury to the road. For this reason we tore up rails *in front* of each train that we feared might be turned back after us, in the expectation that it would be disabled by running upon these places before suspecting danger. Viewing the matter from the standpoint of facts then in possession of Andrews, his decision was right. Had nothing else occurred than the presence of this engine "it would have made no difference."

Leaving the engine and bridge behind, we glided on through Cartersville, a town of considerable size, where there were many disappointed passengers on the platform, and continued without incident until we reached Cass station. The town of Cassville is some distance from the railroad, but the station was important for us as the regular place for taking on wood and water. Here we stopped and began to wood up. William Russel, the tender, was naturally curious about the appearance of such a small train running on the time of the morning mail, with no passengers, and none of the regular hands. Here Andrews told a most adroit and carefully planned story, with enough of foundation to make it probable. He claimed to have been sent by Gen. Beauregard, who was in desperate straits for ammunition, to impress a train, have it loaded with powder, and run it through at lightning speed. Had he been pressed more closely, ne could have produced passes proving himself worthy of belief   But it was not necessary to go so far.   The very appearance of Andrews, tall, commanding and perfectly self possessed, speaking like one who had long been accustomed to authority, was so much like the ideal Southern officer that Russel's credence was won at once.   He knew very well that after such a battle as Pittsburg Landing it was natural that powder should be scarce, and if it did not come at once, what more natural than to send for it?   Seeing the impression that he had made, Andrews, who of course did not work at throwing on wood, but left that to his companions, asked if he could not be supplied with a schedule of the road, as it might be useful.   Russel, in his patriotic fervor, took down and handed out his own schedule, saying that he would " send his shirt to Beauregard " if the latter wanted it !   When asked afterward if he did not suspect a man who made such an unreasonable demand, he answered, " No; I would as soon have suspected Jefferson Davis himself as one who talked with the assur- ance that Andrews did.'

We were now within seven miles of Kingston, resupplied with wood and water, without having met the slightest hindrance, and with a full schedule of the road.   But at Kingston we had more reason to apprehend danger than anywhere else along the route.   A branch road from Rome connected there with the main track, and the morning train from that town would be awaiting our arrival.   This, with the local freight which we hoped to meet, and the complicated arrangement of the switches, would constitute no small obstacle to our onward progress.   The real difficulties surpassed expectations.   Andrews had made himself familiar with the minutest working of the road at this point, as also at Dalton and Chattanooga, and we would soon be able to see how he would overcome the hindrances in his way.

CHAPTER X.

PURSUED!

WE reached Kingston a little ahead of time. A glance showed s that the local freight had not yet arrived. Without the slightest hesitation, Andrews ran a few hundred yards past the station, and ordered the switch-tender to arrange the switch so as to throw us on the side track; then we backed out on it, stopping on the west side of the station, and almost directly alongside of the Rome passenger train, then lying on its own track, which joined the main line still further north. This train was expecting the coming of Fuller's mail, and of course, the arrival of our partial train in the place of the one they were expecting, was a matter of the greatest interest to them. The engineer stepped over to our locomotive and said, with an oath,

"How is this? What's up? Here's their engine with none of their men on board."

Fortunately Andrews was just at hand and promptly replied, "I have taken this train by Government authority to run ammunition through to General Beauregard, who must have it at once."

He waved his hand toward the car in which we were shut up (*representing the powder!*) and they inquired no further in that direction, but simply asked when the passenger train would be along. Andrews responded indifferently, that he could not tell exactly, but supposed it would not be a great while, as they were fitting out another train when he left Atlanta. With this cold comfort they were obliged to be contented; and Andrews, leaving the engine in care of his three comrades, went into the telegraph office, which was on the side of the depot next us and asked, "What is the matter with the local freight that it is not here?" He was shown a telegraph dispatch for Fuller, ordering him to wait at this point for its coming—an indication that it was not very far away. This was the only information vouchsafed to us by the management of the road during the whole of that eventful day!

Andrews returned to his engine, and stood there, or walked about on the end of the platform near by during the tedious moments of waiting. He did not seek to enter into conversation with any one, but quietly answered any question asked. He appeared abstracted and a little anxious, as was natural for one running an express ammunition train, on which the safety of an army might depend! It was fortunate that his real and assumed characters were so much in harmony.

Brown, Knight, and Wilson attended to their engine, seeing that all was in good order with a reasonable head of steam, and refrained, as far as they could, from any kind of conversation, answering all demands in monosyllables. Their position during this enforced stop was embarrassing, but far less painful than ours in the box-car. We could hear low murmurs outside, we knew that we were at a station, and alongside another train, and could hear the tread of feet; but we could not learn why we did not press on. A thousand conjectures will spring up at such times; and the possibilities of our situation were ample enough for all kinds of imagining. We had a tolerably high estimate of our fighting power, and did not doubt that we could capture any ordinary train, or the usual crowd around a village station. But to be shut up in the dark, while—for aught we knew —the enemy might be concentrating an overwhelming force against us, was exceedingly trying, and put the implicit confidence we had in our leader to a very severe test. There was one precaution Andrews had neglected—probably because he trusted so fully in his own marvelous genius—but the need of which was felt keenly afterwards. No lieutenant was appointed. One who could have taken charge of the men, leaving Andrews free to plan and give general directions, would have been a support to us now, and a help to all of us later. With George D. Wilson or some other of the soldiers as authorized second, the force would have been in better fighting trim, and what is of still more importance, Andrews would have felt more free to order the capture of any pursuing train. But at present we had nothing to do but wait till the road was cleared for us.

Before suspense became intolerable the whistle of an approaching train was heard, and the local freight rumbled up to the eastern side of the depot, and stopped on the main track. Andrews made haste to begin the inevitable conversation. He went over and spoke to the conductor, telling him to pull his train, which was quite long, on down the road so that we might get out of the switch and proceed on our way; adding the same powder story. This conductor saw that Andrews was treated with marked deference by the people about the station, and did not hesitate to believe his story and obey the order. But before he had moved his train, Andrews noticed a red flag on the hind car, and at once exclaimed; "What does this mean? I am ordered to get this powder through to Beauregard at the earliest possible moment and now you are signalling for another train on the track!" No doubt Andrews felt all the vexation he expressed. The man said he was very sorry, but it could not be helped; and then he gave the reason, which was a startling piece of intelligence. Mitchel had captured Huntsville and was said to be advancing eastward toward Chattanooga by forced marches; and as they had no force to resist him, they were running everything out of Chattanooga, and had put a large extra train on the track to get the rolling stock, as well as the goods, out of the way. Andrews thanked him for the information and told him to go a long way down the road so that the extra would have room enough to get by, adding, "I must be off the very first minute that is possible." The conductor made no objection, but asked, "What will you do about Mitchel at Huntsville?" Andrews replied; "I do not believe the story. Mitchel would not be fool enough to run down there, but if he is, Beauregard will soon sweep him out of the road. At any rate I have my orders." The train was pulled down the road, and the tedious process of waiting continued.

While the moments are dragging their leaden weight along as the three trains rest on separate tracks at Kingston, it may be well to narrate the experiences of those whose engine was so unceremoniously wrested from them at Big Shanty. We had counted on a great commotion and excitement following the seizure, and in this we were fully warranted; but we also believed that for some little time no one would know what to do. In this we were mistaken. The absence of a telegraph office or engine, or even of any horses, did utterly confound the great majority; but a fortunate mistake on the part of one man served the Confederate cause better than the deepest calculation could have done; in fact, gave rise to the thrilling railroad chase that followed. Conductor Fuller, Engineer Cain, and the foreman of the road machine shops, Mr. Anthony Murphy, sat down to the breakfast table not far apart. Before they had tasted a mouthful, however, the sound of escaping steam, the loud whirr of the wheels on the track, and the outcry that rose in a moment from guards and camp, brought them and all the breakfasters to their feet. By this time the locomotive had started, and Fuller and Murphy, with loud exclamations about the robbery of the train, rushed pellmell with everybody else out on the platform, the passengers who had been uncoupled and left on the road, not being behind others in their complaints and uproar. The whole camp also was in a turmoil. A single glance around showed Fuller that there was no chance for help there; and being a man of quick thought his mind fastened on an idea—utterly wrong as it proved—but which had the merit of putting him vigorously to work. The nearest guard declared that only four men were engaged in the capture—he had only seen those that mounted the engine—and others corroborated him. Fuller remembered the conscripts he had been warned to watch for, and at once the thought flashed across his mind, "Some of those men, one of whom happens to know enough of an engine to pull open the throttle, have jumped on my train to get out of camp, and as soon as they are outside they will leave the engine and run into the mountains. I must follow as fast as possible and try to get it back before I get very badly out of time." Ah! there was no more regular time made on that road for several days!

The presence of Mr. Anthony Murphy that morning was purely accidental. He was going to examine an engine at Allatoona reported out of order. As an officer of high authority on the road, commanding all engineers and firemen, knowing all the engines and everything about the road perfectly, his presence at that time was most unfortunate for us. He was a man of great coolness and good judgment. His first action was far-sighted. He sent Mr. William Kendrick on horseback to Marietta, to notify the superintendent at Atlanta by wire. Mr. Kendrick arrived in time to hold a freight train there till orders were flashed back to drop all cars but one, run up to Big Shanty, load on soldiers and pursue with all speed. This was the first train in chase. A message was also sent from Marietta to Richmond, but no result followed that. My present opinion is that Andrews had in some way arranged for cutting the wire between Chattanooga and Knoxvillle, which was the only route by which the message could have gone around and got on the line of the Western and Atlantic railroad ahead of us.

Had these been the only measures of pursuit—wise and judicious as they were—our task

ANTHONY MURPHY, Supt. of W. & A. machine-shops.

would have been easy. But Fuller's error and his ardent temperament prompted to another course. He called to Murphy and Cain, "Come on with me;" they promptly followed *and the three of them started at a dead run up the track.* The spectacle of three men running vigorously after a flying engine, as if they expected to catch it, instantly restored the mob to good humor, and they cheered, and shouted with laughter! What would have been the fate of these runners if they had overtaken us at the first stop, where we cut wire and obstructed the track, it is needless to inquire. They would have hardly begun such a chase had they not entirely underestimated our number, as well as mistaken our purpose. The Atlanta *Southern Confederacy,* of the next day, in an article full of panic, written before the issue was determined, speaks of us as " some four men yet unknown." The whole article is so full of interest that we give it entire on a subsequent page.

The different running powers of the pursuers were soon made evident in this apparently hopeless chase. Fuller was extraordinarily fleet and of great endurance. His companions were equally zealous but less able for running. They were, therefore, soon spread out for a considerable distance. While putting in his best efforts, Fuller shouted back encouragement to his comrades, but did not wait. The hope of getting his train soon was too strong, and he also feared that the reckless men who had taken it might do some injury to the engine before he could come up. The idea that they might offer any resistance did not enter his mind.

But the chase could not long have continued in this manner for human muscles cannot be pitted successfully against steam. The labored breath and the decreasing pace of the runners showed that they were well nigh exhausted; and as curve after curve was rounded,—for with the instinct of railroad men they clung to the track,—they grew discouraged; but just before their "second wind" was exhausted, they received both help and renewed excitement. They came to Moon's station, some two miles from the place of starting, and have never been able to make even a plausible conjecture of the time consumed in this first stage of the journey; but it is certain that it was just as little as straining muscles and iron will could make it.

Here they learned from the track-laying party that some of their tools had been "borrowed," and a short distance beyond some ties placed on the track, and the telegraph cut. This was our first halt, and the track had been obstructed that a train coming from Atlanta might be hindered. The pursuers here found a hand-car—not one of the elaborate machine cars, which may be driven at great speed, but what Fuller termed a "pole-car." It was at once pressed into service and gave a welcome rest. Fuller ran it backward a little way and picked up his companions who were behind, then drove forward as fast as the construction of the car would permit. Pushing with a long pole in flat-boat fashion, quite rapid time could be made on the level, and on down-grades, while on the steep up-grades, two would jump off and push at a full run.

At Ackworth they got a reinforcement of two men, Mr. Smith and Mr. Stokely, and hurried forward. All idea that they were following conscripts had now been given up. They had learned that the captors had been seen oiling their engine, as if preparing for a long run, and seeming to perfectly understand their work. Fuller and Murphy were now able, for the first time, to consult about their plans. We were running on regular time—a circumstance that they had noted in the reports received from the different stations,—and that time was only sixteen miles per hour, which meant two hours to Kingston. They were making seven or eight miles on the pole-car, and that, if kept up (it was the utmost they could do), would bring them to Etowah in two or three hours from the start. "Then if," Fuller continued, "we can find the old 'Yonah' ready at our end of the branch, we can take her and run up to Kingston in fifteen minutes more. There are to be some extra trains on the road to-day that will bother the scoundrels up there, and the chances are that we will overhaul them at that place, where we will get plenty of help." "But if we do not find the 'Yonah' ready?" was asked. "Why, then, so far as I can see, we are done," was the reply.

They did not stop to speculate, but were all this time pressing on at the very highest speed possible. It might well be that one minute would make all the difference between finding the "Yonah" and her starting back to the iron works, miles away. But there was a sharp interruption just before they came in sight of Etowah, and while they were straining every nerve and looking forward to see if the smoke of the engine was yet visible. We had taken up a rail, and there was a crash, a sense of falling, and they found themselves lying, hand-car and all, in a heap at the bottom of a ditch! If the embankment had been as high at that place as at many others on the road, all our danger would have ended, for no other party that day originated anything against us; but the ground was almost level; and except a few bruises, they were unhurt, and at once placed their car on the track again.

While doing this, they were greatly stimulated and hurried by noticing the smoke and steam of the "Yonah," which they could see across the long bend on their side of Etowah River. If they could only make the distance, a little more than a mile, before the engine went back on the branch! With all their power, like men working for their lives, they drove forward. They were none too soon. The engine was on the main track still, and the tender was just being turned on the half-moon turn-table, preparatory to starting back. But the people there saw the furiously driven hand-car, with the shouting, excited men on it, and at once suspended their work and gathered around, to know the cause of these frantic gestures. Fuller had not much breath left to spend in talking, but managed to say that the Yankees had taken his train, and that he wanted their engine, and all of them with their guns to follow in chase. He, with every man of his party, was well known and there was no stopping to question. Their very appearance, streaming with sweat and almost exhausted, bore witness to the urgency of their haste. A score of strong arms whirled back the tender on the turn-table, and pushed it and an empty coal-car up to the engine, while a number of Confederate soldiers who were waiting to take the next train southward to Big Shanty, piled in also.

Now they were off with a strong, well-armed party, and the chase was on more equal terms. "The 'Yonah,'" in the words of Fuller, "was not a strong engine, but had large wheels, was as active as a cat, and with a light load could run very fast." She was now just in the service adapted to her, and her drivers called out all her powers. As they flew over the ground, it was a refreshing contrast to the exertions on foot or hand-car. The thirteen miles were made in sixteen minutes. If the extra trains at Kingston only entangled us as long as Fuller hoped, the whole affair would soon be brought to a final issue! He did not dread the fight that was likely to follow, for our number was only reported even yet by those who had seen us working as eight, while he had some twenty well-armed men with him at this stage of the journey; and there were the crews of the four or five trains at Kingston. Indeed, Fuller and Murphy might be excused if they rather feared that "the Yankees" might be captured before they arrived.

But where were we while this train was flying toward us? Lying still on the side track at the left hand of Kingston station, enduring those agonies of suspense and intense alternations of hope and fear which were harder to bear than all the exertions of Fuller and his companions! The local freight train came as previously narrated, and had drawn down the road to let the extra follow it, and still give us room to haul out above. Long and tedious was the waiting. But when we almost despaired the extra came. But alas! on this train, also, was a red flag! On being questioned as to the meaning of another train, the conductor said that there were too many cars and too great a load for one engine, and that another section was made up, and would be along shortly. The delay for these two trains had already been little less than an hour; and here was a third train, still blocking the road before us! How Andrews wished that he had taken the risk of running out in the face of the first extra and had tried to make, at least, the station above! We could easily have succeeded But now, in the absence of some telegraphic message—and no report was made to that office, while Andrews hardly thought it prudent to telegraph for instructions!—it would be madness to run out between the two sections of a belated train. It was better to wait, even if that entailed the risk of a fight. For this possibility Andrews made ready; he said to Knight:

"Go back and tell the boys, without attracting attention, that we have to wait for a train that is behind time, and for them to be ready to jump out at the signal, if needed, and fight."

Knight sauntered carelessly along down the train, just as if he was tired to death with waiting and did not know what to do with himself; and leaning against our car, without turning his head or eyes toward us, said in a low tone which we heard perfectly:

"Boys, we have to wait for a train that is a little behind time, and the folks around are getting mighty uneasy and suspicious. Be ready to jump out, if you are called, and let them have it hot and fast."

We did not know how many of the "uneasy folks" there might be about; and so unbearable had become the suspense of being shut up in that dark car, and hearing the sound of voices outside, without being able to distinguish the words or know what was going on, that a command to spring out and begin a deadly strife would have been welcomed as a relief, without much regard to numbers. We said we were ready; we had been ready at any time the last hour! Still we carefully examined the priming of every revolver, and saw that reserve ammunition was in pockets within easy reach. We did not intend, if it came to a fight, to shoot at long range, but to close right in, where every shot would be deadly. That we could, with the surprise of an unexpected assault, and firing each time to kill, have cleared the station of four times our number, I have never doubted. Could any situation be imagined by poet or novelist more trying than that of this carload of Union soldiers shut up in the midst of Rebel trains!

But we were not called upon to quiet the "uneasiness" outside, which was fully as great as Knight had described, in any such summary manner. Andrews played his part with surpassing skill. The people around, and especially the old switch-tender, began to grumble something about being sure that all was not right. A good many questions were asked as to why Fuller with the regular train was not along by this time, and why the superintendent of the road at Atlanta had not sent notice of the powder train. Andrews answered each suggested question very briefly and plausibly, but without appearing at all anxious about their opinion; grumbled a little about the bad management of a road that would allow its track to be blocked at a time like this; gave accounts of himself in the camp of Beauregard, with an air so confident and truthful that no one ventured to question him. I think there was only one thing aside from the dangerous delay, which he really dreaded. He kept very near the telegraph office, and without seeming to do so, closely watched the operator. The attempt to telegraph any kind of a message up the line, would have probably brought on an immediate collision.

Brown narrates a curious little episode as occurring here. He noticed a man who watched Andrews for a short time, and then, when no one else was near, stepped close to him and handed him a large and seemingly well-filled envelope. Andrews smiled, and placed it in his breast pocket. Brown intended to ask about it, but more pressing business put it out of his mind. Probably this was an incident of Andrews's contraband trade, and the package contained an order for goods with the money to pay for them. If Andrews was recognized on this raid by any who knew him, as he seems to have been on more than one instance, he was not at all compromised by such meeting, as long as the train was not known to belong to the Yankees. But, however successful, this day's work must have ended the *rôle* he had been playing, and forever closed the South to him.

Fuller and his party were now not many miles away, and were making more rapid time toward us than had ever been made on that road before. But we knew nothing of that—supposed that we were still an hour ahead of any pursuit that could be imagined. We had been at this place one hour and five minutes! It seemed to those shut up in the box-car nearer half a day! and when the whistle of a train was heard, which fortunately for us was *first from the north*—not the pursuing train from the south—it was about as welcome as the boom of Mitchel's cannon with which we expected to be greeted in the evening when our work was done. This last extra came up to the platform as the others had done, and was at once ordered by Andrews to draw on down the road that we might have room to go out. The conductor obeyed without hesitation, and this obstruction was removed.

It only remained to adjust the switch so that we might again get on the main track. This Andrews directed the old switch-tender to do; but he had been getting in a worse and worse humor for the whole of the last hour; he had hung up his keys, and now roughly declared that he would not take them down again until Andrews showed him by what authority he was ordering everybody about as if he owned the whole road! We who were shut up in the box-car, heard the loud and angry voice, and supposed that the time for us to act had come; yet we waited for our leader's command as we remembered how he had counselled us against being too precipitate. But he only laughed softly as if the anger of the old man amused him, and saying " I have no more time to waste with you," he walked into the station, to the place where he had seen the keys put up, and taking them down, went quietly and swiftly out and made the change himself. The tender's wrath knew no bounds at this; he stormed, declared he would have Andrews arrested, would report him, and many other things. Andrews then waved his hand to the engineer, and as our locomotive came promptly up, he stepped on board, and we glided out on the main track, and were off!

It had been a fearful ordeal, but it was well met. The three men, Brown, Knight, and Wilson, who were outside, declared that they did not see the slightest indication of fear, chagrin, or impatience on the part of Andrews, save what he exhibited when telling how much Beauregard was in need of his ammunition, and what a shame it was that the road should be blocked by any ordinary travel when the fate of their brave soldiers was trembling in the balance. Andrews had explained that it was because he could not get his orders filled without ruinous delays by the ordinary channels, that Beauregard had sent him to bring this powder through by force, if necessary; and declared that if the officers at home did not support the army in the field better, martial law would soon be proclaimed! Such grumbling and threats were applauded by those who wished to be thought especially loyal to the rebel cause. For at least half an hour no distrust was shown.

The hour and five minutes we were at this station added to our two hours run made us now three hours and five minutes from Big Shanty. Fuller was three hours and nine minutes on the way. He came in sight of Kingston just four minutes after we had glided around the sharp northward curve beyond; so near were we to the final collision at this place!

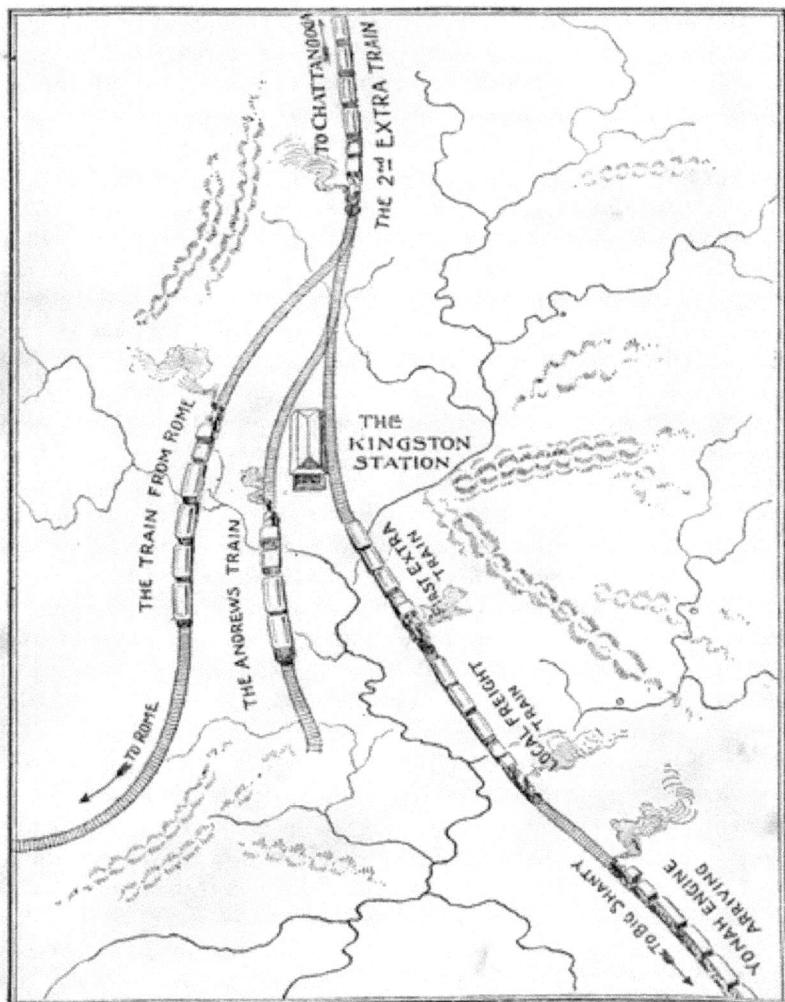

The Gathering of Trains at Kingston.

CHAPTER XI.

AN APPALLING STRUGGLE.

A S soon as we were well out of sight of the station we stopped, and Scott, with a man at each foot to give him a good start, was in a moment at the top of the telegraph pole, the box was knocked off, and the wire cut. We wanted no message of inquiry sent ahead, preferring ourselves to tell the story of the impressed powder train and Beauregard's need of ammunition. While this was being done, others threw a few obstructions on the track. When once more on board we noticed a quickening of speed that after our long rest was delightful. We had been running slowly since leaving Big Shanty, but now Andrews said to his crew, "Push her, boys, push her." Wilson heaped in the wood, and the fire which was but moderate when we left Kingston, was soon roaring, and great clouds of smoke escaping. Our leader's intention was to reach Adairsville in a few minutes, in order to meet two trains there which were now over due. These were the through freight and the Southern passenger trains, and they would wait for us ("Fuller's train") there. Our terrible delay at Kingston was in every way most unfortunate. If there had been no extra trains we could by this time have been at Dalton, forty miles further up the road, with the Oostenaula bridge burned behind us, and, these two trains passed, leaving no further serious obstacle to contend with.

But while so anxious to reach Adairsville, the next station above, where there was a side track, it would never do to leave the way open from Kingston, as the distance is only ten miles, and if the enemy choose to make up a pursuing train at that point, on account of suspicions formed, it might be very embarrassing. So it seemed that we had scarcely got under full headway at a tremendous rate of speed, before the tender-brakes, all that we had on our train, were put sharply down, and we were on the ground almost before the train had stopped, and under the energetic leadership of Andrews were hard at work lifting the track, the readiest mode of effective obstruction in our power. We again cut the wire, and also loaded on a large number of ties and other kinds of wood to be used in burning the bridge. We were the more anxious for abundant fuel as the ceaseless rain, which was now severe, would render kindling a fire without much wood, slow and difficult. No time was lost in these operations, as but a few could work at track-lifting at once, and others were ready for any other useful employment.

Lifting a rail seems easy enough, but it was far from easy in practice. The rail is long and heavy; it is securely bolted to other rails, and fastened with great spikes driven into solid oak ties, which in turn are deeply imbedded in the ground. This was the first place we wished to take up a rail very quickly, and accordingly we were far more sensible of the difficulty than when we had abundance of leisure. We were not excited, for we believed ourselves an hour ahead of any probable pursuit: but to pass the two trains still before us, and hurry to our real work of bridge burning, was an ardent desire. Slowly we drew out spike after spike, battering out the great nails as rapidly as possible with our one iron bar. I cannot tell how many minutes we spent, but time went by swiftly. The large load of ties which were not far away, was a great acquisition, and were all on board before the stubborn rail was half loosened.[1]

The rail was loosened at the southern end, and for perhaps two-thirds of its length, was cleared of spikes. Eight of us, including our strongest men, took hold of it, to try to pull out the remainder by the rail itself. But they were too firmly fixed; and we were about to give up the attempt and wait to batter out a few more, when away in the distance, we heard, faintly but unmistakably, the whistle of a locomotive in pursuit! But faint and far off as it was, no sound more unwelcome ever fell on human ear. Before us, only two or three miles away, there were two trains possibly blocking the track; and behind us a pursuing engine, which in a minute or two more would be upon us. It nearly seemed as if our race was ended! But we did not pause for moralizing; we lifted again and with every particle of strength, as men lift for life. The strong rail bent under the terrible pressure and snapped with a dull *twang!* All of us tumbled in a confused heap down the grade, but in a moment were on our feet, and hurrying towards the car, taking our precious half rail with us. For the time we were saved! No matter who the pursuer was he would be arrested by that break, and give us time, with favoring fortune, to pass the trains above.

Once more we were on board and away! We would soon know whether we were to have a clear track at Adairsville or to repeat the vexatious and

---

[1] Here, so far as I can determine, after a most exhaustive examination, occurred an incident which has given rise to much controversy. In "Daring and Suffering," of 1863, it was narrated substantially as above; but afterward, some Confederate accounts induced me to think that myself and comrades had mistaken the place; for concerning localities our recollections would naturally be less definite than those of our pursuers; but full investigation shows the balance, even of Confederate authority, to be on the side of the story as here given; and it certainly renders all the events that followed more clear and intelligible.

dangerous experience of Kingston. Once more the engine was given full force; we in the box-car were thrown from side to side, sometimes a little roughly; but this did not diminish our joy over the rapid motion which was "devouring" the distance between us and our friends in Tennessee! As we came in sight of the station, there to our great satisfaction lay the freight train, which, indeed, had long been waiting for us, as we were now a half hour behind the time of Fuller's passenger train, and also waiting for the morning passenger train from Chattanooga, which should have overtaken the freight at this point, but which was also late. Indeed, the panic in Chattanooga, and the extra trains on the road had disordered the whole schedule, and enormously increased our difficulties. As we came near the station, speed was slackened, and we stopped on the main track beside the through freight. Andrews at once answered the usual storm of questions and asked others in turn. He heard still more of Mitchel's operations, how he seemed to have captured all their trains on the western road, so that for twenty-four hours not a car had got through, and that the telegrams were being interrupted further and further up the road, so that, from every indication, he was coming to Chattanooga. But Andrews was still more interested in asking news of the down passenger train, which was now half an hour late. No information was received, but the freight conductor had determined to run on south on the arrival of Fuller's train, in harmony with their rule of railroading at that time, by which a following train was to be waited for only a certain length of time, after which the waiting train had the right to proceed. Andrews approved of that intention, saying that Fuller with the regular train would probably wait for him at Kingston. Andrews might have held this train here by giving a message as from Fuller, but he preferred to get rid of it, so that if compelled to back before the belated passenger, it might not be in the way: and if compelled to fight, the fewer of the enemy the better: otherwise, its running down to the place of the broken rail was undesirable. The conductor said to Andrews:

"You of course will remain here until my passenger train comes, and tell them to overhaul me at Kingston!

"No," returned Andrews, "I must go at once! the fate of the army hangs on my getting promptly through with those carloads of ammunition. Suppose the Yankees attack Beauregard! He has not powder enough for three hours' fight."

This was a startling possibility, and forgetting all about Mitchel being in the way, the conductor (the men on both sides had heard the conversation, but had not joined in it) patriotically said :

"Get through by all means; but you will have to run very slow and put a flag-man out on every curve, or you will have a collision."

Andrews answered quickly: "I will attend to that;" stepped on his own engine and motioned to Knight who was still at the throttle. The latter hearing the words about running slowly had put on the steam in a gradual manner, and the engine glided away at a moderate rate of speed

But this was not to last; neither was any flagman to be sent ahead; there had been delays enough. The time had come when it was wise to take a terrible risk. We dared not wait for the passenger train because of the pursuers we had heard, and of the freight which had started toward the break; and we *must* reach the station above before the passenger started out! From Adairsville to Calhoun, the next station that had a side track, is a little more than nine miles. The road runs directly north, is almost straight, and but little removed from level; this is the most favorable stretch for running on the whole line. Andrews said to his comrades, "Make her show how fast she can go; every second saved in getting to Calhoun counts." The effect of giving such orders to men whose nerves had all morning been thrilling with suppressed fire may be imagined! The engine was in the finest running condition. Knight had oiled it carefully during the long waiting at Kingston, and again, in part, at Adairsville, and a heavy pressure of steam had accumulated during the pause at the latter station. Now the full force of the mighty power was turned on at once, while oil was poured on sticks and these fed into the furnace. The three cars and twenty men were no load for the powerful engine, and it sprung to its work with a shock that nearly took every man from his feet! The race against time which followed was grand and terrible. The engine seemed to be not so much running as coursing with great lion-like bounds along the track, and the spectacle from the locomotive as it rose and fell in its ceaseless rapid motion, while houses, fields, and woods rushed by, was wonderful and glorious, almost worth the risk to enjoy! In the box-car, we were thrown from side to side and jerked about in a manner that baffles description. The car was so close to the engine that it felt every impulse of power and there was no following train to steady it. Many times we were startled with the momentary conviction that we were off the track; but there was no cessation of our rapid flight. We hardly knew what it meant, and though we pushed our door partly open, the risk of being thrown out was too great to permit us to open it wide; and gazing at the panorama that flitted by, with lightning-like rapidity, we could gain no clue to this frantic and perilous chase, for there was no indication of a following train that we could perceive. There was no danger of being seen in the opening of our door, for the rapid flight of the train would have attracted all the attention that anything upon the car could. Andrews scarcely looked ahead while making this run. Brown and Knight, however, did keep a sharp lookout, simply for the purpose of seeing when we came near the station that they might shut off steam, and be able to stop there. They had no hope of reversing or stopping, if they saw the belated passenger train approaching. As well try to reverse a cannon ball in its flight! if the train started out from Calhoun before we came in sight, it was simply and inevitably death for every one of us; and the people of the other train would not have fared much better.

Our fireman, J. A. Wilson, gives a very graphic account of this fearful effort to conquer time:[1]

"Our locomotive was under a full head of steam. The engineer stood with his hand on the lever, with the valve wide open. It was frightful to see how the powerful iron monster under us would leap forward under the revolutions of her great wheels. Brown would scream to me ever and anon, 'Give her more wood, Alf,' which command was promptly obeyed. She rocked and reeled like a drunken man, while we tumbled from side to side like grains of popcorn in a hot frying-pan. It was bewildering to look at the ground or objects on the roadside. A constant stream of fire ran from the great wheels, and to this day I shudder as I reflect on that, my first and last locomotive ride. We sped past houses, stations, and fields, and out of sight, almost like a meteor, while the bystanders, who scarcely caught a glimpse of us as we passed, looked on as if in both fear and amazement. It has always been a wonder to me that our locomotive and cars kept the track at all, or how they could possibly stay on the track. At times the iron horse seemed literally to fly over the course, the driving wheels of one side being lifted from the rails much of the distance over which we now sped with a velocity fearful to contemplate. We took little thought of the matter then. Death in a railroad smash-up would have been preferred by us to capture."

Andrews kept his watch in his hand, seeming to notice nothing else, for time was the only element in this part of our problem; and he and Knight, who looked on the same watch, always joined in declaring that the interval of nine miles between the two stations was run in *seven and a half minutes;* and this not upon a magnificent road with steel rails as that road is to-day, but over a poor and neglected track! It must, in candor, however, be allowed that Andrews probably reckoned the interval from losing sight of Adairsville until coming in sight of Calhoun. When near the two stations he would be otherwise engaged; and thus the rate may have been little over a mile a minute—surely enough for all the fear, wonder, and sublimity of motion!

Our escape on this run was exceedingly narrow. The passenger train had begun to move out before we arrived; but it had only just got under way while we were slackening up for the station. A minute earlier in their starting, would have ended the raid. But seeing us coming, and our whistle sounding out loud and peremptory, they backed before us up the track, and the proper officer obligingly opened the switch to let us on the side track. Of course this was done as much in the interest of the passenger train, which could not go on till we were out of the way, as in ours. But they did not go on for some time, and we were obliged to await

[1] Adventures of Alf. Wilson, Toledo, O., 1880.

their movements. In backing they had gone far enough, not only to give us room on the side track, but also, as their train was a long one, to completely block the far end of it, and we could not proceed on the main track until they should pull ahead. Before doing this they naturally wanted some explanation. The lateness of the regular train; our having Fuller's engine, without him or any of his men; and not least, the manner in which we had swooped down upon them like some beast of prey, coming without any signal man ahead at a time when under railroad rules·they were entitled to the road—all this which only some most urgent occasion or public calamity could excuse, called for explanation. Andrews calmly told his story, and the urgent need of ammunition was felt to justify every thing; and all the questions were asked and answered that are common among railroad men on meeting. Yet Andrews would have talked little and would have made a very short stop, had it not been for the manner in which the passenger train bound in his own. We had a good right to be uneasy here, for we had not cut the wires between this station and Adairsville because we had not dared, in the terrible urgency of reaching Calhoun, to delay even for this purpose. A question might come on the wires at any moment which Andrews, with all his adroitness, would not find it easy to answer. Neither had we put any obstructions on the track. This latter omission prepared the way for another race against time, only less swift and fearful than our own.

Thus we were again delayed. Andrews tried gentle and indirect means to persuade the conductor of the passenger train that it was perfectly safe for him to run down and get to Adairsville before Fuller's passenger train. But he was not easily persuaded. The bare escape from collision with our train had shaken his nerves too much for him to wish at once to repeat the experience. Neither did he seem at all in a hurry to move his train ahead and let us out on the main track; but as his train was the only obstacle, it would not have been long, had he continued obstinate, until the reserve force of our party would have been brought into requisition. It may be said here that Andrews was perfectly sincere in telling him that there would be abundant time for him to reach Adairsville before Fuller with his train would be along. We did not think that Fuller would be along that day, and with his own train he was not. But as matters were, if the Calhoun man had allowed himself to be persuaded to start southward, a fearful collision would have ended all possibilities of pursuit, and left us free to burn bridges at our leisure. Here was another of the narrow escapes made by the enemy. To understand this it is necessary to recur to Fuller and Murphy, who were within two or three minutes of Kingston when we left that place.

They were terribly disappointed when they found themselves stopped quite a long distance below Kingston by three heavy freight trains, and learned in a brief conversation with the engineer of the nearest, and the persons who had run down that way on hearing their whistle, that their game had flown. They heard with wonder how long the commander of the captured train had been held there, and how he had succeeded in concealing his real character. The formidable nature of the enemy ahead was now clearly revealed, but it looked for a moment to Fuller as if all his labor had been for nothing, and that he would be able to continue the pursuit only after a ruinous delay. To back all these trains up the heavy grade so that he could get on the side track, and then down again to get off at the upper end, would require an amount of "see-sawing," that would give the captors of his train a hopeless start.

Here arose a difference of opinion between Fuller and Murphy, who up to this time had worked together in perfect accord. Murphy ran ahead and cut loose the "New York," the new and good engine of one of the freights, attaching it to the car which had brought their tools from Etowah. He then called to Fuller to move the Rome engine back out of the way that he might come round on the "Y." But Fuller had different plans.

The Rome engine and train had stood on its own track all this time waiting for his coming; the Rome branch led into the main track above all the impediments. Why not take that engine? No sooner thought than executed. Fuller had taken one foot race that day, and he now took another, shorter but not less important. The engine was headed already toward Chattanooga with only one car attached, and in the most favorable position. There was abundance of volunteers, and no need of explanations, for now everybody was sure that the impressed powder story was false and absurd,—had thought so all the time! Conductor Smith, of the Rome train, gave it for the service at once. All was done so quickly that Murphy saw them start and had to run at his best speed to keep from being left behind! Fuller probably made a mistake in not taking the "New York," as the other engine was much inferior, with small wheels and incapable of great speed. But the distance in which they could use it turned out to be short, and being driven at the height of its power, it is not probable that much was lost; while the time spent in changing the freight trains out of the way might have cost the Oostenaula bridge.

A mile or more from Kingston they found some ties on the track at the place where we cut the wires, and were obliged to stop and throw them off. Of course an effort was made to send a message from Kingston to Chattanooga as soon as Fuller arrived, but we had cut the wire too quickly for them. Continuing on the way, they came in a few minutes to the place where the track had been torn up. A Southern account says that sixty yards had been removed; but this is a gross exaggeration. Track lifting was only intended to make the road temporarily impassable, and one broken rail answered this purpose as well as a dozen. Had there been a regular track layer with the pursuers, a rail would not have caused a great delay; but it was in all cases sufficient for its purpose on this day.

Though we had heard the whistle of the pursuers, they neither heard nor saw us at this point, and came near wreck; but they were on their guard because of the similar break which had caused their fall fro   the hand-car, and by great effort and reversing the engine, they were able to prevent an accident. But their progress seemed to be completely barred. As usual no one but Fuller and Murphy seemed to have the least idea of what to do; in fact during the whole day every hopeful plan of pursuit sprung from their indomitable energy. Too much credit (from the Confederate point of view) cannot be given to them. They were already practiced in foot travel, and once more set out in that manner; all the rest, remaining behind, had no further influence on the fortunes of the day. But at full speed the two pedestrians pushed over the slippery and muddy road and through the driving rain. They felt sure of finding the freight train, or the passenger, either at Adairsville or further on this side. Should they be obliged to take the terribly fatiguing run to the station itself they would probably be too late; but they were determined to do their utmost.

Notice how all things seemed to work against us on this eventful day. If we had not stopped to take up this rail at all, we would have had abundant time to reach the freight and start it south, as we did; and the freight train running south, and Fuller's train running north at full speed would have produced a frightful collision, which could scarcely have been prevented; for the freight man had been induced to set out by the representations of Andrews, and Fuller, on his part, probably believed that Andrews was still running on slow time and had not reached the station above. The stopping to lift this rail, *as it turned out*, was probably the greatest mistake Andrews made. On the other hand, if the freight had waited for Fuller, so great a delay would have ensued that the Oostenaula bridge, which we were now very near, would have been in flames.

But the pursuing pair had scarcely been well breathed in this third foot-race, when they heard the welcome whistle of a locomotive. Fuller, who was ahead, stopped in a place where the view was clear, and gave the signal of danger; the freight was checked up as quickly as possible, and while Fuller told in a few words what had happened, and what he wanted, Murphy who had been distanced, came up, and they sprang on board, and took command. With all the power of the "Texas," which was one of the very best engines on the road, and the best the pursuers had yet obtained, they pushed backward toward Adairsville, and learned that Andrews had left a few minutes before. Fuller took his place on the last freight car which was now the front of the train, and directed their movements. Murphy was the official superior of all the engineers on the road.

He stood by the lever to render assistance when needed and all his orders were cheerfully obeyed.

It was not long till they were back at the station, when Fuller jumped off, threw the switch over to turn the freight cars which were detached at the same time and allowed to run with their own momentum on the side track; and then as the last one passed by, he changed the switch back, sprung on the engine and outran the cars which continued to move parallel with him ! This was quick work. They now had a comparatively small crew, but they were all armed with guns, and loaded on the tender and engine alone. It was true that the engine was reversed, but this, while it is somewhat less handy, does not diminish strength or speed. The first question which confronted the pursuers was whether to risk running up to Calhoun in the face of the delayed passenger train. They did not hesitate, as the way had been made clear for them. It was less than ten minutes since Andrews had left, promising to run slowly and carefully, and if he kept his word, he could be overhauled and enclosed between the two trains before he could reach Calhoun; and even if he did not, the danger of collision would be borne by the train ahead and not by the following one. The marvellous flight which Andrews had made was not, of course, dreamed of. Had Andrews been able to persuade the passenger-conductor to push out, as he did the freight-conductor, Fuller's and Murphy's career would have ended.

These indomitable men now had an excellent engine and ordered full speed. The whole distance of nine miles was made in little more than ten minutes ! There was no obstruction of any kind, and they trusted to the fact of being so close behind Andrews to assure them against any lifted rail.

Before they reached Calhoun, however, Andrews was released from his perilous position. After he had chatted with the conductor and engineer of the down freight for some time and found them indisposed to go on their way, he said in the most matter-of-fact and positive manner: " I must press on without more delay. Pull your engine ahead and let me out." When the order was given in this direct form they were obliged to obey, or give a good reason for refusing; and it may be considered certain that if they had delayed, though Andrews did not threaten violence, yet our engineers would at once have taken control, and executed the order, probably not without bloodshed.

At last we are on the main track with no train between us and Chattanooga ! and if the reports from Huntsville are true there is no obstruction west of that town, as all travel is cut off by Mitchel. There is reason for exultation on our part. An open road ahead and scores of miles of obstructed and broken track behind us ! For the whole morning we have been running with a train right in front of us, or waiting for a belated one. We had passed five trains, all but one either extras or behind time—a wonderful achievement! now the way is clear to our own lines; and the "Y" at Chattanooga is no more difficult of passage than any of the many side tracks we have already successfully encountered. No small amount of the exultation we felt on first taking the train was again ours, as we rushed rapidly on for a mile or more, and then stopped to cut the wire, and to take up a rail (as we hoped) for the last time. The Oostenaula bridge was just ahead, and when that was burned, we would simply run from bridge to bridge, firing them as we passed; and no more of this hard drudgery of track raising and still more terrible work of sitting silent and housed in a dark car waiting for trains to arrive! We had heard the whistle of a following train a dozen miles back; but it probably was one from Kingston, and if not wrecked by the broken rail, would return there for tools. We knew nothing of Fuller's and Murphy's pursuit, and if we had been told the full story, as already narrated, we would have thought it too wild and improbable even for good fiction.

But it was expedient to take up this one rail more, before we finally changed our mode of operations. A piece of torn track had been put before or after every train that we had met. It was well to put a broken road behind this passenger train also, that it might not turn back after us on any sudden suspicious freak, and come upon us while working at the Resaca (Oostenaula) bridge. The crisis of our fate approached, and we believed it would be triumphantly passed. Nothing had as yet been lost but time, and if we were fairly prosperous for fifteen minutes more, all would be regained, and the fulfillment of all our hopes, as far as human prospects could reach, be in our own hands.

No wonder that we worked gladly and cheerfully. Scott climbed the pole with even more than usual agility. Some worked at the taking in of all kinds of combustibles, for we wished to be well provided for the bridge. Every stick and piece of wood we could get hold of was soaking wet, but by breaking and whittling, they could be made to add to a flame, and from the engine, which was kept full of wood for the purpose, we could give a good start to a fire. We had only one iron bar to drive out our spikes; a bent "crow's foot" would have been worth more than its weight in gold; but we hammered away with what we had, and spike after spike was drawn. Here I saw Andrews show real impatience for the first—I am not sure but I may say the only—time. He had altered his dress, throwing off the cape and high hat that he wore while at stations, and had a small cap on, which greatly changed his appearance. The nearing of the time when his plans would all culminate in success seemed to thrill and inspire him. He snatched the iron bar out f the hands of the man who was wielding it, and—though we had strong and practiced workmen in our party—I had not before seen the blows rained down with such precision and force. Some say that he uttered an oath on this occasion, but though standing by I did not hear him: the only words I did hear being directions about the work, given in his mild tones but with quite an emphatic ring of triumph in them. He wanted that rail up in the fewest number of seconds and then —the bridge ! There were several using a lever of green wood and trying to tear up the end of a rail from which the spikes had not yet been drawn; but the lever bent too much, and a fence rail was added and we lifted again. At that instant, loud and clear from the South, came the whistle of the engine in pursuit ! It was near by and running at lightning speed. The roll of a thousand thunders could not have startled us more.

What could we do ? At the end where we had been prying the rail it was bent, but it was still too firmly fixed for us to hope to lift it, or break it like the last. But we did the best in our power; we bent the loose end up still further, and put the fence rail carefully under it with the hope that it would compel the pursuers either to stop and adjust it, or throw them from the track, and then piled into the car and engine with a celerity born of long practice, and with one of its old bounds that jerked us from our feet,—for Brown and Knight threw the valve wide open,—the "General" bore us rapidly on. The impatience of Andrews to reach the bridge had not been diminished by the appearance of this new element in the situation.

Here our pursuers were greatly startled. Their story had been swiftly told when they reached Calhoun, and the engine and tender of the passenger car with a reinforcement of armed men followed them up the road. Fuller stood on the tender of his own train, which was in front, gazing intently forward to see if there was any dangerous obstacle or break in the track, such as they had already many times encountered. Soon he beheld us at work with feelings which cannot be described. Before getting near enough to see our number, we had mounted and sped away, and he saw with exultation that we had not broken the track, and that there seemed to be no obstruction. With full speed he ran on till too close to stop, and then beheld what he believed, at first, to be a broken rail and gave himself up for lost; but it was on *the inside of a curve*, and as an engine running rapidly, throws most of its weight on the outside, when he ran on it, the bent rail was only straitened down, and they were safely on the other side of this danger. The next train which followed almost immediately after, did not notice the obstruction at all There is scarcely a doubt that two minutes more, enabling us to finish getting a rail up at this point, would have given the control of the day into our hands, for there were no more trains on the road either to delay us or to be turned back after us. But as it was, the "Texas" pressed on after us without the slightest loss of headway.

The coming of this train before the track was torn up was by far the most serious misfortune that we had yet encountered. But might it not still be overcome? The plan which first presented itself to the undismayed spirit of Andrews was to use two of our cars as projectiles and hurl them back at the enemy. This was more in accordance with his genius, which delighted in strategy, than the plain course which most of the soldiers would have preferred; that is, a straight out-and-out fight with the pursuing train. Accordingly our engine was reversed—could we have selected a down grade the chance of success would have been better, but we were coming so near the bridge that we could not delay to choose—and when the speed in this way had been checked and the pursuing train was quite close and still going fast, we uncoupled, and bounded on again. But the skillful pursuers were not thus to be beaten. They saw what we were about, and checking their headway when the car was dropped, they also reversed, and coming up to it with moderate force, coupled on, which was easier because their tender was in front.

The bridge was now just at hand. What should we do? To leave it intact was to be thought of only in the direst necessity. We had carried our ammunition—the fuel we had gathered—into our last car, and while it was not as good as we would have liked, yet in a little time we could make a fire. We now punched a hole in the back end of our car—in fact we had done this in passing from one car to another previous to dropping the last one,— and now began to let ties fall out on the track while we ran. They followed us "end over end" and showed a most perverse disposition to get off the track, but a few remained. This moderated the speed of the pursuing engine, which was a help that we sorely needed, for it was now evident that they either had a faster engine than ours, or better fuel. The latter was certainly the case, for we had been using wood very rapidly without any opportunity for a long time past to replenish it.

The first feeling of despondency of the whole route took possession of us as we approached the bridge with our pursuer close behind. The situation was in every way unfavorable for us. If we passed by without leaving it in ashes we felt that one important part of our business would be undone even if we were completely successful afterwards in evading pursuit and destroying the Chickamauga bridges.

Murphy expresses the opinion that we made a great mistake at this point. There was at this time a long and high wooden trestle by which the Resaca bridge is approached. As we came near, we "slowed up," and right in the middle of this trestle we dropped our last car. Murphy says that if we had but thrown it across the track the bridge would have been at our mercy, as well as all the bridges above. He is right, but the difficulties in doing this were greater than he thinks. To pull or push the car off by means of the engine, involved some risk of getting the engine itself off, which would have been fatal. We had no good means of moving it in any other way, and the element of time was all important. The pursuers were right behind, and while they could not have fired on us at effective range, they could very quickly have alarmed the town ahead of us, and then the track could have been obstructed to prevent our passage. It is easier to imagine what might have been done, than actually to do it, even if the circumstances were repeated !

There was no opportunity to turn and fight at this point. The town of Resaca was within a few hundred yards of the bridge, and any noise would bring help from that quarter. Besides our pursuers were armed with guns, and our only chance of getting at close quarters was by an

OOSTENAULA BRIDGE.

ambuscade. Had the day been dry, we could have flung faggots from the engine upon the roof, but now a fire even on the inside of the large frame bridge, would require careful nursing. With a station only a few hundred feet ahead, where the track might be so easily obstructed, and with the guns of the pursuers behind, we could not give time for this; so we slowly and reluctantly passed over the bridge, after dropping the car, and on through the village of Resaca. The pursuers "took up" this car as they had done the other, and pushed them both through the bridge, and left them on the Resaca side track.

It may be well to notice here how our ignorance of the enemy and his ignorance of us both inured to his advantage. There had been already many intimations among us that it would be well to turn and fight rather than to be chased any further. Had the real weakness of the enemy on the first train been known, Andrews would have certainly ordered the attack. On the other hand, if Fuller's party had known how strong we were he could not have induced them to continue the chase, even if the resolute conductor himself had not been willing to wait for help. It was believed at first that we were but four—the number on the engine. The estimate was never raised higher than eight, Murphy suggesting to Fuller even then, that it would be better to wait for the train behind and take on more men. But Fuller resolved to persevere and at least delay us at the risk of his own life. Had it been known that we were twenty, he and his slender band would not have been guilty of the madness of crowding on nearly twice their number, even if better armed, and sure of help at every station. But this madness, *this unreasonable pursuit*, the result of imperfect knowledge, served them well.

After passing Resaca, we again forced our pursuers back by dropping ties on the track, and not knowing whether it was a telegraph station or not, we again cut the wires. No obstructions were placed on the track at this point, but it was on a curve, and taking a rail which had been bent in lifting it I placed one end under the rail at one side and the other projecting diagonally toward the train on the other side. The pursuers saw us start, but seeing no obstructions they ran at a good rate of speed right over this rail! Their escape was marvellous. Persons on the tender jumped a foot high, and one of Fuller's staunchest helpers demanded that the train be stopped to let him off! he wanted no more such running as that! But Fuller, though considering this the greatest of their dangers, would not stop; and it was impossible for him to keep a closer lookout than he had done.

But what conjectures did we form to account for the unexpected appearance of this pursuing train? The story as given to the reader was totally unknown then, and we were greatly perplexed. The matter had great practical importance. Was this engine started after us by an authority which had also alarmed the whole road ahead of us? If so, we would do well to abandon our efforts for the destruction of bridges, and seek our own safety. Of one thing we felt sure: it must have been one of the trains that we had passed at Calhoun or Adairsville that was following; but why? There were three possibilities only to choose from. The first and least serious was that the suspicious conductor at Calhoun, who had been so unwilling to let us pass, had determined that we were impostors, and at his own motion had set out to follow us. If so, we would have to deal only with him, and might yet accomplish a part of our work. Or it might be that the freight had run to where we had broken up the track, had escaped wreck, and, turning back, had telegraphed ahead before we had cut the wire. In this case all the road ahead would be alarmed, and this was probable indeed. Or, once more, a messenger might have been sent down to Marietta from Big Shanty, and a dispatch sent to Atlanta and around the whole circuit of the Confederacy back to Chattanooga, and, before the wires had been cut, to one of the trains we had passed, with orders to follow us closely and prevent us from damaging the road until a train could be sent out from Chattanooga to secure our capture. If either of these latter conjectures were true—and they were the most probable—our race was almost run! We would be obliged to leave the road, and essay the far more difficult task of escaping on foot. If Andrews thought either of these probably true, it would fully account for his reluctance in ordering the capture of a pursuing train; for such a capture could do no permanent good, while every one of his party wounded in the fight would be disabled for the inevitable and terrible land journey ahead, and would surely be lost. In view of the almost hopeless situation as it appeared to us then—far worse than the reality, for the road ahead had not been warned as yet—the heroic constancy of Andrews, who continued to put forth every possible effort as coolly and quietly as if success had been within his grasp, is made brightly conspicuous. There were now three chasing trains; first, Fuller and his men with the locomotive of the down freight; second, the Calhoun passenger, which had immediately followed him and was not very far behind; and last, the train started from Marietta, and loaded with soldiers.

For a time after leaving Resaca we did not run very fast. It was evident that we could not get away from the engine behind us by mere speed; the only hope was in some way to disable them, or to obstruct the track; and we were obliged to be saving of our fuel. But now we were approaching Green's, a wood station near Tilton, and we were determined to have a fresh supply at any cost. So the last wood in the box, with a little of our precious oil, was shoved into the furnace, and Brown, who had now taken the throttle, turned on a full head of steam, and we once more flew along the track. At the same time, we who were in the box-car, put a line of ties along its floor and kept them moving to the hole in the end, and let them drop, as fast as possible on the track. This was rapidly exhausting our ammunition, but it was effectual in enforcing slowness and caution upon the pursuers. Fuller could not run rapidly in the face of such a succession of obstacles. He did the best he could, giving the signal to reverse whenever he saw a tie on the track, jumping off and removing it, and on again, when the engineer would start with a full head of steam, and reduce speed, as the engine gathered headway to such a rate as would admit of stopping in time, when another tie was seen. It was fearfully perilous; and the only wonder is that he was not wrecked long before the chase was done. But he probably understood that we were racing for the wood-yard ahead.

In Sight.—Ties thrown from the Car.

ed against him at this place. The facts are these: the writer and the man in the box-car had come to feel that there was no need of running so long before the pursuing train, which we could see to be a short one, with probably not much if any greater force than our own. Now while as many were at the rail as could find places to work—the process of lifting it with our imperfect tools was very slow, requiring more than five minutes—I said to Andrews, " We can capture that train, if you are willing."

" How ?" he asked. I answered:

"Find a good place on a curve where there are plenty of bushes"—(as the road had numberless curves, and ran mostly through woods, this was easy); " then let us put on some obstructions and hide; one of our engineers can run ahead a mile or two, and come back after us; when the enemy stop to clear the track, we will rush on them, and when we have captured them, our other engineer can reverse their engine and send it in a hurry down the track to clear the road of any more trains that may be following."

Andrews said, in his quiet way, " It is a good plan. It is worth trying," and looked around in a meditative manner as if weighing the chances. Then the enemy's whistle sounded, we saw them rush up to the obstructions we had placed on the track, stop by reversing, and labor as frantically to clear the road as we were doing in trying to raise the rail.

But our efforts were in vain. The stubborn spikes still held, and as they were ready to move on again, Andrews called out, " All aboard," and we dashed away. That was not the place to make a fight as we all knew, for revolvers against shot guns and rifles would have had no chance at long range; but from an ambush we could have been climbing into their engine and cars before they could pick up their guns, and the conflict would not have been many minutes doubtful. This was the nearest we came to what a Southern account called " open mutiny "—a mere respectful suggestion in the line of our work. No officer was ever more heartily obeyed than was Andrews during the whole of this day, and none of us said anything more about this plan for the time, partly because we felt that our leader was better able to judge what was to be done than we, and partly also, I must confess, because we thought he was only waiting for the best place to turn on our foes, and that we would soon have all the fighting we wanted.

The full speed of our engine was again called into requisition as we neared Dalton, and by the aid of a few ties dropped on the track, we were once more a respectable distance ahead. We needed this interval badly, for it was by no means certain that the switches at this point would be properly adjusted for our immediate passage through; and if not, serious difficulty might arise. We might have a battle with forces in front as well as in th   Dalton was the largest town we had reached since leav-

ing Marietta. Here a roa.   ' to Cleveland in Tennessee, where it
connects with the main line from ....hmond to Chattanooga (see map),
thus making a large triangle, or as a railroad man would say a great " Y."
At that time no telegraph wires were on this cross road; they were not
put up till 1877. There were also nu .rous side tracks, and a probability that cars might be left standing on some of them; and as we had
more than made up our hour's delay at Kingston, and were now much
ahead of time, there was no certainty of the road being rightly adjusted
for us. It was therefore necessary to stop at the opening of the switch,
which was fortunately a little way down from the large passenger depot,
which had a shed over all the tracks, and through which we had to pass.

Here the coolness and adroitness of Andrews shone out with pre-
eminent lustre. It is likely that when we had spoken of fighting a little
way back, his mind was occupied rather with the problem of passing Dalton, and of judging by what took place there whether the enemy was
warned. The train was stopped, he ran forward, observed that the track
was clear, spoke to one or two bystanders, and was back to his post in an
exceedingly short time. To one or two who had come up even in these
few seconds, he said, " I am running this train through to Corinth, and
have no time to spare;" and nodded to Knight, who once more put on
the full force of the engine—there was nothing to be gained by care in
avoiding alarm any longer, for the distant whistle of the pursuer was
heard—and we *rushed at the depot*, which then stood right across the
double track, and passed with fearful speed under its roof. Here Knight
got his most terrible fright. The darting into the partial darkness of the
shed was bad enough; but just at the far end, the main track bends
sharply to the left, and the swerve was so sudden, and the speed already
so high, that Knight believed that he was rushing on another side track
and that in a moment would come the awful crash. But instead, the engine instantly righted and he again saw the track straight before him. But
so quickly had we passed that we could not certainly determine whether
the people at the station had been warned or not !

CHAPTER XII.

THE FINAL RACE.

A MILE above Dalton, which was about as soon as the headlong rush of the engine could be checked, we stopped again, just opposite to where Col. Jesse Glenn's regiment of conscripts was encamped in a field. Their position, which was within two or three hundred yards of us, was probably not seen until we were close to them, and it was better to take the risk of their interference than to lose time by seeking another place for most pressing work. Again the wire was cut; but it was a second too late; for a message had just been flashed through, no doubt even as Scott was bringing it down. The usual obstructions were here piled on the track, and we again essayed to take up a rail, for the Chickamauga bridges were just above, and we wanted time enough to get them on fire; hoping that Fuller would stop long enough at Dalton for the purpose of getting his telegram ready, to allow us to finish the track lifting No men ever worked with more desperate energy, but all in vain; long before the rail was loose the pursuers were again upon us.

The race recommenced with all its speed and fury. The great tunnel was a short distance ahead—a glorious place for an ambush, where, in the darkness, the guns of the enemy would be of little value. If Andrews was disposed to fight, there would be the place of all others to do it. With the smoke of our train filling the space, with our party in ambush along the sides, success would be comparatively sure, if they had twice our number, for of course we could not tell how much of a reinforcement they might pick up at Dalton. But we kept right on through the tunnel and the village of Tunnel Hill beyond, where we carefully drew down to conceal our number from the curious eyes of any who might be about the station.

At Calhoun Fuller had received a small but very effective reinforcement—only a boy thirteen years old, but worth a dozen of ordinary men—by means of one of those apparently small circumstances which often influence the course of great events. At Chattanooga the chief officers of the road had become alarmed by receiving no dispatches from Atlanta, or the stations below Kingston. They therefore directed the young assistant operator at Dalton, to jump on the passenger train just then leaving that station and go south, sending them back word at each station passed till the cause of the trouble was found. He had only got as far as Calhoun when Andrews passed, and Fuller in a moment after. The latter hardly came to a stop before he saw the operator, and called him, and without a word of explanation, seized his hand and dragged him on the train. In the run up, however, he made all the necessary explanations, and wrote out the following dispatch:

FULLER'S TELEGRAM.

" To Gen. Leadbetter, Commander at Chattanooga:

" My train was captured this A.M. at Big Shanty, evidently by Federal soldiers in disguise. They are making rapidly for Chattanooga, possibly with the idea of burning the railroad bridges in their rear. If I do not capture them in the meantime, see that they do not pass Chattanooga.

WILLIAM A. FULLER."

This he gave to the operator, saying, " Don't speak to anybody or lose a second till you put that through to Chattanooga. Jump for the platform when I slow up, for I must push on and keep those Yankees from getting up a rail, or burning the bridges."

It was terribly quick work. The operator was at home in the office, and almost before Fuller had cleared the shed he was at the desk and the first words were over the wires. Whether they had time to get the whole message over before the cut is very doubtful, and not material, for the first two lines would answer every purpose. Had Fuller stopped at this point, and himself went into the office long enough to set the operator at work, it is almost certain that we would have had the rail up, and then all the bridges above that point would have been burned; though it is still possible that enough of the message might have been pushed through to secure our arrest in Chattanooga. This was another striking instance of the many narrow margins on which this day hinged.

Just north of Dalton an incident occurred which well illustrates the spirit that possessed the Confederates on this occasion. Benjamin B. Flynn, who resided in the immediate vicinity of Dalton, happened to cross the track just in advance of our engine. Noticing the fearful speed, he raised his hand and said to his companion, " That engineer will be discharged for reckless running." The gesture was seen and assumed to be a signal. He joined in the chase afterward, but that night when he returned home, a detachment from the regiment where we had torn up the track, dragged him from his bed, and tying him up, whipped him almost to death ! '

' See a full account of the affair, with Flynn's own letter, in the Atlanta *Southern Confederacy*, of April 20th, 1862.

As Fuller pressed on toward the great tunnel, even his resolute heart almost died within him, while all his party began to blame him for fool-hardiness. He feared to plunge into its dark depths. It was still filled with smoke from our engine; and he well knew that if we jumped off at the far end and hurled back our locomotive at him, it meant a horrible death to every one on his train; and he was by no means sure that we would not do it. Mr. Murphy, who had so ably stood by him all the while, here counselled prudence, pointing out all the perils of an ambush. But Fuller realized as apparently no one else did the desperate need of pressing on to save the road; and he had made so many escapes and been so marvellously favored, that a kind of fatalism took hold of him. He determined not to lose a minute, no matter what the danger might be. It may as well be said here that no prudent and common-sense kind of pursuit, such as possibly any other man would have employed, could have had the slightest chance of success. But even Fuller quailed as they dived into the cloud of smoke that hung around the entrance of the tunnel, and held his breath for a few seconds (they were still at full speed), till he saw with a sigh of relief a gleam of light ahead and knew that there was no other engine now in the tunnel! On he pressed, for he knew the value of the Chickamauga bridges ahead as well as we did.

But for the wetness of the day all his efforts even yet would have been foiled. We now did what had been in the mind of Andrews, doubtless, for some time past—what he might have tried even at the Oostenaula bridge had not the interval between that and Calhoun been so fearfully short. He ordered us to fire our last car, while we were running. It was said easily but was much harder to do. Everything about the car was as wet as it well could be. The rain fell in torrents, and the wood was drenched in the tender. It was by no small effort and skillful firing that the engine fire could be kept at the heat required for fast running. But desperate fingers tore everything combustible loose from the car, and smashed it into kindling. Some blazing faggots were stolen from the engine and the fire made to burn. The rapid motion with driving rain was an obstacle at first, but as we fed up the blaze and sheltered it as well as possible, it grew rapidly till soon but one could stay on the car and watch it, and all the others crowded on the tender and locomotive. The steam was now gradually shut off that we might come slowly upon the bridge and be able to leave the burning car just at the right place. We came to a full stop at this first Chickamauga bridge, a large one, and well covered.[1] Inside it was at least dryer than on the outside, and we doubted not that with time it would burn well. The only question was, "Will that time be given?" We added almost the last of our oil and nearly

---

[1] The bridges are of different pattern now and would be much more difficult to burn.

the last stick of wood—knowing that a wood station was not far ahead, and if this bridge could be made to burn well, we could have all the time we wanted to get wood and everything else. In fact we put life itself on this last throw, and left ourselves, in case of failure, hopelessly bankrupt. For a considerable time, as it seemed to us, though it must have been measured by seconds rather than minutes, we remained on the other side

Kindling a Fire in the Box-Car.

of the fire watching. Then the inexorable smoke of the foe was seen; the pin connecting the burning car with our engine was pulled out and we slowly moved on. Too clearly we saw the ruin of all our hopes! To wait the coming of our foes was vain. They were now near at hand, and we could see their guns, with which they would be able to fight us at long range. The car which, if the day had been dry, would long before this have filled the bridge with a mass of roaring flame, was burning faster than the bridge. To take it to another bridge was useless, for the drenching rain would have given it little chance to burn away from the shelter of the bridge. Very sadly we left the tall column of smoke behind. The pursuers saw the car, and realizing how serious their loss would be if it was permitted to consume the bridge, they pushed right into the smoke and shoved the burning car on to Ringgold but a short distance ahead, where it was left to smoke and sputter in the rain on the side track.

We were now on what proved to be our last run. I have often been asked if this day was not one of great fear and terror on the part of those who were engaged in the race. For my own part, I cannot honestly lay claim to any greater fear than I had often felt in ordinary military service. No matter what happened, there was the assurance that we still had one resource—the power to turn around and attack the pursuing foe. From the beginning, such a conflict had been present to my mind as a matter of course. Before leaving camp, this had been reckoned a natural consequence of our position. It had been frequently talked of among the men, and not one of them seemed to regard it with any more dread than an ordinary battle. We had been careful to select large revolvers for use, and not for show, and when we found the enemy gaining upon us, or our leader's plans for their destruction failing, we only felt or said that our time to strike would soon come. We did not have the boastful feeling that we were an overmatch for a large body of southern soldiers, for we all knew how desperately they could and often did fight; but of the ordinary citizens gathered up as we presumed our pursuers were, or even of conscripts, we had no great fear. That we had not our accustomed arms was a serious disadvantage, but this could be remedied by getting into close quarters; and we trusted that our leader, who had shown such wonderful skill in management, would be able to put us within short range of the pursuing train where we felt sure that we could quickly give a good account of it.

Probably the fact of Andrews having never been in battle, but always engaged in schemes where his own cool daring and sagacious planning counted for everything, and mere force for nothing, made him hesitate to order an attack which would throw aside all these qualities and determine the issue by simple fighting. A time was near when we would firmly have disputed our leader's command if there had been an officer of any authority among us who could have been substituted for him; but not until Andrews himself had definitely abandoned his authority.

Many times the question has been asked, "Why did you not reverse your engine, and, jumping off, let it drive back at the enemy?" What good could that have done? If their engine and our own had been destroyed, as was very probable, together with a considerable number of lives, we would only have been where we were before we captured the engine at all, except that the whole country would have been aroused, and our disguise thrown off. The second train would have been on the ground in a few minutes and the power of pursuit would have been undiminished. We had no wish to sacrifice our own engine until the last effort possible had been made. To merely destroy had no charm for us, when that destruction could neither promote our escape nor serve a military purpose.

We crouched down as well as we could in the tender while passing Ringgold, that the enemy might not see our number; and when beyond the town we arose and looked about us. The country was mostly wooded and rough, being much cut up by the branches of the swollen Chickamauga creek. We had no fuel, though we might have taken on a few water-soaked fence rails and broken them to burn; but what would have been the use? Every combustible scrap was carefully gathered up and thrown into the engine. Worst symptom of all, a large pair of saddle-bags which we had never seen Andrews without from the time of the midnight conference, together with his cap and some other pieces of clothing that he did not need for immediate use,were flung remorselessly into the furnace. Various papers went along. These were probably documents that he feared would compromise himself or others in case of capture. Such preparations were indeed ominous. But his next command—the last he ever gave to us as a party—was more dreadful still, and for the first time that day there shot a pang of mortal terror to my heart. Not the crash of the engine down an enbankment, nor the coming of another train of the enemy from the north shutting us between two fires, would have caused such a sense of despair and hopeless misery to steal over me. This was the order, which, as intimated before, our party, had they been properly organized, would not have obeyed.

For our situation was still far from desperate. Aside from the capture of the pursuing train, which would now have been very difficult from the fact that we had neither fuel for rapid running, nor the obstructions on board that were necessary to place us far enough ahead for an ambuscade, there was another plan to which our leader was virtually pledged, which presented every prospect of saving our own lives, though it was now too late to accomplish our original purpose. We were some five miles beyond Ringgold, within a mile of Graysville, or nineteen miles by the longest railway course from Chattanooga. From that city westward to Bridgeport was twenty-eight miles further. But the nearest way to Bridgeport was not through Chattanooga, but further south, and by that route it was not distant more than thirty-five or forty miles. The direct course was at right angles with the numerous mountain ranges which here run almost north and south, a route over which cavalry could not be used, and which was known to more than one of our party. Two comrades had pocket compasses which would have guided us in thick woods or in cloudy weather by day or night. Now to have left our train in a body, and without delaying to seek concealment, to have struck over the streams and mountains at right angles, as rapidly as we could go, would have been our most hopeful course. Long before night of the next day we would have been safe within Mitchel's lines! Why not? How could the enemy have captured us? If they sent cavalry, these would necessarily have made long circuits and have been obliged to adhere to the lines of the road, and thus could not have come near us while clinging to the valleys and mountain sides. Even in thick woods they could not have overtaken us. If they followed us with a strong party on foot, we fleeing for our lives, would not have deserved to escape, if we could not have held our distance for forty miles or more. If they had ridden ahead and raised the whole country for a general man-hunt, they would have had only twenty-four hours or less to organize it, and no small party then could have arrested twenty armed men. In fine, this plan of escape through a mountainous and densely wooded country did not appear to me to be more dangerous than a cavalry dash on the lines of the enemy's communications—an every-day military affair. Even if Mitchel did not prove to be in the neighborhood of Bridgeport when we arrived, we would then have been in the loyal mountainous district where we would have met as many friends as foes. All that we needed in the way of provisions and guides, our force would have enabled us to command, and even guns and ammunition could readily have been gathered on our way.

But all these advantages depended on our *keeping together under one head*. An army scattered and disorganized is lost; and our little army was no exception. The fatal command which Andrews now gave as we were huddled together in the wood-box of the tender, was to jump off, one by one, scatter in the woods, and each man strive to work his own way back to the Union army! We hesitated, but had no concert of action, no leader, no time for council, and the instinct of obedience was still strong upon us; but it was a fatal order, and led directly to the calamities that followed. It transformed us in a moment from a formidable body of picked soldiers ready to fight to the death, into a scattered mass of fugitive boys, bewildered and hopeless in an enemy's country!

Yet no one of us felt like censuring our leader for this order, which every one at the moment believed to be a terrible mistake. He was not a trained soldier, and had never learned the power of disciplined men. All that he had achieved heretofore had been by the force of individual effort; and he never seemed so much at home as when thrown on his own resources. In a most pathetic letter, printed on a subsequent page, he expressed perfect confidence in his power, in the absence of a certain contingency, to have eluded the enemy and secured his own escape; and what he felt confident of doing he thought others could also do; or that if they were captured he would be in a position to give them very efficient help. Probably he thought that each man of the party would find the same relief that he did, in being cast entirely on his own resources.

Then it must further be remembered, in explanation of this mistaken order, that Andrews had slept none the night before, that he had been nearly twenty-four hours without food, and that he had spent nearly two days and a night in the most exhausting labors, both mental and physical, that it is possible to conceive. He had seen his cherished plans, when on the brink of success, overthrown by what seemed the remorseless hand of destiny. To the many failures and sorrows of his past life had been added the crowning misfortune of this defeat. It had ever been his sad and tantalizing lot, to almost, but not quite achieve; to succeed grandly up to a certain point, and then to fail through some cause too strong for human power, and too obscure for human foresight. Perhaps under his calm brow he realized this with an intensity of anguish, and felt that the greatest favor he could do those he had led within sight of a horrible death, and into the presence of an enraged and triumphant foe, was to separate them at once from his own dark and shadowed destiny. If so, that was the most fearful mistake of all; and as this order was given, we could almost, as we looked southward through the driving rain and the storm-clouds, behold already the dark outline of the Atlanta scaffolds!

It was pitiful! The "General" had served us well ever since the morning hour in fearful speed and patient waiting, in exulting raptures and in almost despair. It was hard to abandon her now. She was substantially uninjured. The engineers, Brown and Knight, had taken good care of her, and, with wood and oil in abundance, there would have been no difficulty on her part in completing the run to Huntsville. She was still jogging along at the rate of eight or ten miles an hour, and could maintain that pace a little longer. The pursuers had also diminished their speed, so as to just keep us in sight, having apparently no wish to press upon what may have seemed to them like a wounded and dying lion. The command to "jump off and scatter" was repeated with the injunction to be quick about it, as the engineer wished to reverse the engine and drive it back upon the enemy. With such a reason, there could be no more hesitation. It is said that some three or four had already got off at the first word of command; but the most of us had hesitated, not on account of the still rapid motion of the train, but in the idle hope that in some way, this terrible parting might be averted. Now one after another clambered down on the step and swung off. I was neither among the first nor the last; and jumping unskillfully out from the step, instead of for-

ward, whirled over and over on hands and feet for several revolutions.
Rising in a dazed condition, though unhurt with the exception of a few
scratches from the briers with which the place abounded, I looked over
the animated scene, with the deepest interest. The men who had jumped
off were, according to instructions, flying in different directions; a few
others were just coming off the engine in almost the same way that I had
done, while the engineers were attempting to carry out their scheme of
reversing the engine, which could do no good now, except possibly to

Leaving the Locomotive.

delay the inevitable pursuit a little, and give us a better opportunity to
organize our plans. The brakes of the tender were put on still more to ·
diminish speed, and the reversal was made. Here is a slight conflict of
authority. The pursuers say that the brakes were not loosed again; but our
engineers are equally positive that they were. It is not material, for the
result is the same. The steam power was so low, that though the engine
moved back, it was with moderate velocity and I saw the pursuers reverse

also, and coming to a full stop, whistle two or three times as it approached —a seeming whistle of alarm, though there was little in the approach of our poor "General" to fear; and then they moved slowly before it for a short distance till the two were in contact, when the weaker stopped, and the steam was shut off. The great Railroad Chase was over!

CHAPTER XIII.

WHAT WAS ACTUALLY ACCOMPLISHED.

WHOEVER has attentively read the preceding pages will understand that the enterprise described was not a mere hair-brained and reckless raid which could scarcely hope for success and uld have achieved no solid result. On the contrary he will realize that promised much, and came very near success; while its ultimate failure is only one of those chances of war which no human wisdom can avert, d which give to this game of life and death its terrible fascination.

Why did it fail? Four causes may be assigned, three of which are but anches of one.

1. *The delay of one day.* This is the first and far the most important. ad Mitchel been equally delayed, no harm would have resulted; but was not, and in consequence the road was crowded with extra trains. he responsibility for this delay must rest upon Andrews himself.

2. *The rain on Saturday.* This made the use of fire slow and difficult. it it may be regarded as a part of the first cause, for on the day assigned, is difficulty would not have existed. On Friday there was a clear day and oaring wind. We would not have spent much time then in the lifting rails, for a bridge fired and fanned by such a breeze as delayed us in ossing the Tennessee, would in a few minutes have made an impassable rrier. But on Saturday the rain fell in torrents, and everything was most insured against fire.

3. *Andrews's reluctance to fight.* He took the men along as laborers her than warriors; he did not inspect their arms, but simply advised it each be furnished with a good revolver. I was asked by a person to om I gave this reason whether I thought Andrews cowardly! The m might as well be applied to Julius Cæsar! But his plans were based strategy rather than force; and the failure to turn and fight at the pro r time was disastrous. A second in command who would have stood him, and at the critical time have advised him to turn on the foe, uld have remedied this evil. It is the opinion of the Confederates, as ll as our own, that if we had attacked them with half the determi- ion shown afterward at Atlanta when weaker and fewer, we would ve been sure of success. The fault for this neglect was in General

Mitchel, who should have put some determined fighter party, under the direction of Andrews. If he had no other man there were members of the party fully com

4. *The pursuit of Fuller and Murphy.* It was met latter with his coolness and knowledge of all repairs, a on the road, was present that morning. The day befor been. Fuller was probably the only conductor on the r been equal to the terrible vigor of this pursuit; and it w he could have carried it through without the help o it is reflected that a difference of one minute or less at three distinct points, would have given complete succe drances, it will be seen how narrow was the margin, help of Murphy could have been spared.

It required all of these four to defeat us. No on three, would have been sufficient. Looking back over give it as my candid opinion that when we set out fror of success were as good as those of Mitchel in his mar But in war there are no certainties.

It may also be maintained that even in what was score of men engaged rendered to their army, at what ing to themselves, a greater service than ten times their wrought in the ordinary line of military duty, and co share to the overthrow of the rebellon. Three con this claim.

The superintendent of the Western and Atlantic I graphed a request for strong guards at all impo E. Kirby Smith from Knoxville, under date of Apri Leadbetter to send troops from Chattanooga for the were twelve or fifteen bridges on this road alone. all i insure them against the repetition of an attempt such to these is added all other bridges in exposed positior enemy was made uneasy in consequence of our raid, ployed cannot be estimated at less than several hundre active service at a time when every man was urgently i ple diversion of force, this raid was eminently successf

But another result was in the end not less importan increased the rigor of the war-pressure inured in tl advantage of the Union cause. From the time wher Federal soldiers in disguise could penetrate to the ver federacy, a much more stringent passport system was ao at ferries, as well as bridges, were strengthened; arri trains were rigidly examined; and a general atmosph

[1] See also the message of Gov. Brown, of Georgia. Supplen

or many hours the result was doubtful. The terrible passions evoked on the part of all engaged were shared with but little diminution by those who were interested so directly in the result. The Southern people had from the beginning of the war been taught that the coming of the "Yankees," as the Federals were generally called, would bring with it the most terrible calamities to which a people could be subjected. The loss of property, the burning of houses, and even the murder of women and children were not the worst evils threatened. Reports of Southern victories had hitherto been so exaggerated and frequent that these startling dangers had seemed distant; but now in a single day they were brought terribly near. The news that Mitchel had taken Huntsville in Alabama, had followed directly the reports of a great battle on the borders of Mississippi, and the same day the enemy were in their own midst, in the heart of Georgia! It is still possible by the aid of files of Confederate newspapers to enter into this natural, even if exaggerated feeling, which had the most practical importance to us, as it led to the terrible pursuit and probably to much of the ill-treatment that followed.

A circumstance that added to the speed with which the intelligence was diffused, and made the pursuit to be more prompt and better organized, while increasing the general excitement, was that Saturday was a general muster day. As the trains passed they found people already gathered in their towns, with their guns. These were of rude pattern, but very efficient for the use now demanded. They waited with breathless interest, as many of them as could not crowd on the pursuing trains, to learn the result. The telegraph wires were restored as fast as possible and the news sent back. Each message, whether by this channel or by couriers on horseback, was listened to with the greatest avidity. The simple facts were distorted in a hundred ways.

An amusing account is given by a correspondent of the Atlanta "*Southern Confederacy*" of April 18 (written of course after the panic had subsided) of the reception of the first intelligence at Chattanooga. The correspondent says that "the news reached here by telegraph from Dalton to the agent at this end of the line. Quick as thought it was heralded over town, penetrating to every part of this diversified city, that the Yankees had possession of the road; that all the bridges from here to Dalton were burned; that Cleveland, Tennessee, had fallen into the hands of the enemy; and that they were within a few miles of Chattanooga." He declares that while no flag of truce was sent out to surrender the town, no other absurdity was omitted. The whole able-bodied population were enrolled for defense, and some of them were exceedingly unwilling to venture into danger; that streams of people flocked away as fast as they could go; and that for a time panic reigned supreme. It may safely be assumed that the same state of excitement existed in every other town near the line of the road.

But the accounts of the matter as given in Atlanta are still more interesting. This city was in itself very important; it was now made a great storehouse for the South, much of the provision, bacon especially, on which the armies of both Lee and Beauregard depended for subsistence, had been gathered out of Kentucky and Tennessee, and sent here for safe keeping. Camp McDonald was on the line of the railroad between this town and Chattanooga, and troops were continually coming and going over the line. This seizure was at once felt in Atlanta to be a vital matter; and while there was not the same panic from the supposed immediate presence of an invading army with all its horrors of rapine and bloodshed, yet the danger was felt to be imminent. We give an extract from one or two numbers of *"The Southern Confederacy,"* one of the most important journals in the South. For some days a large proportion of its space was given to this affair; on the 15th of April especially, a full account was given embracing a whole page. This is given in the Supplement,[1] as it is a deliberate Southern estimate of the nature of the expedition. But a still more interesting account is that in the issue of the 13th, which gives a view of the feelings of the people while the contest was still in progress. Parts only of the long article are cited, though for the sake of clearness we must repeat some matters already narrated, and give others which have the natural incorrectness of a first report.

"The startling intelligence was received on yesterday morning, that while the regular mail and passenger train was stopped at Camp McDonald, or Big Shanty, and the engineer, conductor, and passengers were at breakfast, some four men, as yet unknown, after having cut loose all but the three foremost cars, got on the engine, put on steam, and shot away like an arrow, leaving the baggage and passenger cars, passengers, engineer, and train hands, and conductor lost in amazement at this unparalleled and daring outrage. Some distance above they tore up the track and cut down the telegraph wires." * * * * *

(It will be noticed that so quickly were the men placed on the cars that only the four on the engine were seen. In one sense this was our misfortune as it rendered the pursuit the less fearful and cautious.)

"They arrived at Kingston where they met the down freight train, and went upon the turnout, showing that they understood the schedule, and the minutest working of the road. * * * * * As soon as the down train passed they shot away with all their speed and mystery.

We learn that a train has been put in pursuit of them, having repaired the track, and hopes to overtake them before they reach any of the bridges over the Chickamauga and other streams. No doubt they are Lincoln enemies, sent down among us to destroy these bridges, to retard our movement of troops, and the thought is a very serious one to us.

[1] Chapter XXXIX.

For cool impudence and reckless daring, this beats anything we ever heard or read of. We are in an agony of suspense to hear the denoucement of the strange and daring achievement." * * * * *

P. S.—(Fuller's pursuit) Arriving at Kingston he got the Rome engine with its engineer, all in fine condition, with perhaps forty armed men, and pressed on. * * * * *

" At Adairsville, the regular passenger trains up and down meet, and the thieves would have to pass them there. The down train due here at four P. M. has not yet (at nine P. M.) arrived, and it is feared that there has been a collision with the engine, though the torn-up track may be the cause of the delay.

"Various surmises looking to a solution of the mystery are indulged here. Everybody at first concluded it was a most daring effort of some Lincolnites to burn the bridges to stop the transportation of troops over the State road. It is reported that the whole of the troops at Camp McDonald were going off yesterday morning and a large number of troops from " * * * * * (these are the Charleston troops already mentioned.—W. P.) "came through here last night on their way to the scene of action. Some said there was really ammunition in the three box-cars they carried off, and the object was to take it to the enemy at Huntsville. We however learn officially that the cars attached to the engine were empty.

"Another solution which has gained credence, and is not at all improbable, is that they were simply thieves on a large scale, and took this method of escape."— *The Southern Confederacy*, April 13, 1862.

A few words may be of interest in regard to those who will appear no more in this history. The locomotive captured, the " General," is itself a famous war relic, being regarded by the Western and Atlantic Railroad Co. with peculiar affection, and carefully preserved. The chase did not substantially injure the grand machine, and she was soon made as good as new. Before the close of the war she had other adventures, being under Federal fire at the battle of Kenesaw Mountain, and in great danger, for she had taken a train-load of ammunition up to the present station of Elizabeth, just south of the Confederate intrenchments, where the Federal shells exploded all about her. For many years she hauled passenger trains over the familiar road, and I had the pleasure, in 1880, of again riding behind her from Atlanta to Chattanooga. Since then she has been used in light work only, and receives the attention of many visitors, who like to look on the most famous locomotive in America. The accompanying photograph has been furnished by the kindness of the Western and Atlantic Co.

Most of those who were prominent in the chase reside still in Atlanta (1887), Anthony Murphy as a prosperous lumberman, and Conductor Fuller engaged in the management of his plantation and town property. Immediately after the railroad raid a number of prominent citizens petitioned Gov. Brown of Georgia to bestow a suitable reward on Capt. Fuller for his indomitable energy and pluck, who in turn recommended[1] him to the Georgia Legislature. He was voted a gold medal and the thanks of the Legislature but had to be contented with the latter only, as gold became

[1] See extract from message in Chap. XLI.

exceedingly scarce in the South and enough for the purpose was not available. It was sad for him that his brilliant services were given to a "lost cause," and could therefore bring no Governmental reward. He has always, since the close of the contest, been on friendly terms with the survivors on the Union side, as they respected bravery, and believed that he did simply what he regarded as his duty.

Capt. Fuller continued to serve as conductor on the same road while it was under Confederate control; but in the summer of 1864 the route began to shorten rapidly. As he graphically said, "Sherman bit a piece

"THE GENERAL."—From a photograph taken in Atlanta in 1887, by the W. and A. R. R. Co. Conductor Fuller and Captain Parrott are shown.

off the end of it almost every day till it was all gone." Then he was made a captain in the Confederate service, and afterwards given charge of all the rolling stock of the road—Sherman was never as successful as Mitchel in capturing such spoils—and kept it out of Federal hands, until the final collapse of the Confederacy. After the restoration of the Union, he continued for ten years more in his old position as conductor, then for seven or eight years was a merchant of Atlanta, when he retired from active business.

CHAPTER XIV.

HUNTED IN THE WOODS.

THE following narratives of that strangest and most thrilling of human experiences, where men were hunted as wild animals, are left substantially unchanged, except where a better knowledge of the country leads to some slight corrections. The fear and horror, the faint hope fading into despair, the numberless devices, pathetic in their insufficiency, for eluding a remorseless pursuit,—all these are better given in their first freshness as they were told twenty-five years ago, than they could now be rendered by the reflections of more mature years. Abridgements have been made in some of the accounts to prevent that sameness which would otherwise be inevitable, because the obstacles each of the party had to surmount, and the objects in view were the same. It will be seen that the means also employed by us were so nearly identical as to afford our pursuers a ready clue. The result of the whole chase is little less than marvellous when considered in itself; but the marvel is greatly increased when looked at in contrast with the exactly opposite issue of another attempt at escape seven months later. If my own adventures are given first, and with more minuteness than those of others, it is not that they are longer or more important, but simply because it is better to present one fully drawn picture, and then give only what is peculiar in other stories; and this complete picture can be best given by any writer of his own experiences.

When we left the train, I confess that for a moment my heart sank within me. I had the ordinary bravery of disciplined soldiers; and while on the train at no time felt afraid to perform any duty that was ordered; but this position was totally new, and of all earthly situations seemed most dreadful. The conviction, very strongly felt, that it was unnecessary, did not add the least comfort. I could see the soldiers pouring out from the pursuing trains, (for a second one was almost immediately on the ground), and hear their loud shouts, and very soon the firing of guns. That a vast hunt was to be organized was only too evident; and the result of capture would only lie between (so I thought) a gun-shot and a rope. The prospect of escape seemed well nigh desperate. I did not know where I was. The descriptions given on succeeding pages are the result of studies along the line, which would leave me in a far different position for a similar attempt to escape now. I only knew that to go far enough north or north-west would bring me to our own lines; but the sun did not shine, and I had no compass, and no knowledge of those arts of the woodman by which the want of guides is sometimes supplied.

But there was no time for meditation—to simply stand still would bring ruin As I saw the fugitives a short distance ahead or to one side of me, two thoughts arose; first, it would be a great advantage if I could reach those three who had been my companions day and night for a week, and from whom I had become separated in the confused tumbling from the cars. It was a pity that I thought of this, for in the attempt to find them I lost the opportunity of finding some others whose companionship would have answered as well. I might even have succeeded in rallying a small band and partially carrying out the plan already sketched—that of rapidly, with no halt either for rest or sleep, hurrying to our own lines. What I accomplished alone, and what was done by Wilson and Wood, has always disposed me to think that with the help of a few comrades I might have penetrated to our lines in a short time. But in searching for Wilson, Campbell, and Shadrack I got away from the others who might have been reached.

The second thought was of a more practical character, and was immediately acted upon. It was that getting away from the immediate scene of the abandoned train would be far better than to attempt concealment. I knew too well the means for discovering the hiding places of fugitives; though in the rain and in the small oak timber, with the attention of the pursuers called in other directions, it was possible that for a short time one or two might have succeeded in concealing themselves; but there would be no gain in this—nothing but a simple loss of time. Travelling never would be more easy or safe for us than that afternoon, if we but kept in advance of the enemy. By the next day it was sure that horsemen starting out from every telegraphic point would alarm the whole country, and all the panic and uncertainty on the part of the enemy would give place to relentless pursuit. The country people would be filled with horror and resentment at the thought of " Yankees " having penetrated their country in disguise, while the military authorities would count no effort too great to get hold of us, if for no other purpose than to learn our character. So the chances of escape could only grow worse by waiting.

If the description of the country given by different members of the party at starting differs somewhat, it is to be remembered that we were " drilled " along the road for the space of half a mile or more, which may have somewhat changed the surroundings. I climbed up the small bank which I found myself facing as soon as I recovered from the slight dizziness produced by the mode of getting off the train, and found a strip of woodland before me separated from the railroad by a fence. From the top of the fence I looked about, striving to get as good an idea of the situation as my bad eyes would permit. Our own train, which had gone ahead, was now coming back, while the pursuing trains were very near.

They had stopped and men were getting off. Our boys were flying across the woods. No time was to be lost, and jumping down I ran in the same direction. I crossed a little brook which ran by the edge of the wood, and came out into a large wheat-field that sloped up a moderate hill. There were three of our party a little ahead and it seemed as if I could overtake them. I put forth my best exertions, calling as I ran, not knowing their names, but assured that any of our own party would be valuable company. They did not hear me, and I labored in vain to overtake them. Never did I find running so difficult. The wheat-field was very soft, and the mud in great masses clung to my feet so that I could scarcely lift them. I was weak, faint, and tired, and it was like trying to run in a nightmare. Hurrying up hill under such circumstances seemed an impossibility.

Convinced that I could not overtake the men before me, or make them hear, I dropped into a walk, and plodding on, reached the other side of the field, and as I left it, the first of the pursuers, Fuller at their head, mounted the fence and entered behind me. But he was as nearly exhausted as we were, and his chase in the woods was faint and useless as compared with that in his own element. When on solid ground once more I made better progress; yet in the woods beyond, I found with inexpressible anguish that the last one of my comrades had vanished, and I was alone in the heart of the enemy's country !

But I continued on, keeping the noise of firing and shouting,—a dismal accompaniment when every shot might mean the death of a comrade, —directly behind me. This was an excellent guide as long as it lasted.

For some time I continued in this manner, crossing low hills and gentle valleys mostly covered with thin woods, until going down a longer slope than usual, I found myself in a bend of Chickamauga Creek,—a little river that empties into the Tennessee a short distance above Chattanooga. It was now swollen by the continuous rains into a formidable stream, and as I looked on its swift current boiling among the rocks it was far from inviting. The enemy in beginning their pursuit, had counted largely on this stream preventing a flight directly westward; and as the east branch of the same stream joins it at Graysville, but a short distance from the starting-point they had good reason for hoping that we would try to follow down the stream and find ourselves in their trap. This happened in the case of four who were captured within a few hours. They came to this river, and fearing to tempt death in the torrent followed its current until they came to the junction of the stream where they were

surrounded and taken—the first capture in this man-hunt. But as I looked at the torrent I realized that to go either up or down stream would carry
me the wrong way,—that is, at right angles with the noise still made by
the enemy instead of directly away from it. So, holding revolver and
ammunition high over my head with my left hand, leaving the right free
for clinging to the rocks or
swimming, I committed my-
self to the angry tide, and
after being t h o r o u g h l y
soaked and almost washed
away, I succeeded in reaching the o p p o s i t e shore.
There I found that only the
form of danger had changed.
The bank rose in an almost
perpendicular precipice high
over my head—more than a
hundred feet, as nearly as I
could judge, —and extend-
ing further on either hand
than I could see. I dared
not recross the stream, and
therefore clambered slowly
and painfully up the face
of the precipice. Several
times when near the top did
I feel my grasp giving way;
but as often some bush or
projecting rock afforded me
a hold by which I could
save myself; and I could not
help thinking, as I climbed,
what a fine mark I would
present to a rifleman on the
other side of the stream!
At length, after imminent

Climbing the Chickamauga Precipice.

danger, I reached the top, utterly exhausted, and fell at the root of a
tree, where I lay to recover breath for a while.

I had been here but a very short time musing on the unenviable position in which I was placed, when suddenly a sound reached my ears which
brought me to my feet with every nerve straining to its highest tension, and
the blood leaping wildly through my veins. It was the distant baying of a
bloodhound !—to a fugitive, the most doleful and repulsive of earthly

sounds! I was to be hunted not by men only, but, like other game, by dogs as well! never can I read the story of human beings pursued by these most revolting instruments of man's savage "inhumanity to man" with indifference!

A few moments listening confirmed my first impression. It was true that they were after us with their bloodhounds! not one pack alone but many, as the different directions in which I heard them—all, however, from the other side of the Chickamauga,—revealed but too plainly. There was no longer safety in idleness, and I hurried off as rapidly as I could directly away from the river, for I knew not how soon similar packs of dogs and men might be in the woods on the westward side of the stream. I was in but poor condition for a rapid journey, having had neither dinner nor breakfast, and having spent all the morning hours in labor and excitement of the most exhausting character. But for a time the urgent need gave artificial strength, and I rapidly placed a considerable interval between myself and the barking of the dogs, which at least indicated the course which judgment and inclination alike urged me to travel—that is, directly away from them.

Some controversy has arisen as to whether bloodhounds were employed in this pursuit. One Confederate in denying the charge urges that very few if any Spanish bloodhounds of pure breed were in the South, and especially in the state of Georgia, at the time of the war. In this he is probably right. No one of our party at any rate is in a position to disprove the assertion. What we do know is that tracking dogs were kept in the districts over which our experience extends, and that these were remorselessly used in hunting us down. It seems to us a small matter whether the dogs which are kept in large numbers by the Southern planters and by the hunters of the mountains, and which were trained to track negroes and savagely attack them, which we saw and heard baying on our track, were of one breed or another. Their employment in this manner added an element of horror to the chase; but it was almost inevitable that they should be so used. They were accustomed to tracking fugitive men; without the employment of tracking dogs it would have been very difficult to prevent slaves from eluding their masters; and just the same advantages were to be found in using them to chase soldiers. Had we remained together, however, they could have done us no harm, as our passage over the country would have been too rapid and in too great force. Even when alone, I did not greatly fear the dogs after the first thrill of horror had subsided, as I did not think they would of themselves attack an armed man, and I hoped, if the sun and stars would only shine, to be able soon to make my way to our own lines.

I sped rapidly over hills and streams for I know not how far. The noise of the dogs grew fainter in the distance. I was outside as I had hoped to be, of the circle which yet bounded the efforts of the enemy in seeking for hidden foes. But as the sense of immediate danger from pursuers relaxed, the wish to have some guide, or at least to know where I was, grew stronger. I had a conviction that I was going west, but I might be totally mistaken. It was also possible that I would run directly into an enemy's camp.

As I descended the long slope of a wooded hill into a wild, solitary valley, I saw a rude hut and a man in the garden beside it. Here was an opportunity for inquiries. With but one man at hand I did not fear arrest, and I thought it in the highest degree improbable that he had yet received warning of the chase. I approached him and asked the distance to Chattanooga. That was the last place to which I wished to go, but it was the best point to get my bearings from. The answer was "eight miles." This shows that I had already passed over a long distance, as we were nineteen miles distant by rail when I started. I also got the direction of the town, so that I might have a point of comparison in any future inquiries. Then I left him. It would have been wise to have also obtained some food here, but I did not wish to risk the delay. I started along the road for Chattanooga, but as soon as the house was out of sight, I climbed the hill to my left, for if I could, I would have liked to circle around Chattanooga at about the same distance, keeping the town on my right hand, till I reached the Tennessee river, which I rightly judged to be directly in my way. The sight of the river, if at a little distance below Chattanooga, would have been as welcome as was the first view of the sea to the fugitive Greeks.

I had now started in a way that, if I could keep it, would be sure to bring me in time to our own lines. I had not been closely followed and was making good speed. No one opposed my progress, and I felt a strength that under the circumstances was wonderful. In fact I had always possessed a great degree of endurance. In the few hours of daylight that still remained, I could add eight or ten miles more to the distance separating me from the pursuers, and while travelling at night would be more difficult, if it continued cloudy, it would not be impossible, and I had the resolution to go on with slight rest all the night. But everything was conditioned on my being able to keep a straight course. The absence of company, and the uncertainty of the fate of my comrades, weighed on my spirits heavily; and I longed for any one of the number to talk and plan with, feeling that then the way would not be so long; and if it became necessary to seize food by force, two would be much more able to do it than one. But I had no hope of meeting any of them now, and was obliged to rely on my own resources.

At this stage, a simple incident occurred which would have been ludicrous under other circumstances, but which went far toward driving away all my new-found hope. After passing down a short wooded bank I came to a road, and in crossing looked both ways to see if there was any sign of travellers; and thus I became well acquainted with its appearance. I then passed on and walked fast for perhaps an hour, when to my great surprise I came to the same place again. I had passed other roads in the meantime, but the appearance of this one was unmistakable. I was greatly annoyed thus to lose my labor when I was hoping to be far on the way to the Tennessee before nightfall; but struck again vigorously over the hill in what seemed to be the right direction. Judge of my astonishment and despair when an hour of hard walking brought me to the same place again! So much time had been lost that I could hear the bloodhounds once more and not far away! I was perplexed beyond measure. A very short time brought me either to the Chickamauga which I had crossed hours before, or to one of its branches, so much swollen that I could not distinguish it from the other. I seemed to be in the mazes of a labyrinth, where all my labor was worse than thrown away. In sheer desperation I took the first road I came to and followed it a long time at a rapid rate, almost regardless of where it led or whom I should meet. The one thing that I felt utterly unwilling to do was to stop and rest, while deadly foes were tracking me out with their dogs. Anything was better than that. Neither did I like to be only fleeing before one pack of hounds after another, without any definite aim of my own. I would rather travel in the road with the certainty of meeting some persons who would attempt my arrest.

In the twilight, while thus pressing on, I met a negro driving a team. From him I learned that I was within *four miles* of Chattanooga! Words cannot describe the tide of disappointment, vexation, and anger that swept over me as I thus realized that in spite of all my efforts I was approaching that town which I had learned to look upon as the lion's mouth! But there was no profit in giving way to despair.

Learning from the negro the direction both of Ringgold and Chattanooga, that I might have two points by which to set my course, I carefully noted the direction of the wind, which was, however, light and changeable, and once more set out. I did not suspect the negro, and it was little that he could have told. Six or seven miles would bring me to the Tennessee if I was able to keep my course. I crossed some terribly steep hills and went down into deep valleys. This did not dismay me, for I supposed that just such country must be passed over on my way. Though I did not know it, the formidable Lookout range was still before me, and to have passed that would have brought me very nearly where I wished to be. I was very hungry when I thought of food, but had a far stronger desire to be on the Tennessee and engaged in devising measures to cross it than I had for eating. Neither did I now feel much weariness.

The need of getting beyond the circle of pursuit before morning, supplied an intensity of will-power which banished fatigue. If my safety had depended on a straight march of fifty miles across the country without food, I think I would have felt equal to the task, if I had only known that I was going in the right direction.

About this time, the moon, which was just past the first quarter, for a moment shone out. I instantly took my bearings and found that I was heading right. Soon it was obscured, but I had hopes that after a little the clouds would pass away, and then all the long night through, I would be sure of the way and fearless of arrest; for at night in that rough country, I did not think I could be taken easily either by men or dogs—the one I could shoot, and the other elude in the woods. But the rain again began to fall and soon became severe. I now crossed a very large tract of deadened timber, and when I reached the other side in the storm I had no idea whatever of the direction, and was again almost in despair. However, I walked on till I came to a large road. This I resolved to follow till I could by some means ascertain my position. But it was very hard to tell which end of it to take. Deciding almost at random, I followed its leadings for several miles, hoping, for the evening was yet early, to meet some traveller from whom I could inquire the way. The mud was fearfully deep and adhesive; but I had gone some miles, when I met three men on horseback. It was too dark to tell whether they were negroes or white men; but I knew they could not ride after me in the woods and therefore ventured to ask them:

"How far is it to Chattanooga?"

"Three miles!"

"Is this the road?"

"Yes, sah! right ahead."

It is extremely probable that these men were sent out to search for members of our party, and that they did not attempt to seize me because of the bold manner in which I accosted them, and because I was so near Chattanooga, and going in that direction. They could have done nothing, however, but fire after me, and a shot from horseback, in the dark, is not usually serious.

To say that I was disgusted and discouraged to find after many miles of travelling, a gain of but one mile since my last inquiry, and that in the wrong direction, very faintly expresses my feelings. It seemed as if I was so hopelessly bewildered that it was impossible for me to travel any but the wrong road! There appeared to be some kind of deadly magnetism, which persistently drew me, in spite of all efforts, toward Chattanooga, the headquarters of the enemy, and the town for the possession of which we had been striving ever since the ill-fated hour when we left Mitchel's camp. Yet the matter, mysterious as it seemed then, is easily explained with the knowledge of the country I have since obtained. Nearly all the roads for many miles in every direction run down the mountain vallies, and centre at Chattanooga. I would be obliged either to cross them all, in which case I was in danger of travelling in circles, or if I adhered to them as I had done for a considerable part of the time, I was more likely to travel north than south, and this took me nearer and nearer the enemy's camp. How easy it would have been on this stormy night to have seized a countryman and compelled him to act as guide had we but remained together! In this case an equal amount of travel would have taken us out of danger.

But the great mistake of the whole flight I made just here. With all my wanderings, I was very near the point I wished to reach had I but known it. I had always thought of Chattanooga as directly between me and the Tennessee. But in fact I was within half a mile of the bank of that stream three miles further down than Chattanooga,—and in daylight would have seen it, or at least the hills beyond, without the slightest difficulty. I was almost at the northern point of Lookout Mountain and could have left the road on the western side, and gone round the spur of the mountain, and before morning have been in a comparatively safe position—all of this with less travelling than I afterwards did. But in the absence of any such knowledge I took a course which carried me far back into the enemy's country. When the horsemen had gone by I followed them some three or four miles so as to put that much ground between myself and Chattanooga. Then I came to a large road running at right angles with my own which terminated where it joined the other. Again I was perplexed, for my old road had wound among the hills in such a crooked manner that it did not even give me the direction of Chattanooga.

Many a time, in old astronomical days, had I wished for the breaking away of the clouds to show me the moon and stars, but never with such anxiety as at this moment; again and again I looked upward. The sky must clear at some time, but when? The opportunity of travel in comparative safety which I enjoyed this night would not last; and my own strength which I was using so freely, would not endure always. At length I made a choice and hurried forward; but, as usual, I chose wrong; for after I had travelled many miles, the moon broke again through a rift in the clouds and poured her welcome light over the dark pine-covered hills among which I was passing. That one glance was most disheartening; I was heading eastward—directly toward the railroad, which it seemed I had left an age ago. I had taken the highway to Ringgold! Wearily I turned and retraced my steps. If the moon would only continue to shine for the rest of the night, I would forget the wanderings of the past, and keeping to the roads that ran west and north, so far as I found them suitable, would still be far on the way by daybreak.

But alas ! the moonshine was brief. In less than half an hour a dark cloud swept up from the west directly ahead, and the moon was once more hidden. As her light died away from the landscape, hope almost died from my heart ! One of my feet, which had been injured a few months before, now pained me excessively. Still I dragged myself along, impelled by a kind of restless energy or nervous fever. The long-continued tension played strange tricks with my senses, which became more marked as the night advanced. I came again to the place I had made the wrong choice of roads, and taking the other end still toiled on.

Another element of discomfort was added to the cup of misery which was now almost full. I was thinly clad, while the rain fell in torrents, and the cold wind, which had risen with the last storm, drove it with great violence against me, until my teeth chattered and I shivered to the bone. The night seemed to be growing constantly worse. I also passed many houses, for I was following a highway, and feared the barking of the dogs would betray me; but the people were either not apprised of our raid so far away, or they did not care to turn out for the hunt in such a night. The storm which added so much discomfort to our situation at least served us the good turn of making pursuit very difficult and disagreeable. I heard but little of the barking of the hounds and that in the distance. The lonely, fearful, stormy night seemed to be all my own.

This comparative repose, for I was now travelling along a straight road and at only a moderate gait, served to soothe my overwrought nerves, and produce the natural result of indescribable weariness. I must rest somewhere. There was no barn or house which I dared approach for fear of being found in my sleep by dogs or men. But I reeled to a large log that lay only a few hundred yards from the edge of the road in a small patch of woodland, and crawling partially under it, not only for shelter from the driving rain, but still more for concealment from my worse dreaded human foes, I soon slept the sound and dreamless sleep of utter exhaustion.

CHAPTER XV.

A SAD SABBATH MORNING.

WHAT follows is an experience so purely personal, so easily explained by natural laws, but so mysterious to me then, that I would hesitate to relate it, were it not that only by such minute pictures of real experience can any just idea of the cost and misery of the war of the rebellion be obtained. Doubtless most that I then felt has been experienced hundreds or thousands of times by fugitive slaves in the olden days, when trying to escape from unendurable bondage; but few of them have had the opportunity to tell the tale of their sufferings. When the war came the circle of suffering widened, in the providence of God, to include those who had thought that such things had no personal interest for them.

Yet the incidents that follow were on the whole marked by a lessening of conscious suffering. Only indirectly do they bear witness to the agony and the effort that preceded and accompanied them. Up to this time the image of that terrible night is graven on my memory as with a pen of fire. After this a most wonderful change took place, and with the exception of a few real incidents that aroused me from my trance, it floated before me with more than the voluptuous splendor of an opium dream. Probably the nearest analogy to the state in which I found myself is that of the drunkard who experiences the horrors of delirium tremens. But there were clear and well-marked differences. The cause was no doubt purely natural, arising from fatigue, hunger, dampness, and intense physical and mental exertion. But I will state exactly what occurred and permit each one to draw his own conclusions.

How long I slept on the water-soaked leaves I have no idea—probably not a great while; but I seemed in an instant to be wide awake with a perfect realization of my position. In addition to this I seemed to hear some one whisper, as plainly as I ever heard human voice: "Shoot him! shoot him! Let us shoot him before he wakes."

The impression that flashed across me was that a party of rebels had discovered my hiding-place and were about to murder me in my sleep to save themselves further trouble; and I thought with a thrill of wonder and almost of relief, "This is the last of earth for me!" But immediately after this followed another thought only less appalling: "Was I insane, or were my senses playing false." I slowly opened my eyes and looked about to see if I could determine the matter.

Can any man in the quiet of home realize what it is to awaken suddenly from a dreamless sleep, and find himself alone, in a dark wood, in a pelting rain, and know that he is hunted by remorseless foes who would not hesitate to tear him limb from limb—the men scarcely less than the hounds! If so, he can estimate the wonder that, with all this knowledge, I felt at the character of the visions that rose before my eyes.

I was accustomed to the darkness, and the moon behind the clouds gave light enough to see dimly the objects around. Directly before me stood a small tree. The first glance showed a tree and nothing else. But the next moment the tree seemed to break up into an innumerable multitude of the most beautiful forms. There were angels, in softest outlines, their heads nodding with plumes above all beauty, and their wings slowly waving with borders of violet and pearl. In every direction I saw the same splendid assemblage. The whole sombre and awful wood was suddenly transformed into a glorious scene, a Paradise of beauty, in which moved a vast company of lovely beings. They were of all kinds that were pleasing and harmless, and on every hand were brilliant colors and the most melodious sounds. There were ladies, flowers, and children; little cherubs floated around on cloudlets of amber and gold. Indeed everything beautiful and pleasant that the imagination could conceive was comprised in this gorgeous vision, which to my unutterable surprise sprung up in the midnight woods. It was a fine beginning for the Sabbath, which had probably replaced the terrible Saturday while I slept.

The most singular fact of all was that though the brain and eye were thus impressed with what had no existence, I was perfectly calm, knowing the whole thing to be but a pleasing illusion, which I could in a moment distinguish from real objects, and not in the least fearing these figures of the brain. On the contrary, they were excellent company. They did not always present the same characters. Sometimes they were old feudal Knights in glittering armor, either on foot or on horseback, but never hostile to me. The finest landscapes would start up from the cold, dull hills around, like mirages in the desert; even language was not denied to my visitants, whose voices were inexpressibly melodious; every thought that passed through my mind seemed to be spoken aloud.

Another marvel I connected with this singular experience, and which I can explain on no other theory than the kindness of my Heavenly Father, is that all of these visions were pleasant and agreeable. No tearing to pieces by wild beasts, no venemous serpents, no demons, such as the drunkard sees, though my situation might well have suggested any of these. I do not know of any merely natural explanation of any of these peculiar features of my illusions.

I also felt refreshed and endowed with new strength. The sleep enjoyed would have accounted for part of this vigor, but only for a very small part, as I seemed now to be lifted at once above all fatigue and every feeling of discomfort. Even the merciless pelting of the cold rain seemed pleasant and luxurious as a cool bath in the heat of midsummer. To walk or to run seemed far more easy than ever in my life. I had scarcely the sense of distance or effort, but the sensation experienced when going at full speed was more like floating along in a dream. Yet the travelling was no dream but a reality, and mile after mile passed with an ease and swiftness that in my strongest days I could never have rivalled. And beyond all the illusions which seemed somewhat shadowy and gauze-like, another faculty penetrated and showed me, though but dimly, the true face of the country.

Once the real and the ideal became mingled and very nearly involved me in serious difficulty. At a cross road some distance ahead I saw what at first appeared to be some of my spectral friends, standing around a fire, the ruddy blaze of which rendered them clearly visible. They were not so beautiful as others, but still I advanced unsuspectingly toward them, and would soon have been in their midst had not my progress been sharply arrested by a sound of all others the least romantic—the squealing of a pig they were killing preparatory to roasting in the fire! This at once drove away all visions and left me in full possession of my faculties. I listened and became convinced that this was a picket sent out to watch for just such persons as myself! They had also some dogs with them, which were fortunately too much absorbed in the dying agonies of the poor pig to give much attention to me.

I crawled cautiously away and made a long circuit through the fields. A dog—possibly a sentinel cur—followed and made himself exceedingly annoying by continuing to bark after me. I did not fear him on his own account, for I had yet my trusty revolver which I had managed to keep dry all this time; but I feared that he might attract the attention of the guard, who if they could not capture me then would at least be able to inform the people of the country, when daybreak came, that I was somewhere in the neighborhood.

At last he left me and I made my way back to the road, but had not proceeded far until I came to some horses hobbled down, probably belonging to the picket behind, and had to make another circuit to avoid driving them away before me. Returning again to the road, I pressed on as rapidly as practicable, hoping before the morning light, as I was now going on a road that seemed to lead straight, to be beyond the circle of guarded roads and planters hunting for fugitives with their dogs. It was a vain hope, but I knew not the gigantic plan of search that had been organized. The visions which had made the lonely wood almost a paradise now began to fade. For several hours, probably from about midnight till near daybreak, they had been agreeable society in the absence of comrades, but they were becoming more and more dim; if I wished their presence, it required an effort to summon up any special scene. At the same time the sense of weariness came back—not excessive at first, but all the extraordinary power in which I had rejoiced gradually passed away. There was, however, no lessening of the grim determination I felt to do everything in my power to secure my own escape. I had no scruple as to the means. After the suffering I had passed through, to deceive the enemy in any possible manner, to enlist in their army with all pledges of allegiance to their cause, would not have caused a moment's hesitation. Neither did the fear of detection or danger seem especially formidable. When a person is in a position as desperate as possible, he may very safely venture changing it for any other; it does not then require courage to accept risks which would under other circumstances be appalling. If I enlisted in the rebel army and was detected, the result would be death; but I would be in no more peril in risking it than I was here. Slowly my plan changed, and even before the morning broke I resolved, if it still remained cloudy, not to try to avoid capture, but getting away as far from the scene of the raid as I could, permit myself to be stopped and questioned with the view of enlisting. There was but one slight flaw in this plan, the character of which will be revealed further on.

The gloomy dawn of the rainy day drove the chill horror of my situation still deeper into my veins. I would find myself staggering along almost asleep; would sink to the ground at the root of a tree, or in an angle of the fence, and sleep for a moment until awakened by the stiffening cold; and then plod on for a short time again.

Here a little incident of a more pleasant character occurred, which was purposely omitted in the first edition of my book, because it might have wrought injury to one who befriended me, as the rebel power then extended over this ground. Wishing to rest for a little time where I would be warmer, I crawled into a heap of corn fodder in a shed. I was awakened soon after by an approaching footstep. Not fearing a solitary man, I did not try to get away. When he accosted me, I told him the old Kentucky story, used on the way down, but in a very brief form, as I was cold, hungry, and desperate. He only sighed, and said, " This terrible war ! I wish all our boys were at home ! " Then he added the welcome words, " Come and let me give you something to eat." I felt that he was a Union man, but did not wish to commit him by any further help than the provisions he offered. I went with him, and he gave me what happened to be ready. Probably he had not heard of our raid, and took me for a deserter from the Confederate army, or a fugitive from the conscription.

I was not willing to remain long, for the house was by the side of the road, and at any moment a cavalry force might ride up; and he was also quite ready to have me pass on. The food he so generously bestowed imparted strength for a time, and renewed hope. It was the only morsel of anything I had eaten for thirty-six hours. He said, as I parted from him, "Don't mention to any one that you have been here," a charge I faithfully observed, until making the present record, which can do no harm.

Once more travelling on I determined to keep a short distance from the road that I might not be discovered by those who were on it. This made travelling very slow but it was more prudent—at least till that time in the day when I should feel willing to travel openly as a person intending to enlist at a convenient point.

As the morning advanced I saw the people going to church. I kept well away from the road, but did not hide, reflecting that I was already so far away that they would not readily connect me with the railroad adventure even if they had heard of it themselves; and if questioned, they would hardly judge me to be anything worse than some poor recruit trying to run away from the terrible conscript law. The sacred day and the church-going people did bring some influences of a religious nature to bear on my mind, and from this time may be traced a series of providences which led to conversion and ultimate entrance upon the work of the ministry. But the beginnings were very faint. I thought of God as the all-powerful Ruler, and with a sense of consolation and rest realized that my life and destiny were in his hands, and not in those of the deadly enemies by whom I was surrounded. I thought of my own father and mother in the far-off Ohio home, probably going to church at that very hour, and remembered with a thrill of regret how little value I had placed on such privileges. I thought how good the Lord had been to me during all my army life, and how poor a return I was making. At this time I was a believer in the Bible, but had not tried to make any practical use of it. In fact, I had a vague notion that I would have to wait until I had passed through some great change, with which I had but little to do, before I could take any steps toward a religious life; and until that came, the utmost in my power was to remain sober, be a good citizen and soldier, and live a tolerably moral life. To some it may seem incredible, but it is absolutely true, that all the falsehoods we had told the enemy, and all the many more we meant to tell, were not thought of as in the least interfering with the highest standard of morality. Indeed, if any one had proposed the view that we were committing sin in deceiving the enemy, I am not sure that we would have understood his meaning, or have believed in his sincerity. Religion, as the sense of a conscious dependence on God and a consistent and sincere effort to walk according to His revealed will, was yet in the future; and before I could seek earnestly to realize it, sufferings were to be met of which I had little idea.

But I turned resolutely away from all thoughts of home or of Heaven or the sacred associations of the Sabbath as if they were mere weaknesses, and concentrated my attention in devising the best methods of answering the questions that would follow arrest. A line of pretense as to my Southern sympathies and history from the time I left Kentucky till the present was outlined, and left to be filled up as occasion rendered expedient. Unfortunately the basis of the whole continued to be the same we had all told on entering the enemy's territory. In arranging such lines of defense, my partial legal education was not without its value; but my best preparation I felt was inadequate to resist a really searching cross-examination, for I lacked the means to build upon.

Such thoughts, and constant plodding on, brought me about noon to the little town of Lafayette, in Georgia. It is a mere country village, though a county-seat, and is in one of the loneliest and most inaccessible parts of the South, being twenty miles away from Ringgold, the nearest railway station, and twenty-seven from Chattanooga. With all my wanderings this was not bad progress for twenty-four hours of storm! Several roads met here, and as I wished to get within following distance of one that would lead south or south-west, I selected what I judged to be a favorable place and asked the way to Rome, and to Corinth in Mississippi. The direction was given, but either there, or somewhere else in the vicinity of the town, I was observed, and a party of pursuit immediately made up. I knew nothing of my danger till a mile or two from the village, when I heard a loud voice behind calling; " Stop there! stop! we want to speak with you!"

Turning, I saw a party of more than twenty men, armed with various kinds of guns, and the foremost of them perhaps fifty yards distant. As some mounted men were among them, and the country around quite open, there did not appear to be any chance either to fight or run. Now was the time to put in execution the plans I had been meditating all the morning. I had money, and had as far as possible removed all trace of the night in the woods from my clothes. Could I induce the rebels themselves to believe my story, and let me travel toward some point where there was a Confederate regiment for the purpose of enlisting? This would be a patriotic errand, and give me an unsuspected point to aim for, until the sky cleared, and I could take a better start northward, either before or after enlisting under the rebel flag.

Accordingly I halted and asked in as bold and frank a tone as I could command, " What do you want, Gentlemen?"

They gathered around quite close before making any reply, and then a conceited little fellow who had the shoulder-straps of a Lieutenant, but who was called Major, became spokesman. He said that they wanted to talk with me for a short time.

I said, "Very well; I will be glad to hear anything you have to say."

There was an awkward pause for a few minutes, and then the Major continued:

"We have no doubt you are all right, but strange things have happened about here, and it is our duty to question every stranger that passes through the country."

"What strange things?" I inquired. It was easier to ask than to answer questions.

"The Yankees have captured a train on the railroad and we are looking for them everywhere."

I was exceedingly surprised; said I had not heard of it; and asked if they were sure the news was true. He said that there was no doubt of it, but that they would like to know who I was, where I came from, and where I was going—the usual formula of questions. I gave my name as John Thompson, of Fleming Co., in Kentucky. I had left home because I could no longer endure the terrible oppressions of the Lincoln Government, and meant, as soon as I had looked about me a little, to enlist in the army and fight for our common rights. He said that this sounded well and was no doubt all true, but it would be necessary for them to search me and see if all my possessions were in harmony with my claims; and he was sure that as a good Southern man I would not mind that slight inconvenience for the country's good. I begged him to proceed, assuring him that he would find nothing wrong. I felt safe enough in this particular, as I had been careful before leaving camp to divest myself of all Federal marks. In fact the South was so full of Northern manufactures, that in the absence of papers or army equipments it is hard to name any article of general use that would have been compromising. He found a very good revolver with plenty of ammunition, but that was in keeping for a Kentuckian. My money was all Confederate, and everything else right. One matter gave me some uneasiness; but it was needless. I always wore my spectacles, keeping them tied on my head for safety. Had they been taken I would have been well nigh helpless. But no reference was made to them either at this or at any other time. The frames were of silver, but had suffered so much by exposure that they showed little traces of it. I have always wished that I might think that they were spared from motives of humanity; but some things are in the way of that explanation.

The Major asked me where I came from the day before. I told him that I had been at Chattanooga looking about for several days. Then he wished to know why I did not enlist there; I was again on familiar ground, for I had heard much of the troops while on the cars two days before. I had also heard the soldiers boasting of the First Georgia as a splendid regiment, and now told the Major that I did not like any of these raw and

12

conscript regiments at Chattanooga, but wished to go on to Corinth and try to get into the First Georgia. This flattered his State pride, and he asked the next question as if he wished a favorable answer:

"Why did you not go on directly to Corinth instead of making a circuit out here?"

I had little idea how wide a circuit I had made, but answered confidently.

"Because Mitchel has Huntsville and I want to keep out of his reach."

This seemed so clear to the little major that he turned to the crowd and said quite heartily.

"We may as well let this man go, for he seems to be all right."

These words rejoiced me, but my joy was premature. I think I would have proposed going back to town with them of my own accord, and hiring a guide on my way to Corinth, which I had money enough to do. But a dark-complexioned horseman, with his hat over his brows, who had not yet spoken a word, raised his eyes slowly and drawled out;

"Well! Y-e-s! perhaps we'd as well take him back to town, and if all's right, maybe we can help him on to Corinth."

I neither liked his tone nor manner, but pretended to be glad of the opportunity, and went promptly along. They took me down the single, long, muddy street which composes almost the whole of Lafayette, and brought me to the *Goree House*, the largest hotel of the place. My reception was kind enough, and, indeed, I had no reason to complain of my treatment in any particular up to this period. I have no doubt that a dinner would have been given if I had asked for it, and I felt a raging hunger; but I did not want to admit that I was fasting. The people of the town gathered rapidly, and apparently many from the country also. The Sabbath quiet was considerably disturbed, though up to this time there was no noise or disorder. The lawyers of the village came in and commenced asking all kinds of questions,—many of them much harder than I had previously answered. When they confined themselves to generalities, I did well enough; but, lawyer-like, they wanted to know all about a host of minute matters. They asked me from what County in Kentucky I came; I said, "Fleming." They asked the county-seat. I gave the name, "Flemingsburg," without difficulty; then they capped the climax by asking me to bound the county! I mentioned a few counties at random that I thought were in that part of the State! They procured a map, and laughing said it looked rather suspicious to find a man who could not bound his own county. I offered to bet that none of their neighbors standing around the door could bound the county we were now in. They would not make the bet, but tried the experiment. I listened most intently, for the words uttered might prove to be of great practical use. Not a man was able to do it entirely; and as they were comparing the map with the answers, I received a better idea of the geography of that part of the State than I ever had before.

Then these inquisitive people wanted to know all about my journey from Kentucky down to that very day and with what people I had lodged. The first part of the story was easy, for there was nothing to prevent imagination from having free rein; but it was very different, especially from the time I claimed to have left Chattanooga on the preceding day. The story became very perilous, as all my auditors knew the country around, which I was fancifully populating; but there was no alternative. If I refused to answer at all, silence would be set down as a confession of guilt and myself as a stubborn and morose fellow; while there might even be no small danger of the lash and the halter! As it was, my assurance puzzled them somewhat, and caused them to hold numerous private consultations in which they could only agree that the matter needed further investigation. There was something not right; but what it was, they could not yet tell. I think from some of their questions that they imagined me to be a rebel deserter, and that my real home was not very far away. Apparently they did not suspect me of being from the North. In the meantime all of them kept in a fairly good humor. Jests passed frequently; if I found the chance to ask a question that would turn the laugh on any of the inquisitors, I used it, and the others would laugh as heartily as when I was on the losing side. I feared no violence while they were in this mood, and, aside from my hunger and fearful fatigue, and the assurance that in a moment the thin varnish of good humor might be turned into hate and revenge, I might also have enjoyed the badinage. But the situation was really terrible. My only hope was dim and far off —that of being permitted to go on my way and meet similar perils, or of being enlisted in the rebel army to face Union bullets, and possibly die in an attempt to desert. This was the best; while a word or an outside incident might in a moment break up our truce, and leave me at the mercy of a howling mob. The latter did happen not long afterward.

For four hours the interminable catechism went on. I saw the hands of the large clock in the room move round and round, and was thinking that I would take advantage of the next temporary pause to demand supper, which would probably not be refused, with the promise that if they would only give me enough to eat, I would answer questions all day Sunday, and Monday too. But I was saved further trouble in that line.

A noise was heard outside; there was a buzz of excitement, and a man dashed up to the door on a horse covered with foam. He shouted,

"*They have caught the bridge burners!*"

In an instant I was almost forsaken, while question after question was hurled at the messenger.

" Where? Who are they? What did they do? What were they after?"—are only a few specimens.

I was as deeply interested as any. My own comrades! perhaps they had already been shot or hung! and I almost felt my heart stand still as I waited for the answer. But it came in a form that I had never for a moment dreamed of.

" *They said at first that they were citizens of Fleming County, Kentucky,*

News from Ringgold. [The copy of the Goree House, Lafayette, Ga., is exact.]

and afterwards owned that they were United States soldiers sent South to burn the bridges on the State road!"

I could hear exclamations of " Ah!" " Oh!" all over the room, and a cry outside, "We've got one of them here!" Fierce eyes gleamed on me, and there were no more jests or questions. Even yet no word of threat or insult was spoken, but stern looks were on every hand. There was a good deal of noise and shouting on the street. Some of the leading men went into another room, but came back in a few minutes, and said to me simply: "We will have to take you to jail." Without a word, I rose from the chair by the window on which I had been sitting so long, and went with them.

CHAPTER XVI.

THE ROLL OF THE CAPTIVES.

THE certainty with which members of our party were identified when arrested arose from two causes. The first captures were so near the place of the abandoned train that the men could be traced almost directly back to it. A party of three who were the first taken were hardly lost sight of from the time they jumped from the "General." To have denied their connection with the raid was useless, and would proba-

bly have led to frightful beatings, as in another c se, or to a speedy death. The captives expected the latter; and they only revealed their names and regiments at the prompting of that deep feeling which makes it hard to die under a false name and character. One man of the most dauntless courage, who never did reveal anything until a council of the prisoners decided to abandon concealment, told me that he had fully resolved, if put to death in the woods, that just before he died he would give his own name and character, and defy them to do their worst. None of those first captured betrayed their com-

SAMUEL SLAVENS. From a war-time photograph.

rades, and no one of them could have given names or descriptions of more than two or three of his more intimate associates, so little had we been thrown together. Not the slightest blame can be attached to these first captives, therefore, for revealing themselves; but unfortunately, they first told the old Kentucky story which had already served us so well that we forgot that it might wear out! After that, whenever a man was found hailing from this state and county, he was set down as one of us, and no denial would even be listened to. It would have gone hard with any

genuine resident of Fleming county who happened to be travelling in the South at that time !

Campbell, the Hercules of the party, Slavens, also a man of massive proportions, and Shadrack, were the first to leave the fagging " General." It has even been said that they jumped off under the command of Andrews before we reached Ringgold, and that as we came near the town the de-

Slavens, Campbell, and Shadrack in chains.

parture of the others was suspended for a little time, so that these were widely separated from the rest. This was well adapted to the idea which Andrews entertained of dividing the pursuit, but the idea itself was not good. They were seen to leave the train and a large party with dogs was qu ckly gathered; having hidden to wait for night they were soon discovered and arrested. They were examined, told at first the Kentucky story, and afterward, finding that circumstances had fixed their connection with the flying train, they gave their names and positions in the army—Campbell claiming, as it had been agreed, that he belonged to Co. K, of the 2nd Ohio. They did this as they thought their fate as soldiers might possibly be better and could not be worse than that of citizens. They were loaded with heavy irons, a long log chain being found, and Campbell, as the largest, having the middle of the chain locked around his neck and an end around the necks of each of the others. They were indeed *strongly united* in misfortune; and in this barbarous manner they were conducted amid a howling mob to Ringgold and afterward to Dalton, in which place they remained for two or three days, when they were forwarded to what proved to be a general rendezvous—Chattanooga.

Parrott and Robinson went into the woods about the same time as the others, and succeeded in eluding the pursuit for some hours; but finally they either got turned around as I did, or finding the Chickamauga an obstacle in their path, they faced eastward and came clear back to the railroad. Here they were observed and captured. Being not far from Ringgold they were at once taken to that point, where the excitement was still at fever heat. Then occurred one of those dreadful scenes we would gladly pass over, but they are a part of the record, and to omit them

JACOB PARROTT. From a war-time photograph.

would give a false impression of the whole story. The two men were questioned and maintained an obstinate silence. Parrott was the younger, being at that time barely eighteen. His educational advantages had been limited, but he had a clear head and a resolute character He was separated from his comrade, and every device used to open his lips. He was threatened with hanging; a rope was procured, and they were apparently about to proceed in their work when a colonel, at least addressed as such—possibly Col. Jesse Glenn, whose regiment was camped near Dalton, and whose men at any rate were engaged in this terrible affair—interfered and saved his life. A lieutenant and four men then took him and stripping him naked, held him down over a stone, while they inflicted over a hundred lashes on his bare back! Several times the whipping was suspended and he was let up to see if he was ready to tell all they wished to know, and especially who was the engineer of the party. From the very first they were anxious to ascertain who had run the engine, probably supposing him to be a Southern man in league with us. Finally the crowd themselves grew sick of the terrible work and it ceased, without the poor boy having uttered one word. His companion, however, believing that they both would be obliged to die, and wishing to die under his own name, gave it, with his regiment. He had also first tried the same Kentucky story in vain. They were then ironed and taken off to jail.

These were all that were captured on the day of leaving the train. In the night following none were taken but Sunday was more fruitful to the enemy.

None of the party had apparently a better opportunity of escape than Hawkins and Porter. They were left at Marietta, and felt wonderfully disappointed and chagrined when, on reaching the station, they saw the train drawing out. To come all that distance to join in a daring expedition, to have all the danger, and then at the last moment to be deprived of the chance of thus striking a blow for their country, from the prosaic cause of oversleeping, was terrible. They strolled around town for some time listening for the news of the seizure of the train. Soon it came with a wild burst of excitement. When they heard this they were almost ready to envy their comrades, and to believe themselves in far greater danger. After a little consultation they resolved that it would be best to do what we all had in mind from the very first as a possible refuge. They knew that any persons attempting to go north would be closely watched, and that the chances of reaching our lines would be but slight. So they walked the eight miles up to Big Shanty and made due application to join one of the regiments there. They were promptly assigned to the Ninth Georgia battalion, and a company in it took a vote in the fashion of southern volunteers as to whether they should be admitted. They were unanimously elected, and all formalities being complied with, they were enrolled under the southern flag. But they did not serve long. In the kindness with which they were received, much story-telling was evoked, and these two gave *their Kentucky story* in all its minuteness. When the news came the next day that all the "bridge burners" who had been captured claimed at first to be from Fleming Co., Kentucky, it was soon recalled that these strangers professed to be from the same place. They were lost! A little separate 'cross-examination was sufficient to involve them in hopeless contradictions, and then they were ironed down and taken back to Marietta, where others of the party were also brought.

The next party in order of capture was Dorsey, Bensinger, Buffum, and Geo. D. Wilson I give their story substantially as told by Dorsey, with a few additions made by Bensinger. Dorsey was among the last to leave the exhausted "General," and as they were running at their highest

speed in a north-eastern direction, Buffum, who had
on a large gray coat, spread open his arms, and ran
right out of it, leaving it on the top of the weeds
through which he was passing. They were soon in
the woods, and there struggled with the same difficulties from the clouds and the want of direction,
that I had found so distressing. All afternoon and
evening they either hid or wandered about, making
slow progress. Late at night they came to a small
log hut. They knocked in vain, and when they
went in, they found the head of the family suffering

under a sickness
that Dorsey be-
lieved to be
feigned, so that their
application for food
here did not avail.
Then they went on
but slowly, for Wil-
son had received a
severe injury, and
was not able to walk
rapidly, while the
others were too loyal
to leave him. Before
morning they tried
again at a better-look-
ing house for food,
against the wishes of
Bensinger, but failed
to get admission, the
people not wishing
under such circum-
stances to open their

Bloodhounds seen on the trail.

doors to strangers. But on the porch they found a bucket of milk, which
they drank to their great refreshment. Again they pushed on, till day-

break. There was no sun to guide them and they resolved to hide till night. From their place of concealment on a hillside, they saw pursuers at a distance nearly all the morning, but these did not discover their game for some hours. At length our comrades beheld a sight which no man could look upon for the first time without feeling his flesh creep with horror. *Three blood-hounds in plain sight* were descending the opposite hill, down which they had recently come, *on their trail!* There were four armed men following close and not allowing the dogs to get far ahead of them. The odds were not unreasonably great for men fighting for life, and our comrades, knowing it impossible to escape the dogs, prepared for the fray; but as soon as the pursuers discovered them, and before coming within striking reach, they set up a shout and were answered in several directions, and other parties closed up, till soon not less than fifty men were at hand. There was no longer hope in fighting, and nothing remained but to try duplicity. Geo. D. Wilson had suggested that it might be well to discard the Kentucky story and try something new, a wise precaution which all of us would have profited by observing. But he did not have time to get his companions drilled into it, and while he spoke with his usual confidence, pretending to be hunting negroes and telling quite a plausible story, yet when separated and pressed closely they soon entangled themselves, and the enemy became convinced that these were none other than the men they were in search of. Indeed, they were found only nine miles away from Ringgold, and in hiding in the woods, and it would therefore be very difficult for them to break the force of circumstantial evidence. Bensinger at once determined that their only chance was to acknowledge their membership in the Union army, and claim protection under the laws of war. He feels sure even now that this was the wisest thing that could have been done. Otherwise they would probably never have been taken to the regular military authorities at all. Several of the captors, as it was, wished to shoot or hang them at once. One young fellow clapped his pistol to Wilson's head and was about to fire, when his hand was turned away by a Major. It is wonderful that some at least of our comrades were not killed when captured. This would only have been what we had expected from the first. Possibly the undaunted bearing of the men, and the readiness with which they joked in the presence of death, had a restraining influence. In view of the succeeding history the whole credit can scarcely be given to humanity on the part of the captors. It may also be that curiosity as to the nature of the expedition led them to take as many alive as possible in the hope that some light might be afforded. A rope had been prepared and brought along for the avowed purpose of hanging all who were captured, and they were making ready to do it, when Buffum said in his most nasal tone, " Well, captain, if you are going to send us over the river, can't you give us one good square meal before we start?"

The leader swore a great oath, with a laugh, that Buffum should have all that he could eat before he was hung!

A very rough fellow pressed the muzzle of a double-barrelled shot-gun against Dorsey's breast, and with finger on the trigger declared he would blow him through in another minute if he did not "make a clean breast of it." Without moving a muscle, Dorsey said, "If you want to shoot, just shoot." He really cared little at that moment, as he had fully made up his mind that they would be hung at any rate. Thus defied, the fellow dropped the muzzle of the gun. They, however, tied the hands of the four and conducted them to a house a mile away, where they redeemed their promise to Buffum by giving them a good dinner. Then they were marched to Ringgold on foot and put into jail—about the same hour that I had found a similar harbor at Lafayette twenty miles distant. On this Sunday evening just one half of those who had been on the train were captured, and also the two who had not been on it. We were not succeed-ing very well in our attempt to escape by scattering!

The number captured on Monday was not large but relatively very important. It has been assumed that when Andrews gave the command to disperse, he wished to go alone with the view of making his own escape the more easily. Those who think this do not blame him, as they judge that by escaping quickly, he would have been in a position to give help to those less fortunate, and they even find some fault with those who kept with him. But Andrews never said anything to indicate that he wished to be alone, seeming to think that as the party had come down in small squads they would succeed best by returning in the same manner.

Andrews, Ross and Wollam left the engine together. They did not make very rapid progress but kept themselves carefully concealed in the day-time. With a good knowledge of the country, they made quite satis-factory progress in a strait line, and by Monday had reached the slope of Lookout Mountain. It seems to have been Andrews's purpose to slowly and carefully work his way to a point where he could avail himself of his passes, some of which he retained to the last. He had also a large sum of money, and, with his skill and address, there is little doubt that, once clear of the immediate surroundings of the expedition, he would be well able to provide for his safety. But the employment of dogs in following the trail, which he seems not to have expected, completely disarranged his plans. On Monday he and his companions were followed in this way. They doubled on their track, and tried all practicable expe-dients for throwing the hounds off the trail, but in vain. They prepared to shoot the dogs, but men armed with guns kept close up. Andrews did
ne shot was not effectual. When finally surrounded, it
ickly at the last as to be something of a surprise, and
lerstood not to have completed the destruction of some

papers which he had been unable to get out of the way. These, with his large amount of money,—more than two thousand dollars,—at once marked him out as a person of consequence, and, in connection with his striking personal appearance, suggested the commanding stranger who had acted as conductor of the pretended powder train. In a day or two after his capture, a number of persons were found who had known him previously, and his identification was complete. Too many persons were acquainted with him in the South for him to have long remained unknown in any event. The same evening, Andrews and his two companions were heavily ironed and borne to Chattanooga.

But two more bands were now at large, taking almost opposite courses. The larger of these was composed of Brown, Knight, Reddick, Scott, and Mason. Scott lost his hat in the first flight, but when they came to a house where a Frenchman lived, he ran in and tried to beg a cap; but not being able to make himself understood, he took down one hanging near, and ran off with it, the good wife chasing him and scolding in French! Knight thus, in substance, describes their adventures:

"When I jumped off the engine the pursuing train was in plain sight. I reversed the engine, but the pressure of steam was too low to carry it back with any force. All the rest of the boys had got quite a start, but I pushed ahead and soon overtook them. We ran southwest till we came to the Chickamauga, where we all plunged in and pulled for the other shore, holding our pistols up out of the water to keep our powder dry, and expecting a shot from the rear every minute ; but we got over and hid in the brush before the enemy came to the other bank. We here kept secreted till dark, when we moved on, but it being rainy, made travelling tedious. We continued to hide by day and travel in the night for some time, but in the dark nights and over the rough country we did not make very rapid progress.

"After six days we came out of the woods, and had to cross quite a slope of clear country, which we got over safely and were going up the mountain to the timber beyond. Mason and Scott were some distance ahead and off to the right, when all at once a dog jumped up right in front of them and a man called out, ' Halt !' They found themselves surrounded and at once surrendered. The rest of us dropped down in the bushes and crawled away around and got clear. Then in going up the mountain we came to where the rocks stood perpendicular." [This was the great rock parapet of Lookout Mountain, W.P.] "We noticed a big hole in the top of this ledge, and we managed to clamber up and get into that. It was quite a natural cave and as it was now getting daylight, we put in the day there. We could look clear over the tree tops, and see into a little town. By their movements we could soon see that they had our comrades there. This was sad, but we had no means of helping them.

"When it became dark, we crawled out, and went up the slope to the top, and then travelled along the mountain. That night was clear, and the mountain top quite level, so we made good time till the morning. At daylight we were at the lower end of Lookout Mountain. We were in sight of a log house off to the left of us, and Brown and Reddick wanted to go and get their breakfast ; but I wanted to hide till night and then go and get our supper. The difference was that if anything happened we would then have darkness to escape in, and the night before us. We took a vote, and as two outvoted one, we went to breakfast. The owner of the house promised us something to eat if we would wait for it to be cooked, but I noticed that he was very uneasy. He finally got a horse out and I told our boys that I thought there was mischief in the wind. We went out to talk to him, with the notion, if all did not seem right, of stopping him till we got our breakfast ; but he said he was going to send his son over to a neighbor's to plough his garden for him. This was so innocent that we let him go, and got our breakfast. Then we asked him the road and distance to Bridgeport, but calculated not to go there ; (but we did !). We started up a hill on the main road, but meant to hide in the woods till dark. But lo ! there was a party of cavalry coming up the other side and we met at the top. That was the kind of ploughing that boy was after ! I was as hungry as any one, and the breakfast good, but I wished it, with the boy and men, and the plough on top of them, very far off. They halted us and commenced their questioning. We told them we were going to Chandler's Springs, Alabama ; had come from Fleming Co., Ky. They asked us for passes, as now no body was allowed to travel through the country without a pass. We had none ; did not know they were needed. They asked us to come along with them to Bridgeport, where, if everything was straight we could get the passes. We went to Bridgeport after a pass, but found that they passed us up to Chattanoga."

The capture of this party is thus noticed in the Chattanoooga correspondence of *The Southern Confederacy*, April 19th. As these men were among the best looking of us, the reader may judge of the others!

"Four rather villainous-looking Yankees were arrested at some point on the Willis Valley Railroad, a few miles from this place this morning. They confessed themselves to be spies: said they were part of the party that had stolen Mr. Fuller's mail train on the W. & A. railroad ; that they were members of an Indiana regiment ; that they had been sent to burn the bridges on the State railroad ; and that nearly every regiment in the western department of the Federal army was represented in a similar way. They are now here in jail awaiting further trial."

The last of all these narratives is intensely interesting, because of its wide variety of adventures, the light it throws upon the loyal East Tennessee Mountaineers—some of the noblest people in the world—and their modes of life in the midst of enemies; and because these men alone of all the band actually reached the Union lines, though there was a startling sequel even to that. These two were Wilson, our fireman, and Wood, our only Englishman. The account is extracted from Wilson's book, though with his consent I have slightly abridged it. The instinct for the water shown by Wilson in his attempts to escape is curious. No others of the party made such use of the streams. After describing the abandonment of the train, Wilson begins:[1]

"After running some distance, Wood and I came to a large, open field, on the slope of a mountain just in front of us. To attempt to cross this space would expose us too much, and we were nearly out of breath. We could hear the enemy shouting, and the constant report of fire-arms warned us that the remorseless crowd was waging a war of extermination. There was no time to be lost. A fortuitous circumstance saved us. The woods were too open for a man to hide in, but as I glanced about I saw where a tree had been cut down, probably the summer previous, and the brush which had been trimmed off

---

[1] Adventures of Lieut. J. A. Wilson (Toledo, O., 1880), page 53.

lay scattered around with the dried leaves still clinging to it. My plan was formed instantly and I told Wood to lay down. I hastily laid a few leafy boughs on him in such a manner as not to show that they had been displaced. Mark was soon out of sight in a little flat, unpretentious pile, that would scarcely be noticed among the other rubbish, and with almost the quickness of a rabbit I slipped out of sight under the heap by Mark's side. I now drew my revolver, and told Mark to do likewise. I felt a sense of desperation which I had never felt before. We were in a high state of excitement, and realized that the frenzied crowd of man-hunters, then deploying all over the woods, would show us no mercy. From the constant report of fire-arms which rang in our ears we had reason to believe that some of our unfortunate comrades were being shot down like cattle, perhaps all except ourselves had been killed.

"In my concealment I felt all the desperation and anguish of mind that a man could feel. We had failed and were disappointed. We had been run down and had gone to the last extremity of human endurance to make our escape. Our enemies were infuriated. We had made such superhuman exertions within the last half hour, ending in our up-hill race to the brush-heap, that we were almost breathless. It did not at that time seem to me that it would require many rebel bullets to finish my part of the story. Several times, as they passed us so close that I could have touched their legs with my hand, I was on the point of springing up, and, with a loud yell, beginning the work of death at close range with my revolver. I could not, even in a whisper, communicate my wishes to Wood, without betraying our place of concealment. But our pursuers made so much noise themselves that they could hear nothing else. They were all yelling, swearing, cursing and shouting. We could hear much of the conversation as they passed us. Two men, each with a musket, as they passed near us. spied two of our comrades going across a distant part of the great open field.

" 'There goes two of them,' said one of the pursuers. 'Come on, let's go for them.'

" 'Let us get some more help,' said the other.

" 'But, you see, they have no guns,' said the first.

"And thus they passed out of hearing, halting and debating, but evidently distrusting the policy of tackling the train-robbers even-handed.

"It was some time in the afternoon when we took refuge in the brush-heap, and in that spot we were compelled to remain far into the night, before we dared venture forth. The night was one of terrible anxiety to us. Our condition was perilous in the extreme. The whole country was aroused and swarmed with soldiers and citizens. Every road and cross-road was watched night and day that none of the 'rascals' might escape. We could hear the deep baying of bloodhounds, as they scoured through woods and fields, but luckily for us, so many men had tramped over the ground in the vicinity of the place where we jumped from the train that the dogs could not work. Still, men and dogs were scouring the woods in every direction and it was unsafe to make *tracks*. To add still more to the wretchedness of our condition, the rain was almost incessant. The place of our concealment was a little lower than the ground surrounding, and much of the time the water was three or four inches deep where we lay. This, with hunger and wet clothing, made us extremely uncomfortable.

"After darkness had closed in for some time we were compelled to come out, capture or no capture. We could stand it no longer. On crawling out, our limbs were so stiff and sore that it was with the utmost difficulty that we could move, and it was only by rubbing and working them vigorously that we could begin to use them. It did not seem that we could travel very far, do our best, with such stiffened limbs. After looking about, we decided to take an opposite course from that which our comrades had taken, thinking there would be less vigilance on the part of the hunters in that direction. We desired also to get into the mountains, thinking we would there have a better chance for our lives. I suppose at this time we were less than twenty miles from Chattanooga. The rain still fell in torrents, but as we went on and our stiffened limbs got limbered up, we began to make good time. Our desire was, as soon as we could get beyond the immediate reach of our enemies, to bend our course in the direction of the Federal lines. But we must by all means avoid Chattanooga. We knew that.

"We traveled as rapidly as we could that night, and about daybreak of Monday morning we saw an old log hut off by itself some distance from any road. We wished very much to get shelter from the cold rain, which had chilled us almost to the point of freezing. We found the hut to be a sort of barn, the mow of which was full of bundles of corn-fodder. We made a hole down in the mow and covered ourselves out of sight and went to sleep.

"About one o'clock in the day, as we slumbered, we were awakened by somebody in the mow, and soon found out that two women were looking there for eggs. One of them nearest us said: ' Here is a hole ; I wouldn't wonder if there is a nest in here ;' and at the same time she thrust her hand down, and, as bad luck would have it, she touched one of my hands and started back with a scream, which brought up the other woman, and they threw off the bundles and there we were. They were both badly frightened and ran for the house with all their might. We hastily crawled out and brushed some of the chaff from our clothes, and after a moment's thought concluded that the best thing for us to do would be to go to the house and apologize to them and in addition try to get something to eat.

"Who is the man who has ever in his life vainly appealed, in a becoming and respect-ful manner, for food, when hungry, to a woman? If man excels in the brutal art of war and killing his fellow-beings with successful and unsparing hand, or being himself killed without a murmur, all of which passes for bravery, noble woman excels in those higher and more Godlike attributes of sympathy for the distressed and charity to the needy. I believe this to be true the world over, where woman is treated as the equal of man.

We went to the door, bowed politely and apologized for the unintentional scare we had caused them. We then told them we had been in pursuit of the train-robbers, and that wet, cold, and sleepy, we preferred to take shelter in the barn, rather than disturb any one at the dead hour of night. This story seemed to be satisfactory to them, when we told them we were hungry and asked them for something to eat. They had just had their dinner and the table still stood out on the floor. They gave us a pitcher of butter-milk and some corn-bread, all they had unless we would wait for them to cook us something, which we did not wish to do. We paid them and left much refreshed and strengthened by our food and rest. We started away on a road, but as soon as we got well out of sight of the house we changed our course, and soon after concealed ourselves in a dense thicket and there awaited the shades of night.

" We had not been in the thicket long before we saw a squad of mounted soldiers pass down the road we had previously left. From their loud talk and their manner of march we concluded they were a party of man-hunters. Whether they had gained any information at the house where we had been we could not tell, but we laid down and kept quiet. When night came we shaped our course as near as we could, without following any road, toward the Tennessee River, east of Chattanooga. Dur-ing the night march we narrowly missed running into a guard post at the crossing of a road, but fortunately heard them in time. We went around them and on our way undis-turbed. At the dawn of Tuesday we had just arrived at the foot of the mountains and breathed easier, for we felt more secure than we had in the open country. We concealed ourselves in a comfortable place and witnessed the rising of the sun. Its loveliness and genial warmth never before cheered me so much as then. But we soon fell asleep from weariness and did not wake until nearly night. We had a toilsome night march, feeling our way over rocks, climbing precipitous places and at other times descending the steep mountain side through bushes and among rocks.

" When Wednesday morning came we found that we were still surrounded by mountains on all sides, with no signs of a habitation or a human being in sight. When the sun got well up and it was comfortably warm, we lay down and took a nap. The pangs of hunger were, by this time, pressing us distressingly. We had only tasted food once since the raid began, and that was the scanty meal we made on buttermilk and corn-bread, Monday afternoon. In this starving extremity we decided that there was no great risk run in this lonely region if we should travel by day, and after so deciding, we pushed on with our utmost energy, as a hungry man will do when he hopes soon to find food.

" We were guiding our course by the sun, and during the afternoon we came out on the brow of a high mountain, overlooking a beautiful little valley, thickly dotted with houses. From our elevated position we could see everything the valley contained. I thought it one of the loveliest sights I had ever seen—that quiet, peaceful little valley. I looked at each house and wished that I could go into even the humblest and ask for a piece of corn-bread. I pictured in my hungry imagination the good things to eat in each little cottage, and wondered how we could safely manage to get a morsel of their stores of abundance to satisfy our great hunger. The more I looked at that little valley the hungrier I became.

" Near the foot of the mountain was a small log house, a little separated from the rest, and we knew it was inhabited from the smoke that curled up from the chimney. We concluded to venture down and apply for food. A young woman appeared at the door, and we told her that we had been lost in the mountains and were in need of something to eat. She invited us to seats, and at once set about getting us a meal. We inquired the way to the next town, the name of which I pretended I could not just speak, but she helped me out by mentioning the name—Cleveland. We learned from her that the town was only a short distance away and that there were no soldiers there. This was gratifying, but not near as much so as the savory odors of the ham, eggs and rye coffee she was preparing for us. We could hardly wait until the corn-bread was cooked, and when she invited us to take seats at her table, we soon gave her satisfactory evidence that we had told the truth about being hungry, although we had stretched the facts a little about being lost. We paid the woman, and, without delay, took leave.

" We felt very much the need of a map, and after a near approach to the little town of Cleveland, and a careful survey of the surroundings, I left Wood in a secluded spot to wait while I walked boldly in and went to a book-store and asked for a school atlas. They had Mitchell's Geography and Atlas. As the author was none other than my command-ing General,[1] I had no reason to doubt that through the aid of his map I might reach his camp, if he had not moved too far since I left I had to buy the geography too if I took the atlas, and taking the books under my arm, like some countryman who lived near by in the mountains, no one seemed to pay any attention to me. We were soon in the woods again, when we tore out such portions of the atlas as we needed and hid the rest under a log, after which we took our course and pushed on, making good progress. We knew that we must, by this time, be in the vicinity of the Tennessee River. Our plan was to reach the river as soon as possible and secure a boat of some kind, after which we would drift down the river to Bridgeport, Stevenson, or some point nearest the Federal lines.

" Towards evening of this day we came to the terminus of the mountain in this direc-tion, and from its great height we had a commanding view of the valley below, which, though beautiful in scenery, was sparsely settled. We descended the mountain and felt our way cautiously across the valley. After a time we came to a log house. There seemed to be no stir about the premises, and, as we were still hungry, we concluded to apply for

---

[1] A natural mistake. Mitchel's men generally believed that he had done or could do everything !

something more to eat. We had been so hungry that we had not dared to eat all our appetites craved at the last place.

"There was no man to be seen about the house, but the woman, who was a noble, dignified-looking lady, plainly dressed, told us to be seated. I noticed her looking at us with that scrutinizing, inquiring gaze of a woman in doubt, and I could read her thoughts as plainly as if she had spoken them. I knew enough about woman, too, to know that whatever her first impressions were they would be unchangeable, so I said, without further hesitation, 'We are in need of something to eat.' She said if we could put up with such fare as she had we were welcome. We told her that we were quite hungry and any kind of food would be welcome. As she proceeded about her work, I noticed that on every opportunity she scrutinized us very sharply, and I became a little uneasy.

"Presently she asked us if we were traveling, to which I replied that we were on our way to Harrison, which was a small village a few miles from there. I still noticed that she was eyeing us keenly and closely, and that her mind was not at rest on the subject, when suddenly she turned, looked us squarely in the face, and startled us by saying :

"'You are Union men! You can't fool me! I know a Union man by his look. You need not deny it, nor need you be afraid to own it, either. I am a Union woman, and I am not afraid to own it to anybody. The secessionists around here don't like me a bit, for I say just what I think of them, whether they like it or not. Further, I know that you are Union men trying to get to the Union army and you need not go to the trouble to deny it. I will do anything I can to help you.

"We stoutly denied any such intention, and told her that we had been soldiers in the Confederate army. But that did no good. She seemed to have made up her mind. So we let her have her own way. Soon after her husband came in. He was rather a fine-looking fellow, with a frank, manly face.

"When supper was over we offered this loyal woman pay, but she refused to take our money, saying that anything she could do for a Union man she would do with a glad heart and willing hands. She said she wished the Union army would come—she would give them everything she had.

"As we took our leave, she told her husband to give us all the information that he could as to our route. 'For,' said she to him, "you know that old Snow, with his company of cavalry, is in the neighborhood, and he will be upon them before they know it. He is watching every nook and road in the settlement to prevent Union men from getting away from the rebel conscription.'

"That night we passed in the woods, and continued our journey Thursday morning. The valley through which our course lay was thickly inhabited, and we had observed the greatest possible precaution, as we supposed, in avoiding 'Old Snow's' cavalry. Our surprise was all the greater, when, without the least warning, we heard the stern command :

"'Halt there, you! Halt, or I will blow your brains out.'

"A hasty glance around failed to discover any safe chance of retreat. We were captured, and there was no course for us to pursue but to submit to the unpleasant inevitable.

"The captain seemed to be a pompous and, according to his own account, a bloodthirsty warrior, for he said it was not his custom to take prisoners but to hang and shoot all who fell into his hands. He asked us a great many questions, including, of course, our place of residence and our names, all of which we answered very promptly, although I will not say truthfully. We told him we lived in Harrison, and gave him some names we had picked up, in which we must have struck him just right, for at once he inquired after the 'old men, our fathers,' whom he said he knew. We told him they were in excellent health. He said he was glad to hear it, for he was well acquainted with both of them. 'But,' he continued, looking at us very sternly, 'boys, it's my impression that you are running away from the conscription and you deserve to be shot as traitors for wanting to join the Yankees.' We told him we had not the slightest intention of that. After a moment's silence, and looking at us steadily, during which time, no doubt, he was mentally debating what course to pursue, he said :

" ' For all I know you may belong to those spies and bridge-burners, and if I did not know your folks I would send you to Chattanooga under arrest ; but I will tell you what I will do : if you will take the oath and promise to go back home and stay until I call for you, I will allow you to do so. I have known both your fathers for many years and have great respect for them. They have always been true men to the South, and out of consideration for them I will permit you to go back on the conditions I have named.'

" Now, there may be those with a nice discrimination of conscience who will con. demn me and my comrade in misfortune—who has long since ceased his struggle with the cold charities of the world—for what we did. But, dear comrade, or reader, I pray you before you lightly pass sentence of condemnation, remember that 'circumstances alter cases.' The professional detective or spy lives a life of constant deception. He professes to be what he is not. Whether great exigencies of a public nature justify the practices necessary to the successful pursuit of such a profession, may be a question on which moral philosophers can well disagree, but which I am not competent to discuss.

" But we were in no condition for hair-splitting on minor points. Conscience, where the moral perceptions are to be consulted, and conscience, where a fellow's neck is at stake, are two different things. We were not professional spies or detectives, although, for the time being and for the good of the cause in which we enlisted, we were to all intents and purposes practicing the arts of a spy. Our game had been a desperate one from the start. The players on the other side were as desperate as we were. The stakes on our part were to save our necks from the halter—from the death of felons.

" We had told a plausible story to this officer, by which we had so completely deceived him that he proposed to let us go, conditionally. He had named the conditions, and for us to have rejected them would have refuted the statements we had just made to him, namely, that we were Confederates. Besides this, our detention a single hour might betray the falsity of our story about our living at Harrison. We were liable to be exposed any moment by some of the new troopers who were constantly arriving. We accordingly signified our acceptance of his conditions, and he at once ordered us to follow him, he leading us back to what proved to be the house of a rank old rebel and within a half mile of the house we had left the evening before. Here he went through the cere- mony of what he termed administering the oath, after which he, with the aid of the hot-tempered old woman of the house, gave us the most fiery lecture on the subject of Southern rights and Northern wrongs we had ever heard. After the captain and the old woman had both exhausted their vocabulary, we told the captain that we hoped it would not be long until he would find it convenient to call upon us for our services in the cause. He seemed much pleased at the favorable effect his eloquent harangue had worked upon us, and as we hastily shook hands with him preparatory to leaving, he handed us back our revolvers, which he had previously taken from us.

" This we considered a lucky escape, and we started off in fine spirits after the depress- ing uncertainty occasioned by this capture. It was not long until we were again in the mountains, where we soon after found a place of safety, and rested and slept till near night. After we awoke we talked over the situation. What we desired was to get across the wide, thickly-settled valley to the river and find a boat. How to do it and evade capture was what concerned us most just now. If, by a streak of bad luck, we should again fall into the hands of old Snow or his crowd, we would fare hard, for we had promised him to take the back track. In this state of perplexity we decided to trust ourselves in the hands of the man and woman who had treated us so kindly and professed so much devotion to the Union cause. We knew, however, that if we ventured near this house that we must do it with great caution, otherwise we might be discovered, and thus not only be captured, but compromise our good, kind friends.

' It was, therefore, late on Thursday evening when, having left Wood a few paces from the house to keep a look-out, I went noiselessly to the door and knocked. The family had retired and the house was still as death. I knocked again and again, but finally heard the woman tell her husband there was some one at the door. Soon the man opened the door and seemed to know me at the first glance or by the sound of my voice. He spoke to me kindly and invited me in. While he was speaking to me I observed from some indications, I could not distinctly see, that his wife stood near by to kill me instantly in case any sign of foul play had been noticed. Those were times in Tennessee and Northern Georgia, and other places in the South, when shocking tragedies took place. Men were hunted and shot down in their own door-yards and homes, for their loyalty to the old flag, and these persecuted people were generally ready for the worst and defended themselves to the death. In this defense the women often took a ready hand. The woman I am now speaking of would have been a dangerous one for any rebel to have attacked, if she had been given the least warning or had half a chance. It need then be no matter of surprise that she held a cocked rifle on me as I stood near the door, ready, on the least suspicious movement on my part, to drop me in my tracks.

" I told the man I would like to speak a few words with him privately. He stepped a few paces from the door so that we were sure no person was in hearing distance. I then, in a low tone of voice, asked him if he could, of his own free, voluntary will, assist a Union man in distress, if he had the opportunity. I then paused and watched him intently, and at once noticed that he was embarrassed. He acted like a man who suspected some trick—as if he thought I had been sent to entrap him for the purpose of betraying his loyalty. I was assured by his actions ; for had he been a rebel and had wished to entrap me, he would have unhesitatingly answered, ' Yes,' and encouraged me to reveal myself. I relieved his embarrassment by saying, ' There is no trick in this ; I am a Union man in deep trouble, the nature of which I am not just at liberty to mention now. I need a friend and assistance.' He then answered and said he would render any assistance in his power, not only to us but for the Union cause.

" Wood had by this time come to where we were and I told the stranger to hold up his hand and be sworn, which he did, and I administered to him the following oath :

" ' You do most solemnly swear in our presence and before ALMIGHTY GOD, that you will not betray us to our enemies, but that you will do all that lies in your power to secrete, aid, protect and defend us.'

" To all of which he answered, ' I will.'

" We then shook hands, and after making sure that no ear could hear us, I revealed to him a part of our story and who we were. He was a brave man and a true man, and hearing our story seemed to increase his interest and friendship in our behalf. We watched this man closely to observe if he took such precautions as a man would take who honestly desired our safety, and were gratified to see that he did. About the first thing, he told us that we must not come near, nor be seen about his house. He told us to follow him, and he led the way to an old abandoned house, where he had first lived when he located on the farm, and which stood in a secluded spot, remote from the road. In the centre of the old floor was a trap-door, which opened into a hole about four feet square, which, during the occupancy of the house, had been used as a sort of cellar. Here we took up our quarters. He then went to the house and brought out a bundle of quilts for us to lie on. He next told us to avoid talking loud, and keep out of sight, in which case we would be perfectly safe until he could get an opportunity to pilot us safely out. He told us that no human being would be apprised of our whereabouts, except his wife, who was our friend, and would do as much for our safety as himself. He then left us and went to the house, first telling us that he would visit us in the morning, and bring us rations. We fixed ourselves very comfortably with the quilts, and, although our bedroom would not admit of our stretching our limbs out full length, we doubled up and enjoyed a very comfortable night's rest, something we had not done before for a long time.

"The next morning, Friday, we heard our friend not far off calling and feeding his pigs, and not long after he quietly lifted off a board over the little cellar, when we put out our heads, shook hands with him, and took a sniff of the morning air. He carried a small basket which seemed to contain corn, which he passed down to us, but we found our breakfast underneath the corn, and after taking it out we replaced the corn and gave him back the basket. He spoke a few encouraging words to us, telling us we must not get restive but bide our time, when he replaced the board, scattered some straw over the old floor and left us.

"Some time after, when we were talking in a low tone of voice, we heard footsteps on the ground and the board was lifted and some one spoke to us in a friendly voice. We put out our heads, and there stood before us, with the basket of corn, not our sworn friend of the night before, but his wife, the true Union woman.

"'I knew,' said she, 'that you were Union men all the time, and I am still ready to make good my promise, to not only do all I can for you, but for the Union cause.' She told us that her husband had gone to assist a neighbor about some work and left us in her charge, and that she had brought our dinner. She spoke a few words of encouragement to us, and praised our daring effort, as she termed it, to steal the railroad away from the rebels, at the same time expressing her sorrow that we had not succeeded, and that the Union army could not before that time have taken possession of the country and driven the rebels out. By this time we had taken our rations from the basket and replaced the corn, and she replaced the board over us and scattered straw about, as her husband had done, and left us.

"In this way we remained secreted for several days. This delay was for several reasons. In the first place, we were nearly disabled with sore feet from our night marches in the mountains. In the next place, we knew that the longer the time that elapsed after the raid, the less vigilance would be observed by the rebels, who would tire of the pursuit. Then, most important of all, we had to wait till our friends could find a suitable person to conduct us out to the river safely, for the nights were, at that time, almost as light as day.

"A trusty guide was found in the person of the brother of the loyal woman whose guests we were. This young man, who knew the country well, conducted us by a circuitous night-march to a creek, perhaps McLarimore's, a tributary of the Tennessee. Our great trouble had been, in this mountainous country, to keep the right course. Even if we knew the direction, it was next to impossible to follow it by night travel on account of the unevenness of the country. It was this that made us so anxious to reach the river, which would afford us a sure means of night travel, and guide us to a point near the Federal army. Unfortunately, when we reached the creek the boat was on the opposite side.

"Here our guide took his leave of us, and we set about finding a way to secure the boat. I first thought to swim the creek, which was very high and running driftwood. After considering the matter, however, I adopted a better plan. Mark secreted himself, near the bank below, where I could easily find him. I then went to an open space on the bank and hallooed. It was now daylight, and a man soon answered. I told him I wished to cross over, and he soon came and took me to the other side. He was unable to change a five-dollar Confederate note, and I told him I expected to cross back next morning, and would try to have the change for him, which he said would do. I then walked briskly on the road leading to Harrison, until I came to the first turn in the road, when I went into the woods and hid myself until dark. After dark I went back and cautiously approached the place where the boat was tied. After satisfying myself that the 'coast was all clear,' I hastily paddled over to the other side, took Mark aboard, and we were soon floating toward the Tennessee. After encountering some troublesome blockades of driftwood, and a rebel steamboat or patrol gunboat, we arrived safely in the Tennessee River.

"This patrol boat gave us some concern. She lay in the mouth of the creek with her 'nose' to the shore, while her stern lay not far from the opposite bank of the narrow stream. When we first saw her lights, we supposed it to be a cabin near the banks of the creek, and did not discover our mistake until we were right up to her, for the night was pitch dark, and it was raining. These latter circumstances enabled us, by lying down, and quietly steering our boat close under the stern of the steam craft, to glide by unnoticed. I thought if we only had our crowd of train boys along, and Wilson Brown to man the engine, we might easily have taken possession of the craft, and given the rebels another big scare, and, perhaps, all of us escape. But it might not have been any easier to steal a steamboat and get away with it than a railroad train. We drifted on, and in a few moments after we were happy voyagers in the Tennessee River, going down stream with the swift current." [The river was now in spring flood, very high and swift.—W. P.] "This night was one of the worst I remember. Those comrades who have campaigned in East Tennessee, will not need be told how disagreeable a cold rain storm is there. The incessant rain was accompanied by a high wind, blinding our eyes much of the time, while the dark, rapid, seething waters carried our little boat on with maddening fury. Sometimes we would find ourselves going round and round in a great eddy or swirl, next striking the point of some island, or nearly knocked from the boat by some low-hanging tree from a short turn in the river bank, or getting a startling thump from some on-rushing log or drifting tree. We were in constant apprehension, for in the black darkness we could not see whither we were going, and so benumbed were we with wet and cold, that we had but little control of the boat, and our ears were our only guide for safety.

"When the night was pretty well spent, we began to have a little anxiety as to where daylight would catch us. We knew we had been making good time, and that Chattanooga lay not far ahead. We also knew that it would not do for us to show ourselves in that locality in daylight. We now began to keep a look-out for a safe landing place. After several ineffectual attempts we found that to land along the steep banks, in our benumbed condition, was both difficult and dangerous. We soon discovered that we were passing what seemed to be a small island. We hugged close along the shore until we reached the lower end, and a place where the rapid current did not strike our boat, and by the aid of our paddles and the overhanging tree-branches, we effected a safe landing in the dark, and drew our boat up on the bank. We took shelter under a great forked tree, and wrung the water from our coats.

"The storm by this time had changed to sleet and hail, and it did seem to me that we must perish with cold. We beat our benumbed hands and arms about our bodies to try to keep up the circulation of the blood, but we were chilled to the bone. I have never, not even in the coldest winter of the North, experienced so much suffering from cold as I did on that terrible night. Poor Wood, who afterwards died of consumption, seemed to suffer even more than I did. Never did I see the light of day approach with more gratitude than on that dismal island at the end of that night of terror. The sun brought no warmth, but its welcome light revealed to us a cabin near the shore, from whose stone chimney the smoke was curling up. We at once decided to go there and warm ourselves, even if we had to fight for the privilege, for we might as well perish fighting as with the cold.

"We at once launched our boat and crossed from the island to the shore. As we landed on the bank to go up to the house, Wood, whose teeth were chattering, and who looked both drowned and frozen, said to me : 'Alf, you will have to make up some lie to tell them ; they will ask us a thousand questions.'

"I said, 'I don't know what I can tell them ; I am too cold to speak the truth, though.' But I told Mark to say but little, so that we might not "cross" one another in our story.

"We were admitted to the cabin, and, as I stood before the great fire-place, I noticed the family viewing our bedraggled, drowned, forlorn appearance with some curiosity, especially the man of the house. After I got so that I could talk freely, I inquired if there were any boats about there. He said he knew of none except his own, which the Confederates allowed him to have to cross over to the island to his work. He then asked me if we were looking for boats. I told him we were, and that we had orders to destroy all we found, with the exception of a few owned and in charge of the right kind of men. I told him the object, of course, was to prevent Union men from running away from the conscription.

"'I thought that was your business,' said he. 'There was a lot of soldiers along here a few days ago and destroyed every boat they could find.'

"He asked if we stayed at Chattanooga. I told him that our company was there. I further said : 'Then you don't know of any boats along here, except your own ?' He said he did not. After some further talk, I asked him if we could get some breakfast with them. He said we could. I then told him we were in the condition of most soldiers—that we had no money, but that I did not think it any dishonor for a man in the service of his country to ask for food. He said it was perfectly right.

"We then took off our coats and hung them up to dry while we were at breakfast. After we had become thoroughly warmed we took our leave, telling the man to keep an eye out for any boats that might possibly be lying about loose in the vicinity.

"We now resumed our boat voyage, and did not spend much time hunting for strange boats, but availed ourselves of the first good opportunity to land and secrete ourselves. Our hiding place was in a thicket in a field, near enough where our boat was tied so that we could watch it. The storm had subsided, and during the afternoon the sun shone out bright and warm and a high wind prevailed.

"Sometime before night, a man and a boy passed across the field not far from us, and the boy soon got his eyes on our canoe and cried out, 'There's a canoe, pap !' They went down to it, and, from their actions, we saw that they were going to take it away. I spoke to Wood and told him that it would not do to allow them to do so, and we walked out of the thicket on the further side from them, and leisurely came down to where they were, when I said :

"'Hallo, there ! what are you doing with that boat ?'

"'I thought it had drifted here, and I was going to take care of it,' was the reply.

"'That is a government boat,' said I. 'We tied it up here awhile ago on account of the high wind.'

"I then repeated the boat story which we had before told at our last stopping-place. This seemed to be an entirely satisfactory explanation to him.

"I then said to Mark : 'Do you think the wind will admit of our proceeding on our way to Chattanooga ?'

"The man spoke up before Mark could answer, and said : 'Men, I would not advise you to venture on the river now. It is not safe. You had better go down to the house, and wait till the wind falls.'

"This proposition suited us well enough, under the circumstances, so we accepted his invitation and accompanied him to his cabin. We found his wife a very talkative old lady. She sympathized heart and soul, she said, with soldiers, for she had a son in the army, who sent word home that he had a pretty hard time of it.

"Night came, but the wind still blew a gale. They invited us to stay all night with them, but we told them that it was absolutely necessary that we should be back to camp by the next day, if possible. We learned that we were only five miles above Chattanooga.

"About midnight the wind went down, and we pushed out in our little boat, and long before daylight we were quietly drifting past Chattanooga, that most "ticklish" point. When we had fairly passed, we felt that the greatest part of our task was over. We began to imagine ourselves almost back again among our old comrades of the 21st. We felt encouraged and jubilant. We soon found out, however, that it was not all

Wood and Wilson going down the Tennessee River.

smooth sailing yet. [They were now, after a week's travel, within a half mile of the place I had reached the same day we abandoned the train. But the difference in their favor was immense: They had a boat and knew where they were.—W. P.]

"Some ten or fifteen miles below the city the river runs through a deep gorge, and narrows down to only a small proportion of its former width. The mountains rise abruptly from the water in frowning grandeur, while great rocks, from dizzy heights, project out over the rushing, foaming torrent below. To increase the troubles of navigation here, the river makes a sharp turn to the left, after a long, straight stretch, during which time the water gathers great velocity of motion, and suddenly dashes against the wall of rock at the elbow, recoils, and forms a great, rapid, foaming eddy, after which it rushes on down the gorge in mad fury, as if trying to get revenge for the check it has just received. We perceived, even in the darkness, that there was danger ahead. The great roar and noise caused by the dashing of the angry waters against the rocks warned us. We hugged the left bank with our little boat as closely as possible. As we passed

the angry whirlpool, into which we seemed to be drifting, our boat was struck a tremendous blow by a floating log. We thought we were dashed to pieces. The blow
hoisted us away, however, several yards to the left, and we went flying down the gorge
like the wind. We pulled at our paddles with might and main to keep the water from
swamping our boat, which sank pretty low in the current and was now going at railroad
speed. We soon reached smoother water, and again felt ourselves safe.

" It was now getting light, and, as we drifted on, we saw a man on shore motioning
with all his might for us to approach him. As there seemed to be something unusual
about his actions, we pulled in a little, when he hailed us and said if we went on as we
were then going, we would be drowned in spite of fate. He said, ' You are strangers in
these parts, ain't you?' We told him we had never been down the river before, although
quite familiar with the country. He then said, ' Strangers, whatever you do, don't try
to run down through the " suck." I have lived here all my life, and have known men who
were well acquainted with the river to be drowned there. It is much worse than the
place you have just passed.'

" We tried to persuade him to go with us and pilot us down, but he said he was not
well. At last, however, with much urging and the promise of three dollars, he consented
to go. We rowed to the shore, and, after providing himself with an extra paddle, he
came aboard and took charge of our craft, which we ran as close to the left shore as
possible. The water ran with such great velocity and force, that we found it almost
impossible to control the boat, although we all had paddles and were pulling as if for life.
Our new pilot understood his business well and knew how to man a boat.

" At the place where we apprehended most danger, the river runs through a narrow
gorge. The whole volume of water, thus circumscribed, draws right to the centre of the
channel. After a ride which I never wish to repeat, we passed in safety, with no further
mishap than getting our boat nearly full of water, which we soon bailed out. Our pilot
now gave us careful directions as to the course we should take in the river below, after
which we dismissed him, first paying him three dollars, which we felt had been a good
investment, as we would have doubtless been drowned but for the accidental fact of
meeting this man. Though it had been our practice to travel only in the night, yet we
had been compelled, through the difficulty of navigating this part of the river, to travel in
daylight, which was imprudent, as we were constantly reminded.

" I may state just here a fact, which is well known to all men who in time of war have
tried to escape from prison. The most critical part of a journey is that which lies
immediately between the two contending armies. At such places, between the two
hostile lines, patrols are constantly moving about. Outposts are established on all important roads, while vidette and picket posts, in command of the most active and vigilant
officers, are constantly on the alert for spies, scouts, or prowling bands of cavalrymen
from the enemy's camp. Every stray man is picked up and sent to the officer of the
guard, who either sends him to the guard-house, or to the General at headquarters, and if
the unfortunate fellow does not tell a pretty straight story, or if there is anything suspicious about his appearance, he is put under strict guard, and, perhaps, ordered tried by
a drum-head court-martial, charged with being a spy. It is the worst place in the world
to be caught fooling around—this ground between two hostile armies in camp. A man
is almost certain to be captured, unless he is well posted, and if captured, he must give
a very strict account of himself.

" As before stated, we found it unsafe to travel in day-time, and, shortly after dismiss-
ing our pilot, we spied a squad of rebel cavalry on the right bank of the river. Luckily,
the river was pretty wide at that place, and we chanced to be well to the other side. They
yelled to us to come ashore, but we pretended not to hear them, and acted as if we were
intending to land leisurely on the far side. We were too far away for convenient musket

range, and did not fear them much, but the circumstance caused us to think it best to land a few miles below, and secrete both ourselves and the boat.

"During the voyage of the following night, or rather just before daylight, we passed the Bridgeport railroad crossing. We could see the guards on the bridge, but did not know whether they were rebels or Yankees, so in this uncertainty we let our boat drift quietly with the current, and passed by unnoticed. We supposed confidently that General Mitchel had occupied Bridgeport. So after we had passed below the bridge, out of sight, we landed, and Mark remained with the boat while I stole up to the camp to find out what kind of soldiers were there. It did not take me long, however, to discover that they wore butternut uniforms, and I hurried back to the canoe. Mark's disappointment knew no bounds. I could scarcely convince him that I told the truth.

"About sunrise we stopped and hid our canoe, and feeling somewhat hungry and also anxious to learn something about the Federals, we concluded to skulk off a short distance and see what we could find. It was not long until we found a cabin, where we got breakfast and learned that the Yankees were at Stevenson, or a short distance the other side. Soon after leaving this cabin we met a squad of soldiers in full retreat. They told us that we had better be 'lighting out;' that the roads and woods were 'alive with Yankee cavalry. They are in Stevenson and pushing on this way in heavy force.' We expressed some little apprehension, but went on a little further, when we met more rebel militia, who told the same story. It seemed as if there was a regular stampede among them.

"We now became convinced that if we could get safely to Stevenson we would be all right. So we went back to our canoe and rowed down the river again, until we thought we were about opposite the town, which is about four miles north from the river. Then we tied up the canoe and struck out through the woods for the town. Just before reaching the place, we had to cross a creek, after which we ascended a very long, steep hill.

"When we reached the top of this hill, we were somewhat surprised to find ourselves right in the town, but not half so much astonished as we were to find no blue-coats there, but the streets swarming with rebel soldiers. We had been woefully deceived by the stories of the frightened fugitives we had met in the forenoon, and had unwarily entrapped ourselves." [It was only for a very short time that the Federals were absent from this point, which was vital to Mitchel. The high water had destroyed a bridge, and until it was repaired he had withdrawn his forces.—W. P.]

"Wood proposed that we should start and run, but I saw that course would not answer, so we determined to put on a bold front, and take our chances, though we knew we ran great risk. We met and spoke with a number of soldiers. Some of the officers noticed us carelessly, while others paid no attention to us as we passed them. We went into a store and bought some tobacco, and inquired for some other trifling things, and then started off as unconcernedly as if we were a couple of country fellows, accustomed to visiting the town. We had gone some little distance, when we were met by an officer, who stopped us and said that he would have to inquire our business there, and who we were. These were pointed questions, but we knew it would be necessary to meet them. We told him who we were and all about it, and he appeared well satisfied with our answers and was about to dismiss us, when, unfortunately for us, another man, I think a citizen, came up, and, pointing at me, said :

"'That is one of the rascals that was here last night. He rode through the town, cutting all the flourishes he knew how. I know him. He dare not deny it, either.'

"In explanation of this man's singular, unexpected, and to us fatal accusation, I will say that I afterwards learned that a squad of daring troopers from the Fourth Ohio Cavalry, had, on the previous night, made a dash into the place. This explained the stories told us by the flying fugitives who had, by their fright, beguiled us into the rebel camp.

"As soon as we were thus detained, I directed all my attention to destroying the map in my possession, by tearing it in pieces in my pocket, dropping portions of it whenever opportunity offered, and chewing up much of it, until I finally succeeded, without detection, in disposing of the whole of it. Had this map been discovered in my possession it would have been strong evidence against us

"This man's story ended all hope of our getting away, and we were prisoners a second time. No sooner was attention once directed to us than we were surrounded, and scores of fellows saw in our appearance something suspicious. We told the most plausible story we could invent, but it was of no use. In spite of our protestations of innocence, we were bundled off under guard, put on a hand-car and run up to Bridgeport, where the commanding office was stationed.

"We reached Bridgeport soon after dark, and there we were again stripped and searched. Boots, hats, coats, socks and every under-garment underwent the strictest scrutiny. They could find nothing, and were about on a stand as to what judgment to pass on our cases, when fate again turned against us by interposing a circumstance which ended all hope in our favor. An excited fellow, who came and stuck his head in among the gaping crowd, who were staring at us, declared, in a loud voice, that we belonged to Andrews's spies and train-thieves.

"All eyes were turned on him instantly, my own among them. Of course, he felt bound to back up the assertion, although I believe he lied, at least such were my feelings. The spirit of resentment rose up within me, until I could have killed him without compunction if I had possessed the power, for in the next breath he said, 'I know those fellows! I saw them on the train!'"

It is possible that the man was right in regard to Wilson, for, as fireman, he was on the engine and was seen by a great many people ; and he wore just the same clothing as then.

The Southern Confederacy, of April 26th, thus alludes to this latest prize

"We learn that two more of the bridge-burning party were captured not far from Chattanooga, and brought in there two days ago, and that one of the party is still at large, or has escaped to the Federal lines, having never been caught."

In the latter statement they were mistaken, as they now had every one!

CHAPTER XVII.

FIRST PRISON EXPERIENCES.

THE little Major who had acted as spokesman on my first arrest, now escorted me to jail, accompanied by the whole crowd that had gathered around the hotel. He demanded my pocket-book—which he never returned—and the county jailer in whose hands I next came, finished the work of emptying my pockets. The jail, in front of which this robbery took place, is a little, mean-looking, brick building standing on a side street. The two rooms on the ground floor were for the jailer, and the two above for prisoners. I was taken up an exceedingly

The Iron Cage at Lafayette, Ga.

narrow stairway, and, turning to the right, entered a fair-sized room, with bars across the windows. But the noticeable feature of the room was a large iron cage, nine by twelve feet, which stood in the middle, leaving but a moderate passageway between the cage and the walls. The cage was formed of broad iron slats crossing at right angles at intervals of perhaps six inches. It had the same kind of slats for a roof some feet below the ceiling of the room, and a solid iron floor, a little above the room floor. It had also a very heavy iron door, which the jailor unlocked, and I was bidden enter. One man was in it—a Union man, as he said, though I was rather disposed to believe him a detective, put there for the purpose of finding out as much as he could from me. My reflections when I stepped in, and the jailor secured the heavy lock to the massive iron bars across the door, could not have been more gloomy if the inscription Dante saw over the gate of hell had been written over this door:

'All hope abandon, ye who enter here.'' ·

Never before had I been locked up as a prisoner, and now it was no trivial matter—a few days or weeks. Very slender hope, indeed, was before me. I was there as a criminal, and too well did I realize the character of the Southern people to believe they would be very fastidious about proof. A high value was never set on the life of a stranger, and now with the fierce war fever, and the madness caused by our raid, added to the fear of a Federal advance, the situation was most desperate.

In that hour my most distressing thought was of friends at home, and especially of mother. I thought of their sorrow when they should hear of my ignominious fate—if they ever heard ! The thought of dying unknown,—of simply dropping out of life without any one ever being able to tell what had become of me,—seemed terrible. That all my young hopes and fond dreams of being useful should perish, as I then had every reason to believe, on a Southern scaffold, was almost unbearable. But for a moment only did these thoughts sweep over me; the next they were rejected as not only useless, but tending to unnerve me, and prevent that readiness to take advantage of any possible chance for escape that might offer. At least here was rest and quiet, and the experiences of the past day rendered these most enjoyable. No more questions to answer, no straining to the last limits of physical endurance, but shelter, warmth, and rest ! To these was soon added another great comfort, for a fairly good supper was brought. My companion said he was not hungry, which suited exactly, as I ate the allowance of both without difficulty, and then began to feel sleepy. My cage-mate wanted to ask me some questions; but I had been questioned sufficiently, and gave him very short answers. He had plenty of blankets, and wrapping myself in them, I soon fell into a deep sleep—profound and dreamless—such as only extreme fatigue can produce.

I did not wake until the yellow sunlight was streaming through the barred window the next morning. The air was wonderfully pure and fresh after the long rain. At first I had not the slightest idea of where I was. The bed was drier and the blankets warmer than in camp. The iron floor was unfelt. For a little time I lay in the delicious languor of perfect comfort, looking up at the lattice-work of iron without knowing what it meant. Gradually a sense of my situation stole over me; yet when all the terrible memories of the past two days came trooping back, despair did not come with them. There might yet be the opportunity of escape, if a little time was given. Or the enemy might, by the fear of retaliation, be forced to spare us. Before leaving camp I had read of some rebels captured in Federal uniform while trying to burn bridges in Missouri—an offense very similar to our own ! Morgan with his men had also assumed the Federal uniform, and thus managed to reach Gallatin, Tennessee, and destroy valuable stores.[1] Some of his men had been captured and were now in the hands of our forces. There was little doubt that if these were held for us, life for life, the rebels would be slow in beginning the work of hanging. There was also a large number of irregular privateers and others, who, according to all the laws of civilized war, were as much exposed to summary punishment as we. Indeed, all the rebels were, under a strict construction of the laws of nations, punishable with death for treason and disloyalty, and at this time it looked as if the doom of the evil Confederacy could not be far distant. I did not find it hard, from my outlook of deep personal interest, to hope that something might be done on our behalf. I well knew that being inside the enemy's lines in citizen's dress, gave them the right to put us to death by a sentence that would have at least the color of military law; but no military law *required* the exertion of that power. If they could use us to better advantage by holding us as hostages for their friends in similar condition, or exchanging us for them, all military usages and all considerations of humanity would be satisfied. While, then, the great probability was that the enemy in the heat of passion and vengeance would put me to death either by mob-law or by a drum-head court-martial, yet if they could in any way be induced to let some little time pass, and cool reason to get the upper hand, there was a chance that I might either be able to break away. or be saved by the power of threatened retaliation.

But that there might be the slightest chance for interference on our behalf, our character and position must be clearly known. If we died even under circumstances of the greatest cruelty while claiming to be Southern citizens, there could be no help. It would be an effectual answer, when our government asked why a certain soldier was put to death, to say, "We did not know he was a soldier; he gave a false name and was put to death as one of our own offending citizens." Very quaintly did my prison companion put this thought. He said, "If you are innocent of being one of the bridge-burners, as they call you, they will be sure to hang you. But if you are one, and should claim to be a soldier, your government will not let so many of you perish without making some effort to save you !" Slowly I came to the conviction that I would be better off in my own name and character.

[1] See Morgan's own report, March 19, 1862. War Records. Series I., Vol. X., Part I., page 31.

I had time enough for deliberation. During all the morning hours throngs of both ladies and gentlemen came in to see the caged "Yankee." The place was well adapted for the purpose, and I was viewed much as spectators look at a bear in his den. A guard was employed and the outside doors left open. Visitors came in and passed around the cage and out again—a continual procession. They made a great many odd remarks, criticising me without any regard to my opinions or feelings. Sometimes compliments were thrown in; but I heard a good many things about myself that I had not suspected before. The ladies were particularly free in these comments, which sounded strangely enough to "the beast in the cage." Questions were asked, but as the guard would not allow any one to stop for a regular conversation, I did not care to answer. I was too busy trying to arrange plans that might lead to my getting out of the position in which I was placed. I knew that there was no regularly constituted military authority at Lafayette, and that if I should be summarily dealt with while there, it would only be the act of an irresponsible mob; and in that lonely region such a thing was by no means unlikely. I now felt that Chattanooga, the headquarters of that district, was the best place to be. Besides, it was likely that I might there meet the four of whose capture I had heard; and while I was sorry that they had been taken, yet, since they were in the toils, I longed for their companionship. I thought that by working on the curiosity of my captors I might induce them to send me away.

Accordingly I told the guard, who carried word to the jailer, that I would like to see the vigilance committee. They were informed, and in a little time I was taken before them. It is probable that I would have been sent for at any rate to ask me some more questions of the same nature as those of the day before. They had their lawyers and clerk ready. But I cut the whole matter short by saying:

"Gentlemen, the statements I gave you yesterday were intended to deceive. I will now tell you the truth."

The clerk got his pen ready to take down the information, and there was a general hush of expectation, while the president said, "Go on, sir; go on."

"I am ready," I continued, "to give you my true name, and to tell you why I came into your country."

"Just what we want to know, sir. Go on," said they.

"But," I returned, "I will make no statement whatever till taken before the regular military authority of this department."

This took them completely by surprise, and they used every threat and argument in their power to make me change my resolution. But I told them that it was a purely military matter, and that if they were loyal to the Confederacy, they would send me as soon as possible to the commanding General. This looked so reasonable, and the possibility that they might lose some valuable information by an opposite course was so great, that they decided to send me that very day.

It was a journey of twenty-seven miles, and required that we should start promptly after dinner. It would take me over the same ground in part that I had traversed during that long and terrible night of wandering. We had left the cars near Ringgold, and I had gone nearly twenty miles westward; then many miles east, and back again; then south beyond Lafayette, a distance of at least twenty-five more; so in that twenty-four hours I must have gone, including circles and meanderings, much more than fifty miles,—enough if put in a straight line in the right direction to have carried me safe within our own lines; while here I was still in the enemy's power !

I was remanded to jail to wait for the preparation of a suitable escort. After dinner, a dozen men called, and conducted me to the public square in front of the court-house, where a carriage was in waiting in which I was placed; and then commenced the process of tying and chaining

A great mob had gathered, which was increasing every minute, and becoming very excited and angry. They did not look upon my removal to Chattanooga with any favor, seeming to consider me their own especial victim. They declared that I could be hanged as easily and well there as at Chattanooga, and obviously did not wish their village robbed of a tragedy or a sensation. They questioned me in loud and imperious tones, demanding why I came down there to fight them, what I meant to do with the train—for though I had not yet admitted being among its captors, they took that for granted—and added every possible word of insult.

My position was serious. The committee in whose hands I was had no more real authority than those who were hooting and howling around. I could see that they wished to protect me, if for no other reason than to carry out their own intention; but I did not know how far they would be able. As the mob grew more violent, I tried to help the committee. Anything like the manifestation of fear would greatly increase my danger, but I knew that the typical Southerner admires coolness and courage above everything else; so acting on Shakespeare's advice to " assume a virtue if you have it not, " I selected some of those who seemed most prominent and addressed them. They answered with curses; but even this seemed to relieve their feelings. I spoke very quietly, and watching a chance to get in a joke about the manner in which I was being tied, and affecting to treat the whole matter as hardly serious, I soon had some of the laughers on my side; and then I was less afraid, and could play my part still better. Before long, I had the gratification of hearing one say to another, " Pity he is a Yankee, for he seems to be a good fellow. " The insults and oaths decreased, and something of the qualified good feeling of the day before returned. I wish to put it on record that never during the whole time I was a prisoner did I have five minutes conversation with any Southern man, or men, without all manifestation of angry feeling being suppressed

I had now been secured in such a manner as literally " to make assurance doubly sure." The mode of fastening was exceedingly curious, and gives a clue to the feelings of dread with which the enemy had been inspired by our raid. One end of a heavy chain, eight or nine feet in

Chained in a Carriage.

length, was put around my neck and fastened with a padlock; the other end was carried back of, and under the carriage seat on which I was sitting, and locked to one foot in the same manner; the chain being extended to its full length while I was seated, thus rendering it impossible for me to rise. My hands were locked in hand-cuffs; my elbows were closely pinioned to my side by ropes; and to crown all, I was firmly bound to the carriage seat ? Then my conductor, the little major, took his seat at my side, and, —Oh ! spirit of chivalry,—two men followed on horseback armed with shot-guns ! With all these precautions, I ought to have felt safe, but I did not! All preparations being completed, I took leave of the village, to see it no more for twenty-four years.

As we journeyed, the sky, which after the morning burst of sunlight had been overcast, suddenly cleared. The first dawnings of spring smiled on the hills around and I felt my spirit grow more light as I breathed the fresh air, and listened to the singing of the birds.

My companions were talkative, and though I could not quite forget the indignity they were putting upon me in thus carrying me chained as a criminal, yet I knew it would be unavailing to indulge a surly and vindictive disposition, and therefore talked as fast and as lively as they could.

The guards themselves did not subject me to any insult, and even endeavored to prove that the extraordinary manner in which I was secured was a compliment to me. I could not so view it, and would willingly have excused the tying and the compliment together! The most that I had to complain of in their behavior was their conduct when we passed a house; and I should not perhaps complain of that, for it is hardly in human nature to miss making the most of a great curiosity. They would call out,

"Hallo! we've got a live Yankee here:" then men, women and children would rush to the door and stare as if they saw some monster, calling in turn:

"Whar did you ketch him? Goin' to hang him when you get him to Chattanooga?" and similar questions without end.

This was only slightly annoying at first, but its perpetual recurrence grew terribly wearisome, and was not without its effect in making me think that perhaps they would hang me. In fact, my expectation of escaping was not very bright; but I considered it my duty, for the sake of others as well as of myself, to keep hope alive, that I might be ready to take advantage of any favoring circumstance. The roads were muddy and stony, and our progress slow; but though the carriage in places jolted terribly over the frequent ledges of rock, I was in no danger of being thrown out; while the carriage remained right side up, I was sure to be in it! yet this did not render the ride easier. The afternoon wore slowly away, as we passed amid grand and beautiful scenery that under other circumstances would have been greatly admired; but now my thoughts were elsewhere.

It was not so much death I dreaded—a soldier must get control over that fear;—but to be hanged! Death amid the smoke and excitement and glory of battle was not half so terrible as in the calmness and chill horror of the scaffold! and sadder yet to think of the loved ones at home who would count the weary months as they went by, and wish and long for my return till hope became torturing suspense, and suspense deepened into despair! These thoughts were almost too much for my fortitude but I knew no resource except patience and endurance There was another help, but I had not then learned its power, and did not know how to seek it. Yet I am disposed to believe that I was led nearer Heavenly help that very evening.

The sun went down and night came on—clear, calm, and with wonderful depth of blue One by one the stars came out—my old friends! I gazed upon their beauty with new feelings, as I wondered whether a few more days might not bring me a dweller beyond their shining; and as I thought of the blessed rest on the eternal shore, where war, and hate, and chains are never known, my thoughts took a new direction. I had no earthly hope or support; why not seek that which had been promised to all the weary and heavy-laden? If I must die, why not prepare to enter a better world? The cost was as nothing to me now; and what were all the vanities of the world, were I even free and able to enjoy them, in comparison with the deep peace of God, something of which I felt even now stealing into my soul. Country and home! these were the objects of my fondest affections; why not add another still dearer: the heavenly home with the universal Father there! I had never tried in earnest to be religious by yielding my heart and life to God in the way of His appointment; though like so many others, I had intended at some time to do so; but the great controversies in which the nation's life was involved, and the bustle and excitement of camp, had nearly driven all such thoughts from my head. The idea of country and of God were now brought close together as never before. I felt, with a kind of half pity for myself, that I was sure to die, while God would deliver the country; and people would praise those who had died for their nation's life—that would be little; but if God said it was well done, that would be everything. Why not leave it all to him—life, death, and the future! He who made the earth and the stars could make no mistake, and could not be defeated by the malice of wicked men! The grand form of Lookout Mountain arose on my left; the stars spanned the whole valley; and Chattanooga—which might mean a scaffold,—was but a short distance ahead. Whether an influence came from these things, or from the stillness and quiet so congenial to exhausted nature after the excitement of the past week, or from a still higher source, the theologians may determine—I only know that the memory of that twilight, when I was being carried, chained, to an unknown fate, is one of the sweetest of my life. The babbling guards had subsided into silence—perhaps also feeling something of the soothing influence of the eventide; while lofty thoughts of man's destiny filled my heart, and I felt nerved for any fate. Yet this was but a transient experience. Rude shocks of another character swept it away; and great suffering and bitter anguish of spirit were yet to be passed before I could confidently claim the comforts and joys of our holy religion.

We reached Chattanooga before it was quite dark and at once drove to the Crutchfield House, the headquarters of Gen. Leadbetter, then commanding. While the guards ascended to inform him of our arrival, I remained in the carriage. As soon as we had entered the town the word had been given:

"We've got a live Yankee; one of them that took the train the other day."

I was not the first one of the party captured, as the reader has already learned, but was the first brought to Chattanooga; and the curiosity to see me was on that account the more intense. The people ran together from every quarter, and in a few minutes the street in front of the hotel was completely blocked. They behaved like a mob, jeering and hooting, calling me by every opprobrious name and wanting to know why I came down there to burn their property, murder them and their children, and set their negroes free! I could think of no reason to give for these terrible things that would bear statement just then! I was greatly amused (afterward!) by the variety of their criticisms on my appearance. One would say, "It is a pity so young and clever-looking a man should be caught in such a scrape." Another, of a more penetrating cast, "could tell that he was a rogue by his face; probably came out of prison in his own country." Another was surprised that I could hold up my head and look around on honest men—arguing that such brazen effrontery was proof of enormous depravity. I did not feel called upon to volunteer any opinion on the subject.

There was one man I noticed in particular. He was tall and venerable looking; had gray hair and gray beard, a magnificent forehead, and was altogether commanding and intellectual. He was treated with great deference and seemed to me like some college professor or Doctor of Divinity. As he pressed his way through the crowd, I thought:

"Surely I will receive some sympathy from that venerable-looking man."

His first question was promising. Said he:

"How old are you?"

I answered, "Twenty-two, sir."

Gradually his lip wreathed itself into a curl of unutterable scorn as he slowly enunciated:

"Poor young fool! and I suppose you were a school-teacher or something of that kind in your own land! and you thought you would come down here and rob us and burn our houses and murder us, did you? Now let me give you a little advice; if you ever get home again (but you never will!), do try for God's sake and have a little better sense and stay there."

Then he turned contemptuously on his heel and strode away, while the rabble around rewarded him with a cheer. I never learned who he was. After that I looked no more for sympathy in that crowd!

My conductors were a good while gone, engaged, as I supposed, in letting Gen. Leadbetter know the kind of a person I was, and how I came into their possession. No doubt they also told him what I had promised to reveal on my arrival. I had not forgotten this myself, and had gone very carefully over the ground and decided where the line must be drawn between truth and fiction—how *little* of the former would suffice to establish my standing as a Union soldier, and how *much* of the latter could be administered with probable profit. In spite of the good thoughts I had entertained an hour before, there was no qualm of conscience as to telling him—the enemy of my country, and in rebellion against lawful authority! —anything that might injure him or his cause, or contribute to my own interest. I knew that commanders are always careful to get all the military information they can out of prisoners; and I would give him abundance, if he proved disposed to accept my kind of information.

Gen. Leadbetter was a hard drinker, and a northern man; but as is not unfrequently the case, he was more extreme in his views than the natives with whom he had cast his lot. He afterwards acquired a most unenviable reputation for cowardice and cruelty combined. Such was the man who was to have the direct control over myself and comrades at this time. Had we fully known him, we might indeed have despaired.

As I entered the room I saw at a glance that he had been drinking, though not so much as to render him incapable of business. He treated me with a kind of pretentious politeness, causing me to be seated in front of him, and then said that he had learned that I had a communication to make to him which I had refused to give those people out at Lafayette, in which I was perfectly right. Now I might tell him everything, and would find it to my advantage.

As he spoke I was carefully studying him, and trying to estimate the amount of deceit that might be profitably employed. I was encouraged; for he seemed like one who would deceive himself, and think all the while that he was showing wonderful penetration. I told him that I was a United States soldier, giving my name and regiment; declaring that I was detailed without my consent, that I was ignorant of where I was going, or the work I was to perform, which I only learned as fast as I was to execute it. He listened very attentively, and then asked who the engineer was who had run our train. I refused to tell. The Confederates were exceedingly anxious to find the name of this person, imagining that he was probably some high official of the Western and Atlantic railroad. Any member of the expedition could at any time have purchased the promise of his own life by telling who our engineer was—that is, if he had been believed. Then Leadbetter wanted to know the purpose of our expedition. I professed to be ignorant so far as any direct knowledge was concerned, but he still questioned and I gave him my conjectures as facts.

All that I told him was what any person of judgment would have supposed—just what the Southern editors did conjecture. He acted as if receiving very valuable information, but expressed a doubt whether Mitchel had men enough to follow up such an enterprise; saying that, according to his information from his own spies, Mitchel had not more than ten thousand men. It struck me as ludicrous that a general should be discussing his information thus with a prisoner; but I was determined

Confronted with Gen. Leadbetter.

to mislead him if I could, for his estimates were singularly correct. So I said, that this must refer only to the first division of Mitchel's force, and did not take into account the troops that were ready to come to him by rail from Nashville and the North. He said this was perfectly true, and wanted to know how large this reserve was. I told him that in a month Mitchel would have in all over sixty thousand men! He asked what would be done with such an army? I told him that of course we did not know, but that the general opinion was that Mitchel would soon take Chattanooga, then Atlanta, and that there was no force that could stop him till he had reached the sea and cut the Confederacy in halves ! Lead-better seemed profoundly impressed; said that he had no idea that Mitchel had so many men at his disposal. (Had the Federal Government given Mitchel such an army for such a purpose from Halleck's and McClellan's surplus, how different would have been the story of the war !) I do not know whether this misinformation had anything to do with the disgraceful manner in which Leadbetter ran away from Mitchel at Bridgeport two weeks afterwards ! '

Then the General wanted me to tell just how many men we had on the train, and to describe them so that they might be recognized if captured. The latter I could not have done if it had been to save their lives and mine. The former I would not do, as I could see a bearing it might have on the pursuit. So I answered firmly, "General, I have freely told you whatever concerns only myself, for I want you to know that I am a United States soldier, and under military protection; but I am not yet base enough to describe my comrades, or help you in any way to capture them."

"Oh!" answered he, as if the idea was amusing, "I don't know that I ought to ask that."

"I think not sir," I replied.

"Well," said he, "I know all about it. Your leader's name is Andrews. What kind of a man is he?"

I was thunderstruck! How could he have Andrews's name and know him to be our leader? Perhaps Leadbetter was a deeper man than I thought, and had been playing with me all this time. (For I have given a small part only of our conversation, but all the general drift of it.) I never dreamed of the true reason. I had every confidence that Andrews would get away and try some measures for our relief. So I answered boldly:

"I can tell you only one thing about him; and that is, he is a man you will never catch."

The smile on Leadbetter's face became very broad and self-satisfied as I said this, but he only added:

"That will do for you."

And turning to a captain who stood by, he said:

"Take him to the *hole.* You know where that is."

The subordinate gave the military salute, and took me from the room. *At the door stood Andrews* heavily ironed, and Ross and Wollam with him! I then knew why the General smiled. We did not openly recognize each other; but my heart sank lower than it yet had done. Our very leader a

---

' April 28th, E. Kirby Smith writes Leadbetter in answer to a question that he "has no information of reinforcements reaching the enemy at Nashville ; their force at that point is stated at five regiments." War Records, Series I., Vol. X., Part 2, page 460.

helpless prisoner! Who now could carry the news of our disaster, and arrange for exchange, or stay the enemy's hand by the threat of retaliation! I thought our ruin was utter and irremediable; and that we had sunk as low as possible. But before I slept, I learned that there was a deep yet lower!

CHAPTER XVIII.

THE OLD NEGRO JAIL AT CHATTANOOGA.

[AUTHORITIES.—*Daring and Suffering*, 1863.—*War Records.* Series I., Vol. X., Part 1, page 632.—*Adventures of Alf. Wilson*, Toledo, 1880.—*Key West Chronicle*, Nov. 15th, 1862.—Newspaper accounts published by all survivors in 1862-3.—Sworn Testimony in War Office by Buffum, Reddick, Parrott, Bensinger and Pittenger.]

THE sufferings of Northern prisoners in the South constitutes probably the most terrible chapter in the history of the war. Attempts to soften the fearful story have met with slight success. The lot of the prisoner of war is always deplorable, as accommodations are scanty, and the hardships of camp life greatly aggravated. But the Union prisoners in the South suffered more than is usual in military prisons. The Southern States were slenderly supplied with means for the care of bodies of troops numbered by thousands; the Northern armies were pressing severely, and tightening the blockade by sea and land with the express design of depriving them of necessaries for prosecuting the war; and in case of scarcity it was natural that Northern prisoners would first suffer. But to this was added a terribly bitter feeling, which sometimes found delight in gratuitously embittering the prisoner's lot. The horrors of Andersonville cannot soon be palliated or forgotten.

But the sufferings of ordinary prisoners was far exceeded in the case of the Andrews raiders. Our leader had been trusted by the enemy and had betrayed them. We had inflicted an amount of fright altogether disproportionate to our numbers; and we were now believed to be beyond the protection of the laws of war, and almost beyond the pale of humanity. It was thought that we were selected for our desperate character, and therefore would require an extraordinary amount of guarding to prevent us from escaping or doing further injury. Such considerations no doubt had weight in the minds of our captors.

But these alone are not sufficient to explain the story that follows. I have hesitated in regard to telling it at all; but there is at least one good reason for recording all that the proprieties of language will permit—a reason which also goes far to account for the full horrors experienced. Nothing better shows the spirit of the institution of slavery, and the debasing effect it produces on the master class. Those in whose power we now fell had been used to seeing men, women, and children publicly sold, whipped, hunted with dogs, or shut up like wild beasts in dens. With such experiences they would not be likely to care much for the sufferings of enemies, whom they had come to regard as the friends of the enslaved race. Accordingly it is in the negro prisons that our band found their most fearful experiences.

The story of the little, old, Chattanooga prison cannot be fully told. Terrible hardships which had to be lived through in agonies of shuddering disgust, and in utterly helpless disregard of the decencies of life—a daily and unceasing combination of pain and loathing—can hardly be told by one friend to another, much less spread on the cold printed page. The reader will remember that for every painful thing related, a dozen more are behind, which dare not be named. Let it be understood that there is no exaggeration. Photographic accuracy, within the limits already indicated, is aimed at. This worst of all the prisons has long since been swept away; but its memory will never grow faint while one of its hapless victims survives. The story rests not alone on my evidence, but is established by sworn testimony published in the War Records.[1]

The captain, who was appointed my conductor, called a guard of eight men at the door of Leadbetter's room, and led me for some distance through the streets of Chattanooga. Two of the rebel soldiers linked arms with me, one on each side, two walked in front, and four followed behind. I could not help telling the captain that they took better care of our men than we did of theirs; that I had once guarded a Georgian a long distance without any help, and with no handcuffs on him. He did not resent the implied reproach, only saying that they meant to make sure of me. At length we came to a little brick building, surrounded by a high broad fence. It stood, as I learned long afterward, on Lookout street, between Fourth and Fifth. The ground sloped rapidly upward, so that the back of the jail was built into the hill while the front was level with the surface of the ground. The jail had two stories, with two rooms in each story. It was quite high for its length and breadth. The jailor and family lived in the upper and lower rooms at the north end, and the rooms at the south end were the prisons, the lower being entered only from the upper, and that in turn only from the jailer's room. This prison when built was intended for the accommodation of negroes by their humane owners. Another and much larger prison, in which were confined the great majority of white offenders, and afterward of war prisoners, was situated on Fourth and Market Streets.

Swims, the jailor, was a peculiar character. He was old, perhaps sixty, with abundant white hair, and a dry and withered face. His voice was

---

[1] Series I., Vol. X., Part 1, page 632.

always keyed on a whining tone, except when some great cause such as the requests of prisoners for an extra bucket of water aroused his ire, when it rose to a hoarse scream. Avarice was a strong trait. He seemed to think his accommodations vastly too good for negroes and "Yankees,"

The Swims Jail at Chattanooga.

and that when admitted to his hospitality, they should be thankful and give as little trouble as possible. With such notions it is easily seen how much he could add to the sufferings of prisoners. One thing favorable was that he was fond of a dram, and when indulging, became very talkative, revealing many things that we could not otherwise have learned.

We halted for a moment at the camp-fire of the guard outside the

Entering the Dungeon.

gate; then Swims came out grumbling about being disturbed so much, and unlocking the gate, admitted us. We crossed the yard, ascended the long outside stairway, and from an outside landing entered the bedroom. From this a door opened into the prison. I looked around by the light of a candle the jailor carried, and thought I understood why General called the place a "hole." The room was quite small, and entirely destitute of furniture of any kind except a long which lay on the floor. There were five or six old, miserable-looking in the room, whose clothes hung in tatters, and who presented a starved, dirty, and wretched appearance. It was a dreadful place, I shuddered at the idea of taking up my abode in such a den. But I soon found I was not to be so highly favored, and a little more experience was sufficient to make me look almost with envy upon these old men. Said the jailer to the Captain, "Where shall I put him?"

"Below, of course," was the prompt reply.

The jailer advanced to the middle of the room, and kneeling down, took a large key from his pocket, and applying it to a hole in the floor, gave it a turn, and then with a great effort, raised a ponderous trap-door at my feet. A rush of hot air, and a stifling stench as from the mouth of the pit, smote me in the face, and I involuntarily recoiled backward; but the bayonets of the guard were behind, and there was no escape. The ladder was then thrust down, and long as it was, it no more than penetrated the great depth. The wretches whose voices I could hear confusedly murmuring below were ordered to stand from under, and I was compelled to descend into what seemed more like the infernal regions than any place on earth. It was hard to find the steps of the ladder—for the candle of the jailor gave almost no light, and I had on my handcuffs; but I went down, feeling for each step, to a depth of some thirteen feet. I stepped off the ladder, treading on human beings I could not discern, and wedged in as best I might. Then the ladder was slowly drawn up, and in a moment more the trap fell with a dull and heavy sound that seemed crushing down on my heart, and every ray of light vanished. I was shut into a living tomb—buried alive!

I could feel men around me and hear their breathing in the darkness, so that I knew the den was crowded full. Though it was night and cool outside, the heat here was more than that of a tropic noon, and the perspiration soon oozed from every pore. The fetid air and the stench made me for a time deadly sick, and, worst of all, there was an almost unbearable sense of suffocation. I wondered if it could be possible that they would leave human beings in such a place till death came in this horrible form—death, which could not be long delayed. I thought of "The Black Hole of Calcutta," where a hundred and six years earlier so many Englishmen were suffocated in one terrible night by a savage East Indian.

I had heard of negroes being burned alive or whipped to death in our own
South, but these horrors were always, I supposed, meant as vengeance for
some fiendish outrage. Yet of all the forms of death, that by slow suffocation had always appeared most dreadful, and this now seemed imminent.

As I had been brought to this place in the dark, I knew nothing of
character, and after the first moment of stupefaction, resolved to explore
its size and nature. No one of my companions had yet spoken to me
I to them. Whether they were black or white, soldiers or citizens, chained
like myself, or with the free use of their hands, I could not tell, and
scarcely liked to ask, lest the answer should add new misery. I jambed
my way through the living throng to the wall and felt along it to learn if
there was door or window. There was no door, the only entrance to the
fearful place being by the trap down which I came. Neither were there
any windows, but I found two holes in the wall, opposite each other, each
little more than a foot square, and filled with three rows of iron bars. The
walls, as could be told at the holes, were very thick, being made of an

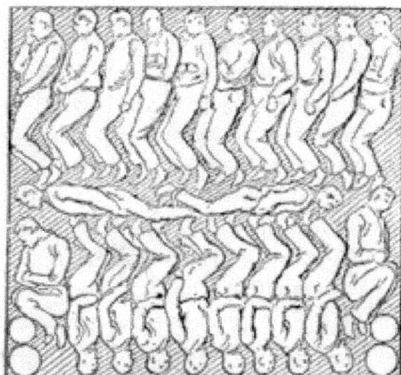

Manner of Sleeping in the Swims Jail at Chattanooga.

inner case of oak logs, and a brick
wall outside. Even in day-time, these
holes gave little light, for one was
close under the outside stairway already described, and the other below
the level of the ground. Yet a little
air could come though the thick-set
bars, and served to revive me,—
making it possible to endure life
here for a short time.

When the first shock had passed
and I became partially inured to the
terrible oppression of the atmosphere, I tried to ascertain something
of the condition of my companions

The most fearful description of this place of torment that can be given is
contained in the plain cold figures,—the number of the prisoners and the
size and manner of their lodging. Before I entered there were fourteen
white men and one negro. This evening the number was increased to nineteen and soon after to twenty-two, at which point it remained for many
days. The room was just thirteen feet square, and about the same in
height. These numbers are not approximations, but are meant to be accepted exactly and literally. The entire furniture of the room consisted of
four buckets for water and slops! And here twenty-two men had to remain day and night, with no respite, and no power to leave the room for
any purpose, for more than two weeks! It was possible, as will be seen by
reference to the accompanying sketch, for all to lie down at once; but it

required the nicest fitting and no small degree of crowding. There were two rows of ten persons each, occupying the space of thirteen feet, and two persons could rest between the feet of the rows. But when one turned, all in his row was obliged to turn likewise; and as all were chained some manner, the crowding, the exclamations, and the clanking of chains in the black darkness of this dungeon presented as good a representation of the realms of the lost as has ever been known!

My prison mates received me very kindly and answered questions freely. I had no hesitation in telling them who I was, and this at once won their confidence. They were Union men from various parts of East Tennessee. Many of them had been in prison for six or eight months, and the offenses charged varied from that of simply preferring the old government to the new, slave-built Confederacy, to that of bridge burning, or of being helpers of the Union army. The latter were called spies. One of them was blind, the rebels accusing him of only feigning blindness; but from all I could observe, I think it was real.

I was greatly interested in the one negro in this miserable place. He was very friendly and anxious to be of service to us in any possible way. Some days after my arrival, he was taken out and brought back again after an hour or so, seeming to be in a good deal of suffering. His story, which he gave as if it were the most ordinary thing in the world, moved me to indignation which I would gladly have expressed in some way more vigorous than words.

He was arrested and imprisoned on suspicion of being a fugitive slave. The law in such cases did not put the burden of proof on the person arresting, but on the negro. Aleck had been treated as law and custom provided. He was first carefully examined, and whipped till he made some kind of confession: then he was put in jail, and advertised in accordance with that confession. If a master appeared and proved property, he was obliged to pay all jail and whipping fees, costs of advertising, and a liberal reward to the person arresting; and then, usually flogging the negro unmercifully for the trouble and expense he had caused, he could take his property. But if no answer came to the advertisement, it was taken for granted that the negro lied, and he was brought out and flogged into a new confession, after which he was remanded to jail and again advertised. Thus they continued, if no master appeared, flogging and advertising, for a year, when the poor fellow was sold at public auction, and the proceeds applied to pay the expenses of all these barbarous inflictions! No trial was allowed by which the negro might prove himself free. When once arrested, unless he happened to have some powerful white friend, his doom was sealed; and in this way, in the old slavery times, many a freed negro found his way back into bondage.

No answer having been received to the advertisement for Aleck, he had been taken out for one of his periodical whippings. He had now been in this prison for seven months, and was to remain five more, with no prospect but that of being sold into perpetual bondage. We pitied him from the bottom of our hearts, and are glad to believe that, if he lived, the triumph of the Union armies relieved him from his dreadful position. These things were not all ascertained on the same evening though several of them were, for I did all I could to get a complete mastery of my surroundings, that I might be ready for any possibility of escape. But the chances were slight indeed. The floor and the walls were of solid oak, many inches thick; a circle of guards was all the time on duty outside; and the only egress was by means of the ladder put down in the presence of the jailor and a strong guard.

As we were talking in the darkness, we heard the tramp of many feet on the outside stairway, with the clank of chains, and listened to learn what next was coming to pass. The noise came overhead, and then the trap-door opened and a stream of comparatively cool air poured down from the room above, and drew in through our narrow windows. We breathed with a sense of indescribable relief—drinking in the air as the desert traveller drinks from the mountain brook !—oh ! what a luxury it would be, if that trap-door could only be kept open ! It might have been if our life or comfort had been valued by those in authority.

A number of men were seen above by the feeble glimmer of the jailer's candle, and the long ladder was thrust down and seized by a man below to prevent it from striking some head, and it was clear that others were being sent down. The Tennesseeans cried out; " Don't put any more down here. We're full ! We'll die if any more are put down."

These most reasonable remonstrances produced no effect, and were not answered. On came the new victims, clambering slowly—there were three of them—as men chained together were obliged to do when on such a road. I stationed myself at the foot of the ladder, and made to them some such safe remark as, "This is a hard place to come to !" and in a moment found my hand caught in a warm strong grasp, and " Pittenger," " Ross," was mutually whispered. It was Andrews, Wollam and Ross ! I had seen them chained in front of Leadbetter's room. Now they were here; and the sense of misfortune seemed lightened by half. To die in the company of friends was better than to die alone. I pitied them and wished them free; but it was far better to be confined together than for us to endure the same suffering in separation. A whisper more to Ross brought me the information that they had given their names and character. I told my story, and the kind of a place into which we had come; while they gave me the history of their adventures since we had parted on leaving the train. There was so much to tell of the past three days' history (this was Monday evening), that we did not do much more in

The Chained Men Descending into the Dungeon.

making the acquaintance of the East Tennesseeans that night. The latter kindly allowed us to take a corner close to one of the window holes, where we could the more readily converse. Ross and Wollam agreed with me that our best course would be to claim, with all our strength, to be detailed Union soldiers, not denying what we really did,—only claiming that we were not volunteers, but were ordered on this expedition with no choice, and simply obeyed orders. Andrews also approved, but said that his case was separate from ours, and much worse.

An hour or two passed in such conversation not altogether unpleasantly and it came time to sleep. A soldier is not particular in such matters, but never had I been placed in such a situation as this. Lying down in the woods unsheltered from the pouring rain was bad enough, but this was far worse. The night before in the Lafayette jail I had abundance of room, air, and blankets. The lattice cage was a palace in comparison with this loathsome "hole." We were warm enough, though we had no bedding whatever; in fact, one of the first things we did was to disrobe as far as our irons would permit. Many of the East Tennesseeans, who were not fettered, were entirely naked, because of the intense heat. But we adjusted ourselves in our corner, so that our chains might cramp us as little as possible, and ceasing to talk, were soon asleep. The others in the room were already unconscious. The one advantage of the terribly close atmosphere was that the great heat and the slow carbonic acid poisoning from the impure air rendered slumber easy, and we found little trouble in sleeping early and late. The arranging for rest required very nice adjustment, and if any one wanted to change his position he was sure to arouse all who were near him; or if he rose up to go for a drink, he had to take along those who were chained to him and was apt to tread on his neighbors, which gave rise to some warm altercations with the result of still further disturbing our slumbers.

We were aroused the next morning early,—as it seemed to us, but really late,—by the opening of the trap-door, and the delicious shower of cool air that fell upon us. As we looked up we saw the white head of our old jailer framed in the opening, and heard him saying, in drawling tones, "Boys, here's your breakfast;" then he lowered, hand over hand, a bucket at the end of a rope, till it was within reach, when it was eagerly secured from below.

There was a strong brotherly feeling among the East Tennesseeans, and not less in our party. So the division of the food—a matter of great importance—was scrupulously fair, the weak getting as much as the strong. Aleck was given as much as any of the rest. There was but a tiny fragment of corn bread and a still smaller slice of unsavory meat for each, not one-fourth as much as I could have eaten with the hearty appetite I still retained. I felt hungrier after devouring it than before. It was now nearly nine o'clock, and we were told by the old citizens of the jail that meals were only served twice a day, and that we would get no more till the middle of the afternoon. They also encouraged us by saying that after we had been in there a month or two, we would not feel so hungry! It was not hard to believe this; in fact, we thought that less time than that would cure our hunger forever!

There was now a feeble glimmer of light in our den, by means of which we could form a better idea of its character than on the preceding evening. But there was nothing reassuring in the survey. Our eyes had become used to the feeble twilight, which was all that ever visited us here, and we could look on each other's pallid faces. The Union men had not been able to change their garments for months; no water was ever given to them or to us for washing; and their faces wore a look of hopeless misery, in the majority of cases, which was terribly pathetic. Under no circumstances did our spirits sink so low, or did we suffer so much from mere imprisonment, as did these men. The very danger we were in, the assurance that our fate, one way or the other, would probably be determined in a short time, the confidence in our own power to make a bold strike for freedom if the faintest chance was allowed, and still more the sense of comradeship, and the interest in telling our stories, and cheering each other,—all these contributed to make us bear up courageously.

During the day G. D. Wilson, Dorsey, Buffum, Bensinger, Porter and Hawkins came in. They told an interesting series of adventures. After capture they were taken to Ringgold, and the same evening to Marietta, where they had been placed in a prison but little inferior in every vile element to our present abode. They were guarded by a heavy force of Cadets from the military school at Marietta and narrowly escaped a mob. They finally heard a clanking of chains, and were called out and chained by the necks in couples, and also handcuffed. Then they were taken north over the line of our fiery chase, and stopped at Dalton. While waiting here, some of the ladies came with their servants and brought them a really first-class supper. This was one of the few pleasant incidents in this part of our history. Dorsey complains that he and his companions were not accustomed to eat with irons on, and that they were not very graceful! The ladies probably excused them, for two of their number drew near and talked with them, weeping freely. One lady gave Dorsey a rose, which he preserved as long as he could. There was a mob outside threatening to attack them, and seemingly the more enraged by the presence of the ladies. But the guards withstood bravely, and they reached Chattanooga in safety, where they were taken to a hotel to await orders, and the landlord at his own cost gave them a good breakfast. Then they were brought to us in the Swims jail. On the way through the crowded street, with their chains, a loud-voiced man called out,

" *Will them hounds hunt ?*" Had they been free, he might soon have learned !

Others of our party joined us, in bands of two or three, at short intervals, and told us the thrilling story of their adventures. After a little, the East Tennesseans were removed from the lower to the upper room as others of our number were brought,—until at last we were all below and they were all taken out, so that we had the dreadful place to ourselves. Buffum, with grim humor, would greet the new arrivals, saying in his nasal way. " Well, boys, they took you in *out of the dew !*"

[ ] brought in chained in some manner, but expected, when [ ] ngeon, that these irons would be removed. In this we were mistaken. Either from an excess of precaution, or from the wish to punish us, the irons were retained. Some were fastened together by neck chains, and others by handcuffs. They economized in the use of the latter by making one pair serve for two men, the right hand of the one and the left of the other being locked together. Reddick was the first to whom I was *strongly attached* for some time, but was afterward glad to exchange him for Buffum, who had a very small hand, and was able by a little painful squeezing to draw it out altogether, leaving me with simply a pair of cuffs dangling from my left hand.

That this terrible place was swarming with vermin, not only rats and mice, but other kinds, smaller and worse, will be understood without further statement, as well as our helplessness in guarding against them while in the dark, and in chains, with not even as much water as we needed for drinking! One of the grievances of the jailer was the great amount of water we drank, or rather wished to drink, for our persuasions shouted up through the trap-door could not induce him to greatly increase the allowance which he deemed right; but the officer of the guard would sometimes order him to bring us more, which he would do with great grumbling. Usually the water-buckets would be empty long before the time at the giving of food when he judged it proper to replenish them.

Hunger was also very pressing, and as one or two of the number had managed when searched to secrete a little money,—all the rest had been taken from us—we resolved to try to buy additional food. Accordingly Swims was asked if he would buy for us. He asked the guard, and finding no objections, said, " Yes; you can buy, if you have the money." We had a very earnest consultation as to the form in which the money would go furthest, and finally settled on wheat bread and molasses, as the latter was very cheap. The money was handed up in the evening, and we made great calculations on having a royal breakfast next morning,—all we could eat ! The time came, and we were more eager than usual for the lowering of the bucket, but when it was seized there was only the starving allowance of corn bread and spoiled pork ! Some one called out, "Swims, how about the wheat bread and molasses?" He leaned over and said in his slowest and most provoking tones: "*B-o-y-s, I lost that money!*" Could we have reached him, I am not sure that he would not have been lost also! It is hard to imagine how angry starving men could get under such circumstances. We called to the officer of the guard, who came to the trap-door, and heard our grievance, but only laughed at us, saying that if we trusted Swims with our money we would have to take the consequences. Fortunately we had not put quite all our funds on this one venture, and the officer tried his hand with better results; but the amount purchased was too small to be of any considerable benefit.

CHAPTER XIX.

GLEAMS OF HOPE.

A very erroneous impression would be given the reader if he imagined that we spent our time here in nothing but hopelessly bemoaning our misery. There were times when our situation seemed overwhelming; but usually we kept our minds busy during waking hours in telling stories, in speculating on the prospects of the war, or planning escape. We had no idea that the rebellion would last much longer; the great armies of McClellan and Halleck, we thought, would soon crush their opponents, and Mitchel would be upon Chattanooga, and thus deliver us from our horrible confinement. Could we live but a little longer, the chances of deliverance were good, and we were human enough to talk of vengeance that would follow. There was but little of a religious character in our conversation; that came later; but we discussed all things relating to the country and its policy with the deepest interest. It now seems scarcely credible that, among us all, there were but two out-and-out abolitionists—Buffum and myself. Many a heated discussion had we on slavery and on the propriety of arming the negroes. But the two of us could always win an advantage by appealing to our own experience. *That* was an irresistible argument! When the others refused to credit some horrible atrocity of slave times, we could retaliate by saying that there were people who would not believe in the reality of such treatment as we were receiving; that this was a natural fruit of slavery, and that we were getting a chance of seeing how the negroes must have liked it! A good deal of temper was evoked by such reasoning, but slowly its force made its way, and slavery could not long have been maintained had its existence depended on a vote taken in our prison. Andrews took no part in our discussions beyond saying that he was no abolitionist, but believed that negroes should be better protected against cruelty.

We also talked much of home-life and friends, and thus became really acquainted for the first time. Hitherto we had known little of each other; but now this interchange of thought and history brought us close together, and caused the hours to pass very much more pleasantly. We talked of future plans when we should be released from this place. We permitted no one to get down-spirited. Had we been confined in solitude, the dread and foreboding would have been more terrible. But we made a league against fear and fretting.

Every precaution was taken to preserve our health, that we might be able when there came a chance for escape. No one was allowed to stand so as to obstruct the ventilation through the wall-holes; the exercise possible in such a limited space was taken; and the help and good will of all was a wonderful relief; but we needed every possible support.

Soon there came a diversion which enlivened our conversation by giving us new themes, and secured for some a brief relief from the " Hole," but which brought the deepest fear with it.

Andrews, our leader, was summoned for trial. He procured able lawyers, who interposed all possible objections, and succeeded in securing considerable delay. Indeed, in all the proceedings there was a slowness and apparent hesitation which bore striking testimony to the importance they attributed to him.

When all of our party were brought into the prison we carefully reviewed the situation and ascertained how much the enemy had learned in regard to us. Most had already given names and regiments, and had claimed to be American soldiers. Nothing could be gained, therefore, by denying what we had done, or our true character. The first could be proved by those who saw us on the train, by our captors, or by our own admissions; the latter, so far from wishing to deny, was our only possible defense. To Geo. D. Wilson and myself fell the main task of outlining our defense and drilling all the party into it. There was no difference of opinion as to what was best. It was only important that we should all tell consistent stories. We therefore resolved that we would say, when examined, that we had been appointed by our officers to serve under a man whom we had never seen, but whom we supposed to be an officer from some other regiment or brigade; that the nature of our service was not clearly made known to us further than that it was to destroy some line of communications in the enemy's country; that we were in citizen's clothing only that we might not alarm the citizens among whom we were to travel; that we did not see any pickets of the enemy or pass them at any point; that we were in no camp; that we supposed our expedition to be within the rules of war and ourselves entitled to protection as prisoners of war; that we had obeyed Andrews as we considered ourselves bound in duty to do; and that if we made false representations, it was only what we were led to do by fear of discovery. We were not to say anything about Campbell being only a citizen; and were to refuse under all circumstances to tell who our engineer was. It is by no means certain that the enemy would have dealt any more severely with the engineer than with others, had they ascertained that he was merely a soldier. But we could not know this, and judged from the persistency with which they sought to discover him (for another purpose), that his fate would be sealed as soon as he was known.

Admitting that we were under the orders of Andrews would not in the least embarrass him, for this much could be proved from what he was actually seen to do on the train; and he had already declared as much when captured. But we never revealed anything that would throw light on his employment as a spy in our service. The story which we thus planned to tell was so closely adhered to that the enemy never learned that we were volunteers and intelligent participants in the enterprise.

All were examined at least once, either on being brought into prison or afterward. My own ordeal was more protracted, probably because I had been the first brought before the commanding General, and also because I had been very willing to communicate what I knew—up to a certain point.

Wilson and Wood, in a narrative published in the Key West, Florida, *New Era*, Nov. 15th, 1862, when they had reached that place on their way home from prison, and while many of us were still in the enemy's power, used the following language:

"A court-martial was ordered for Andrews, and Pittenger of the 2d Ohio was taken out as a witness ; and by alternate offers of pardon and persecution they endeavored to make him testify against him, but he was true to his word and his companions, and the court could gain nothing from him."

The effort to gain additional information was less simple and more protracted than is here indicated. The ladder was thrust down, and I was called out. I went through the street once more, still wearing my handcuffs *on both hands*, attended by eight guards, and was brought before an officer, I presumed either the Judge Advocate or the President of the court martial. Here I was told that they wanted to learn several things, among others, the name of our engineer, who were engaged in this affair, and the relation that the man Andrews bore to us and to the army. The officer said that he thought I could tell if I wanted to; and that if I did, I need have no fear of any prosecution for myself. It would have been easy to refuse to say anything, but I thought it better to answer, "Everything that concerns me alone, I will tell you freely, as I want you to know that I am an American soldier, and that I have done my simple duty; but I will not tell you anything that might tend to injure my companions."

He answered that he could promise me nothing unless I would tell all I knew. I said that I asked no promise, believing that when they understood the case they would only hold us all as prisoners of war. He said that this was very probable, but that he would have me separated from my companions and see me again in a day or two.

Accordingly when taken back I was not put in the jail but kept in the yard outside under the charge of six guards, who stood by me day and night in regular reliefs of two at a time, with orders not to let me, on any account, speak with those inside—which could have been done only through the front window-hole. How foolish they would have felt had they known that for days previously Wilson and I had been drilling the others just in view of making our stories harmonize! There was no need of any further communication. One of the guard, however, with possibly a little Union sentiment or humane feeling in his veins, did tell my comrades just what I had said to the officer, and thus relieved any uneasiness they may have had,—though they had little.

The relief to me in this open-air confinement was indescribable; yet at night it was chilly, and I induced the officer in charge to let me lie down in Swims' kitchen. The next forenoon I saw the old jailer engaged in reading a paper that seemed to interest him very much. I wanted news also, but as he was sober, it was useless to question him. Waiting till he laid the paper down on the window-sill, I edged up to it, and by a quick motion of my manacled hands, I slipped it into my bosom. The guard saw nothing. I glanced at it and beheld—it was wonderful that I should get just this paper—a full account of our expedition, occupying a whole page, with many other interesting items.[1] It was too valuable to be lost. I had expected that it would soon be snatched from me, but had hoped to secure the news first. Now I was determined to save it. I sauntered about the yard,—the guard did not oppose my walking about if I did not go near the fence or the hole in the dungeon wall—but as soon as I got close to the latter, I walked straight to it. A dozen eager questions were asked, but I knew my time was short, so I said, "Take this," and shoved the paper through the triple bars. It was seized instantly, and,—I have that paper now, twenty-five years after—a most precious relic! Then I told them that I was doing well, and that there were rumors of Mitchel's advance. This I had heard as I did many other things, from the guards. I asked them if all were well inside; but before an answer was given, I heard a sharp order, "Come away from there!" and the sergeant of the guard, a tall Georgian who had been rather pleasant to me, was running up, white with passion, and the two guards, who had been seated on the door step, were also coming with fixed bayonets. He upbraided them for their neglect of duty, and then scolded me, winding up by telling them that if they saw me within ten feet of that place again, he wanted them to shoot me down without call or warning. I took this patiently, for I had accomplished my purpose. Soon after the jailer missed his paper, and there was a great search, from which I was not exempt, but all in vain.

While thus kept in the jail-yard I borrowed—either from Mrs. Swims

---

[1] This article is given in Supplement, Chap. V.

or one of the sentinels—a small copy of the New Testament. I had often read and studied it at home and in Sunday-School, but now, when turning the leaves with ironed hands, it read like a new book. I found an intensely practical meaning I had never dreamed of. It took but a little while to read from beginning to end with an attention which often made me forget guards and chains. Amid the doleful scenes and memories of that fearful prison this precious reading was a sweet oasis, and was not without result afterwards.

When taken before the officers again, I told my story in just the lines we had laid down, which made us simply detailed soldiers acting under orders. They questioned and examined a good deal. I refused, directly, to name the engineer, and varied from the truth in other matters just as far as I judged it expedient. At length I overheard one of the officers say to another, in effect, "It is no use. He is either ignorant, or too sharp to tell us anything we want to know." Then I was informed that if I did not know anything more I would have to be put back with my companions. I said, "As you will;" and soon found myself in the "hole" again, where I received a warm welcome. I did not like the place, but was glad of the company.

None of us knew just the line of defense taken by Andrews or upon what he based the hopes he did, certainly, to some extent, entertain. While all his money had been taken from him, he was still believed to be able to command it from outside sources, or could make promises which were believed. He was charged with spying and treason. The line of defense as below indicated I have gathered from scattered hints, from his Flemingsburg letter, and from his own last informal declarations. He seems to have sought to make the work that he did appear as small as possible, and his own motive to be only money-making, with resulting benefits to the South far greater than the loss. What follows is built upon scattered hints, and is therefore offered with no small degree of diffidence, yet comprises the only theory that seems to meet the facts.

He was known to the Southerners as a blockade-runner. This trade, if he could only persuade the Federals to permit it, was of great advantage to the Confederates. He might afford to do a good deal for the Federals and the balance of advantage be still on the Confederate side. Now he claimed that he was offered by General Mitchel, who greatly wanted an engine, the privilege of trading South to the extent of five thousand dollars per month as long as the war lasted, on the simple condition that he would seize an engine and carry it through, he Mitchel supplying the men for the enterprise. Andrews did disclaim to the enemy that he intended to burn any bridges, or do any harm beyond the comparatively trifling one of carrying off this engine. In harmony with the same line of defense he tried to make himself appear very ignorant, and in letters intended for their eye, the spelling is fearfully bad—so bad that it looks as if he over-did his part. As a help to the men who were under him, Andrews declared his belief that they were all detailed without their previous consent, and that they knew nothing about what was to be done, but that Mitchel simply sent them to carry off the engine. He summoned his old partner, Mr. Whiteman, to prove that he was a blockade-runner, and that this was his real business. He was greatly dissatisfied with the conduct of Mr. Whiteman on the trial, thinking that he received far more injury than good from him.

The defense was but feeble, though it was possibly as good as could be made under the circumstances. But the enemy had no notion that a Federal General like Mitchel would send his men away down to the heart of Georgia for an engine, when, as experience showed, he had been able to get them by the score in Kentucky and Alabama. The car left burning on the bridge, and the admissions of the men who were first captured as to their intentions of bridge-burning, also tended to discredit the idea that the running of the engine through was the only object. Neither would the Confederates easily believe that the Federals would permanently establish so large a contraband trade. The facts were that Andrews had gained their confidence, and they had admitted him into their midst to travel and enrich himself as their trusted ally; when suddenly they found him at the head of the most daring enterprise the enemy had yet under-taken. He could not now turn around, even with his marvellous adroitness, and unsurpassed powers of deception, and make them believe that the enterprise after all was but a little affair, and was intended as the means of deceiving the Federals only the more completely. He had played back and forth once too often; and their enmity now was in proportion to their former confidence.

But it is possible that Andrews did not look for an acquittal, only wishing to interpose delays in the hope of a Federal advance, or of find-ing an opportunity to escape. With a man of his boundless fertility of resources, each day brings new possibilities. During the trial the trap door was opened several times, extra, each day, which was no small gain and three or four days were occupied before the end was reached.

The Atlanta *Southern Confederacy*, of April 26th, thus notices the result.

"We learn that the court-martial at Chattanooga have completed the trial of the ring-leader of the bridge-burning party. Their decision in the case will not be made public, however, till their finding is approved by the Secretary of War. We learn that no more of them will be put on trial till this decision is passed upon.

"This leader of the party is named Andrews, and is said to be a partner in a well-known firm in Nashville, and had not heretofore been suspected of hostility to the South. He was hired by the Lincoln authorities to burn the State bridges, and, if possible, to bring through to them an engine. Those accompanying him belonged to the army, ... had been detailed to do the work."

The closing paragraph of the report is hard to understand, even allowing for the usual latitude of misinformation. No member of our party was from Kentucky except Campbell, who carefully concealed the fact. It may be simply a reminiscence of our Kentucky pretensions. Andrews himself had been a member of the Kentucky State-Guard; but it is not likely that he would give that as a reason for being in the Union service. The paper says:

" We are informed that the one who turned State's evidence against them is a Kentuckian. He said he was one of the State Guard in the days when neutrality was in vogue ; that he was entrapped into the service by belonging to this State Guard and accepting arms from the Yankee Government before the Yankees came into the State, and was unable afterwards to get out of the service ; that he was always friendly to the South, and that it was always his determination to fight for the South if forced to take any side ; but that neutrality and the State Guard had deceived him as it had thousands of others. Before he was fully aware of the fact, he was in the Lincoln army and could not escape from it."

On the 28th of April, orders [1] were sent to General Leadbetter in which E. Kirby Smith directs " that the spies be tried at once. " The instructions from Richmond referred to above had no doubt been received. No sentence had yet been awarded Andrews, who remained in prison with us, just as before trial. But other elements outside of Confederate bounds entered into our fate. There is little doubt that if these trials had proceeded while in hot blood we would all have died.

On April 26th, the Union Government, through Gen. Wool, then Commissioner of Exchange, notifies " the Insurgents " that our forces hold many more persons who are liable to be put to death as spies than the Confederates do; and that while no one has yet suffered the extreme penalty on the Union side, this forbearance cannot be counted on in case the work of death is begun. [2] It is probable that the great delay and great care in publishing sentences arose from this cause. The reason for suspending the trials, however, was nearer at hand. On the very day this order was given, Gen. Mitchel had advanced as far as Stevenson, and was cutting out work for the Chattanooga rebels which rendered any further court-martialing for the time out of the question.

---

[1] War Records, Series I., Vol. X., Part 2, page 461.
[2] War Records, Series III.

CHAPTER XX.

GENERAL MITCHEL SAVES THE RAIDERS.

I THINK I never in my life experienced a sweeter dream than in the terrible Swims prison a short time before we left it. I have no superstitions regarding such things, but a dream that fills the mind with inspiration and hope for days after is certainly good, no matter what it may be held to signify

We had been talking about our prospects, and differed, as usual, some maintaining that it was our duty to keep in good heart, for there was hope even yet. J. A. Wilson, Dorsey, and Mason were rather disposed to regard hope as useless and deceptive. They thought that we would have to die at any rate, and the sooner we passed out of such misery the better. When the discussion grew languid, I tried to read from the paper obtained as narrated above,—a very difficult task, for the light was but faint at noonday. Then I leaned back against the wall and sunk into dreams.

I thought I was in a mountain country. The exceeding purity of the air and sky was most grateful. The wide horizon and the lovely valleys made me feel that this was indeed a beautiful world. For a time I knew not where I was; but I saw great snowy peaks, of such spotless, dazzling whiteness, that it was glorious to look upon them; and then it came to me, "This is East Tennessee, and I am not far from Knoxville." (I had never been in that part of the country, and there are no such mountains in reality, as I saw in my dream.) The sky over the peaks was of the most intense blue, and the purity of this and the white was indescribable; my soul rejoiced in it. I seemed to gaze for hours on this beauty and sublimity, and then—opened my eyes and found that I had slept but a few minutes! No doubt the little, dark, dirty, and narrow room was the cause of my dream by the law of opposites; but I found myself lifted up in spirit, toward the altitude of the mountains. I could not believe that we would be left to perish in the darkness of this dungeon.

While we were passing through such terrible scenes, how did those fare whom we had left behind? In the camp deep anxiety was felt in our behalf. When we did not return at the appointed time, all believed we had perished. Every prisoner was closely questioned, but no positive tidings could be gained. Even Gen. Mitchel could not learn where we were confined, or whether we had been captured at all. He finally declared to his son, F. A. Mitchel, that all of us had been hanged. The enemy wished apparently to hide the knowledge of us, probably fearing retaliation on the part of our commander; and it is possible that this was the reason for our confinement in the tomb-like Swims prison, from which no word could reach any one. There was much mourning for us by the camp-fires, and even more in our own homes. I have sometimes wondered whether the sufferings endured by mothers, wives, sisters, and all the helpless loved ones, is not greater than that felt by the soldier. Andrews's betrothed in Flemingsburg was waiting for him with a suspense in which her life was literally bound up. Scott, Slavens, Buffum, and Mason had wives to wonder why no letters came, and why no one in the army would answer questions about them. My own mother said long afterward that for months together she never laid down her sorrow night or day; in sleep there was a vague sense of trouble, and on waking a realization in a moment that something was wrong even before the heartache became distinct.

On the 28th of April, Mitchel[1] began his movement toward Bridgeport with no other design, than to force back the enemy, and throw him on the defensive, while securing his own line of communications to Stevenson. He would gladly have advanced to Chattanooga, and even Knoxville, had a sufficient force been given; but in the absence of this, it was best by a striking demonstration to put the enemy in fear. On the 29th he attacked Bridgeport. The enemy had a considerable force at that point, but as usual Mitchel was able to surprise them completely. This he accomplished by driving in their pickets on the line of the railroad, making the impression that he intended to advance in that direction, and then suddenly moved across the country, dragging two pieces of artillery by hand. There was slight resistance, and the enemy's outpost fled in the direction of Jasper, without giving any warning. A little incident here illustrates the readiness of resources which always distinguished Mitchel. His meagre column was moving swiftly forward to the surprise of the enemy, when a formidable obstacle was encountered—a little creek or gully, a dozen yards across, and quite deep. To build a bridge would not take long, but minutes were precious, and the noise of chopping and other work would be sure to alarm the enemy. Something must be done to get the cannon over. The infantry was moved up to a line of fence and the command given, "Let every man take a rail." It was done in an instant, and all marched by the gap, the rails were thrown in, and the line counter-marched into place again, Mitchel saying to them as they passed, "Quick and silent, my brave boys! We're

---

[1] War Records, Series I., Vol. X., Part 1, pages 655, 656. Report of Gen. Mitchel, April 29th, and of Gen. Leadbetter, May 5th, 1862.

within a squirrel's jump of them!" It was done with such swiftness that there was scarcely a perceptible check in the advance, and the enemy, opened upon with cannon and musketry from an unexpected quarter, did not stay to try conclusions, but fled precipitately.

As they passed the great Tennessee West bridge from the island to the main shore, they made an attempt to blow it up, and not succeeding in doing any great injury, they next tried to burn it. Volunteers were called for from the 2nd Ohio (my regiment), and Captain Sarratt leading, they rushed on foot through the smoke over the burning structure, reached the island and put out the fire, thus saving the most important part of the bridge. Some very important captures resulted, among others some of the men who had been guarding us but a few days before, so that some information of us did reach the Federal army.

Captain Sarratt heard a voice calling, "Don't fire on me; I'm coming over;" and immediately saw a soldier running out on the bridge from the east bank. The rebels did fire on him from their side, and were in turn fired upon by the Second Ohio men. The fugitive succeeded in getting safely over and proved to be one of our original band who had been arrested on the way South and put in the rebel army. His companion was arrested while trying to escape, and for a time was confined in the upper room of Swims prison, where we saw him; but at last he was returned to the army, and got safely through to the Union lines.

More important still for us, a small force of Mitchel's did advance a considerable distance toward Chattanooga, causing a great panic there. No one thought of court-martials for the time being. Indeed there was fear that we might be freed, and all our guards made prisoners. Had Mitchel known how small a force held Chattanooga, though he could not then have maintained the place against Southern reinforcements, he might by a dash have saved all of us, Andrews included, and wrought great distruction of rebel property. But he had no spy like Andrews to reveal to him the exact posture of affairs, and the opportunity, much to our sorrow, was lost.

We were not altogether in ignorance as to what was going forward. Our guards were uneasy. The old ones were taken away and we had entirely new faces. Swims got considerably intoxicated, and while he was lowering our scanty breakfast to us on the morning of the thirtieth, he held his white head over the trap longer than usual, and drawled out: "They say Mitchel is coming." We ate slowly, encouraging him to talk, and managing in a little while to get out of him about all he knew in regard to the fight. He said people blamed General Leadbetter for letting Mitchel get the better of him, and that now they had no men in the town able to defend it. All this was glorious news, and we listened for the booming of cannon, and for a time we heard them, though faintly. We would have been glad to have a few shots in the jail yard and were even willing to risk one or two in the old jail itself—enough to make a breach in the walls !

But our captors had no intention of loosing us. Soon the trap-door was opened in great excitement and hurry, and we were all called up the

The Raiders Seated in the Cars.

ladder. In the upper room our irons were inspected, and new fastenings added. Then we marched away from the old jail, as we hoped forever, and on through the town to the depot, where we waited for the cars. We understood it all. But we knew Mitchel's wonderful speed in movement so well that we hoped he was really coming to Chattanooga, and would get on the railroad in some way, before we could be run further south. The train was late, and we had no wish to hurry it. To simply be out of doors, in the free air, was delicious. We lacked the vigor of three weeks before, but were rapidly reviving. Several actually staggered from weakness when they first came up out of the terrible pit.

The cars came before long and we were once more on board and moving southward. How vividly we remembered leaving that station for our first journey over the State road! It was now the first of May. We had looked into the hollow eyes of death since then; but were now enjoying a short respite, and could afford to revel in the beauty of the outward world!

O, the joy, the gladness of being again under the canopy of heaven, and of looking up to its unfathomable depths with no envious bars to obstruct our view! I have often looked upon romantic scenery in May, but never have I more deeply felt that this is a pleasant world, full of beauty and goodness, than on that balmy evening when the rays of the setting sun streamed over the grass and forests in their path, and poured in yellow radiance through our car window. But a glance at the guard with his musket in the seat beside me and at my own handcuffs had a sobering influence!

Our raid had not been forgotten, and as it became known that we were passing along the road, a mob greeted us at every station. It is not necessary to linger upon the manner in which they accosted us or the questions they asked. Mobs are nearly all alike, and the one in Atlanta will answer as a specimen of the others.

There was a failure of railroad connection in that city, and we were obliged to wait from very early in the morning till late at night. This was the first time that we had been in Atlanta, and our visit now did not impress us in its favor. Little did we think of the awful and hallowed memories it would soon bear for us!

But the mob which now gathered was fierce and bloodthirsty—determined to hang us. Our guards were equally determined to prevent. As soldiers they wished to fulfill their trust, and in addition, had been with us long enough to imbibe some kindly feeling. Several persons were severely injured in the strife, but the guard-prevailed. While the disturbance was at its height, a man seemingly as rude as any of the others succeeded in reaching the window, and, watching his opportunity, he slipped a paper into my hands with the single magic word, "A friend!" There was glorious news in the paper—nothing less than the capture of New Orleans! For a time private sorrow and danger were forgotten in the exhilaration of national triumph.

The cause of secession at this time was far from bright. I took pleasure in talking with the officer in charge of us, and others who were intelligent, and found them discouraged. They would not knowingly give us information, or let us have papers; but by pretending to know we were often able to splice out our slender information. The officer admitted that McClellan was moving with an overwhelming army on Richmond, and that they had no force adequate to resist him. Everything looked bright for the Union cause, and our only uneasiness was whether we would live long enough to enjoy its final triumph

Our old friend, the Atlanta *Southern Confederacy*, the next morning (May 3rd) thus spoke of our passage through the town:

"THE ENGINE THIEVES.—These notorious individuals arrived here yesterday morning on the train from Chattanooga. The leader, Andrews, has often been in our reading-room during his peregrinations in the Confederate States since the fall of Nashville.

"Before that time he was engaged in running the blockade, bringing articles of necessity for merchants, manufacturers, etc., from Cincinnati, Louisville, and other points in the enemies' domain. He made it a business, and was quite successful in it, and he retained the confidence of our people in Nashville, where he resided while so engaged.

"The other prisoners, his compeers in the attempt to burn the bridges, are all sharp, intelligent-looking men, no hard-looking cases like Yankee prisoners and East Tennessee Tories usually are. We learn they will be sent to Milledgeville for confinement."

The last statement was a mistake, for we were taken to Madison, where some six hundred Federal prisoners were confined, and we indulged the hope that we might be put with them; but we soon found that the brand of criminality was not yet effaced. We passed the dilapidated cotton factory where our soldiers were kept, and on to the old county prison which was then unoccupied. It was a gloomy stone building with two rooms, but they were both above ground, and had doors. The party was divided between the rooms. The heavy stone walls rendered it quite damp, and it would have seemed a forlorn place had it not been for experiences in Chattanooga. But we were away from Leadbetter, and our captain talked with us and showed us kindness, though he did not dare to take off our irons.

As a rule we had the most perfect harmony in our own company; but sometimes there was a slight ripple on the current which did not last long. Dorsey sends me the following incident which occurred to himself and his chain-mate in the upper room·

"Porter and myself fell out and tried to fight All were engaged in an argument—a very common thing. Porter wished to cross the room, and being in a hurry, rose to go. I was interested in the argument, and did not move off at once. Porter urged me to rise, and when I still lingered, he jerked the chain on my neck. (They were chained by the necks as well as handcuffed.) I also seized the chain and gave a jerk. There was another jerk or two on both sides, and I sprang to my feet and seized Porter's end of the chain around his neck, and began to twist it trying to choke him! He retaliated. This was all that our handcuffs would allow us to do, and there we stood, twisting our chains, and unable to do more. The other boys set up a laugh, and soon shamed us out of it. We liked each other better after this incident than before."

At certain hours the citizens of the place were freely admitted to see us, and ranged themselves—always in the presence of the guard—along one side of the room and talked over the topics of the day. They expressed great admiration for us and what they were pleased to call our daring exploit. They said they had not expected such things from the cowardly Yankees.

But one visitor did not come for mere curiosity. He was dressed in rebel uniform and talked about as any others would have done; but when all had gone, Andrews informed us that he had recognized in him a former acquaintance and a spy; and told us that by signs he had made known to him the word he wished carried to our forces, and that now we might depend upon our situation being known. We could scarcely credit this, but soon the captain of the guard, who came to bring us our supper, confirmed it all

He said that a most remarkable occurrence had just taken place. The Provost Marshal had learned from some source that a Lincoln spy had been in the town, and had at once sent a guard to seize him. He was found at the depot just as the cars were coming in, and professed to be very indignant at the offer to arrest him, saying, scornfully, that he had papers in his pocket that would prove his character anywhere. A little abashed, they released their hold upon him, asking for the papers. He put his hand in his pocket, as if searching for them, and fumbled about till he noticed that the train, which had started, had got under a good degree of headway, and then, when it was just possible for him to reach the last car, he flung the soldiers aside, and ran for it. He got on board, but they were too late, and as there was no telegraph station here, they were helpless.

The Confederates, on this, stopped all our visiting; but we felt sure that news of us would directly reach our own lines, which was far more important. In this we were disappointed—we never heard from the spy again. Whether he was captured somewhere else, or his information lost in the rush and hurry of other events, we never knew.

Our stay in Madison was only three days, after which the Confederates, relieved of the fear of an immediate advance on the part of Mitchel, ordered us back to Chattanooga. Again we were compelled to run the gauntlet of insulting and jeering mobs that marked our course on the Southward journey. We travelled in rude box-cars, and while wet and filthy, these were not half so hard to endure as the thought of going back to our old quarters. But the journey was rendered easier by the fact that ever since leaving Chattanooga we had been in the hands of one set of guards commanded by Captain Laws, and they had discovered that we were human beings. They talked freely with us, and did all that they safely could to render our condition more endurable. One re-

16

sult of this was, that when brought back to Swims' domain, the commander interceded for us, and we were allowed to remain in the upper room—a priceless indulgence. The poor East Tennesseeans, however, had to go below. Our hearts bled for them, but they were only fourteen now, some of them having been removed, while we were twenty-two. This room was the same size as the under one, but it had three windows instead of two, and these much larger—real windows—with only one row of bars, and so high that they all admitted light as well as air. We could see over the jail fence in two directions. These were immeasurable advantages, and we felt deeply grateful to Capt. Laws and Col. H. L. Claiborne to whom we owed them. Yet these mitigations were only comparative. Our imprisonment was still rigorous beyond anything that could be expected in a civilized country.

It was amusing to see the exaggerated caution with which we were guarded. Even when below, where a man unassisted could scarcely have got out if the locks had all been taken off, the jailor never raised the trap-door unprotected by a strong guard. Now that we were in the upper room, their vigilance increased. They would bring a guard up into the jailer's room, which opened into ours, and array them in two lines with leveled bayonets before our door was unlocked. At the same time the stairway was guarded, and a strong guard in a circle always walked their beats clear around the prison. We were all this time closely chained! Yet Swims would grumble and predict some great trouble from showing the Yankees so much indulgence! How such things provoked us to make an effort to break out!

But this wearing of chains so long seemed useless. Col. Claiborne had made some proposals toward relieving us, but in vain. We tried earlier, and with better success. Knight had concealed his penknife when searched, by putting it up his sleeve, and adroitly turning his arm as they felt for concealed articles. Now from some small bones in our meat he made keys which unlocked the handcuffs. With strings and hairs, the padlocks on the chains had been opened before. We were given a good deal of trouble, for these fastenings had to be put on again when the door was opened. The outside stairway was useful as a warning. As soon as a foot was heard on it, the signal would be given and there was rapid work in "locking up." Had we been detected, there would have been work for the blacksmith in welding us fast, and the dreaded "hole" would probably have again received us. But we were never detected.

Our days were much longer and more pleasant here as we awoke earlier and sought out more employment than below. Mock trials gave us much amusement. We needed some kind of government, and had to try, and punish, offenders. Campbell was made judge, and had usually the sport of carrying out his own sentences—a task for which his immense strength and unfailing good humor well fitted him. The opposing counsel made long and learned speeches—so eloquent and interesting that no hearer ever left the house while they were in progress !

A more refined enjoyment was found in singing. Andrews had been a music teacher, Ross possessed a voice of marvellous sweetness, and several others had talents above the average in this direction. Practicing together, they soon acquired great proficiency. Many of the songs were of a tender and melancholy cast, such as " Twenty years ago," " Nettie More," etc. Three of these songs were invariably sung; and the words wake an echo out of the past more powerfully than almost anything else. The first was appropriate to us all, though it was first taught by Ross:

> "Do they miss me at home, do they miss me?
> 'Twould be an assurance most dear,
> To know at this moment some loved one
> Were saying, I wish he were here:
> To feel that the group at the fireside
> Were thinking of me as I roam.
> Oh yes, 'twould be joy beyond measure
> To know that they missed me at home."

The next was Ross's favorite and was in quite a different vein:

> " Twas in a grove I met my love,
> One soft and balmy night ;
> I owned my flame, she did the same,
> And trembled with delight.
> When at the gate we parted late,
> I blest my lucky stars,
> And stole a kiss to seal our bliss
> Between the wicket bars."

But the " Carrier Dove " was contributed by Andrews and had a melancholy appropriateness to his own condition which often brought tears to our eyes. We knew his history only in part then, but could not help feeling that the song had more than mere melody for him. He was always a little reluctant to sing it: but he had given it once in a sentimental hour, and we could hardly be satisfied any evening afterward without hearing it.

> " Fly away to my native land, sweet dove,
> Fly away to my native land,
> And bear these lines to my lady love,
> That I've traced with a feeble hand.
> She marvels much at my long delay,
> A rumor of death she has heard,
> Or she thinks, perhaps, I falsely stray ;
> Then fly to her bower, sweet bird !

Oh, fly to her bower, and say, the chain
    Of the tyrant is o'er me now ;
That I never shall mount my steed again,
    With helmet upon my brow ;
No friend to my lattice a solace brings,
    Except when your voice is heard,
When you beat the bars with your snowy wings,
    Then fly to her bower, sweet bird !

I shall miss thy visit at dawn, sweet dove !
    I shall miss thy visit at eve !
But bring me a line from my lady love,
    And then I shall cease to grieve !
I can bear in a dungeon to waste away youth,
    I can fall by the conqueror's sword ;
But I cannot endure she should doubt my truth :
    Then fly to her bower, sweet bird !"

Our special time for singing was in the evening twilight. Some one would start a song rather feebly, but others would join, and then for hours in the gathering darkness song after song would pour forth, as glad and free as if not strained through prison bars. The guards liked very much to hear us sing; and frequently citizens of the town would gather outside of the jail fence, where we could be heard with perfect distinctness, to listen to the "caged Yankees." These songs, and the favorable report of all the guards who were brought into contact with us, caused a sentiment in our favor to spread rapidly through the town. This probably was the reason that no further trials, notwithstanding the order of Gen. Smith already quoted, took place in Chattanooga.

We soon had a better opportunity of talking with our guards, and learning a little of what was going on in the outside world, as well as influencing sentiment more directly. When Col. Claiborne first visited us as Provost Marshal, he said boldly that it was a shame and disgrace to keep men in such a condition. After vainly asking permission to remove our irons, he gave us, on his own authority, another wonderful indulgence. He ordered us brought into the jail yard to breathe the fresh air for an hour every afternoon. This was but an ordinary precaution in the case of twenty-two men kept in a thirteen feet room, while warm weather was coming on; but we were deeply grateful. We could, on such occasions, sometimes talk with citizens in the presence of the guards, and gained much interesting information from the guards themselves. Mrs. Swims had a very different spirit from her husband, and did give a little extra food to some sick prisoners, for which they cherished great gratitude. Another lady came from a large mansion on what was called "Brabson's Hill," a little way from the prison, and being in Swims' kitchen, was permitted to talk to Andrews, expressing great compassion for him, and afterward sending a few gifts, with the permission of the officer of the guard, which were highly prized

But I was much more interested in the romantic history of a colored man named Wm. Lewis, whose house—a large two-story frame, still occupied by him—was in plain view only a square away. He asked permission to send us some lettuce, of which he had a large quantity, and this formed quite an addition to our slender rations. From the guard, I learned his story. He was a slave, but being an expert blacksmith, had purchased his time for $350 a year. He was soon able to buy his wife and himself at $1,000 each. Then he set up a shop, hiring other hands, and bought his six-year-old son for $400; his mother and aunt for $150 each, as they were old, came next; two brothers followed for $1000 each. A slave trader bought his sister for him for only $400—the best bargain he had made; then he paid for his house, and laid up a large amount of money besides. Such a man is a genuine hero. He was not able to do business in his own name, under the black laws, and was obliged to pay a white man largely to legalize his transactions. When he saw that the ruin of the Confederacy was inevitable he purchased tobacco with his disposable funds, and, storing it, was able to sell at a handsome profit. He has since lived with the esteem of all men, and presents an example of triumphing over difficulties seldom equaled. His house was the only landmark in the vicinity of the old Swims jail which I was able to recognize in Chattanooga twenty-five years after.

Another reason for the greater humanity now exhibited, was that the word had gone out that we were only detailed men — not volunteers. All of our number had told this story; I had especially insisted on it, and in all our conversations we were sure to make it prominent. Of course in the form we put it the claim was false, but its effects were the same. Andrews, who had nothing to gain in that direction, had been careful to confirm all that we said, and it was accepted as truth, that " those poor Yankees were sent on that terrible raid without being told what risks they ran." Of course our officers were severely blamed for such reckless inhumanity to their men; but this did not hurt the officers, for they were not in rebel hands. It was most likely this strong feeling running in our favor, to which Gen. E. Kirby Smith long afterward bore testimony, that led Col. Reynolds,[1] President of the Court Martial, to make an excuse to get away from that ungrateful post into the field.

But this general and rising tide of sympathy in our behalf produced what seemed to some of us one disastrous effect; and so little can the results of events be foreseen that probably less friendliness would have saved the lives of many of our number. Our best hope was that of es-

---

[1] War Records, Vol. X., Series I., Part 1, page 658.

"orts from the enemy's power   We had surmo'nted
in..... of ...   '-~st, a hopeful attempt.   Our irons could all
be off whe....   '' make formidable weapons in close
quarters; we were ...   . ..   '-~le," and could meet the
enemy on almost equal ter....   ' ~t good wishes of
guards and citizens would not avail ...   ~al power;
and that at any rate it was nobler and more .   .ur fate in
our own hands, and strike for freedom.

But Ross and Geo. D. Wilson had great hopes from what officers had
said to them of an early exchange.   They had been so often told that we
would be held as soldiers simply, that they had come to accept it as un-
doubted.   They pointed to our increasing indulgences as evidence that
the malice of the enemy was relaxing; and although a force of twenty-six
men was constantly on duty, whose officers imagined they were keeping on
our chains, there did seem to be some reason in what was said.   Ross
was the only Freemason of the party, and in that way had received some
trifling favors, and was confident that if any special danger threatened us
he would be given a hint.   But we finally bore down all opposition to attempting an escape by one argument which no generous mind could resist.
It was that Andrews had been tried, and while his sentence had not yet
been given, it was sure that most of the favorable considerations in our
behalf did not exist in his case.   It was our duty to give him a chance, as
well as to avail ourselves of it.   Thus after some delay we came to the
conclusion to strike for liberty.

Two plans were proposed.   The first, which I suggested, had at least
the great merit that it could be tried at once.   Delays are always dangerous, as we had already proved.   It was proposed to have all our irons off
when the guards came up with our supper, and then making a rush upon
the leveled bayonets outside, before the guards had recovered from their
surprise, to have them disarmed, and then to pour down stairs on the guard
below.   When they had been secured and all of us armed, we could have
"double quicked"to the river, crossing it if possible, or if not, into the
mountains to the east of Chattanooga.   We would not again have committed the error of scattering.   The plan was not more difficult than the
first capture of the train, in which we had easily succeeded; and was certainly under more favorable circumstances than a similar attempt made
afterward which had a great measure of success.   Of the first rush, I had
no fear whatever.   It would have been sport to see how Campbell,
Slavens, Brown and Knight would have handled ordinary men in a narrow
room.   Muskets would have been of no avail at first, and before the surprise was over, we should have had more than the enemy.

This plan would have been accepted but for Andrews.   He was always
disinclined to a fight, if the same object could be attained by strategy.

It was then agreed, on his suggestion, that when we were being brought in from our breathing-time in the yard, Wollam should manage to secrete himself under the bed in Swims' room through which we always passed, and remain there till late at night; then come out and open the door from that side, and let us out to proceed as in the first plan. My objections to this were the risk that Wollam would run of detection either when hiding, or when trying to open the door; and that we would be less likely to get a number of guns from the guard. We would probably be pursued by them when ourselves unarmed; and it also required us to wait till the night promised to be dark.

While waiting there was great talk of exchange. Mitchel had captured a younger brother of Gen. Morgan, but his own son had been taken by that chief, and now they, with many more on each side, were to be exchanged. A lieutenant who had been paroled for the purpose of effecting the exchange, visited us, and the most sanguine hopes were raised that we might be included. It is possible that if our authorities had known of our condition and had made a peremptory demand, backed by the threat of retaliation, they might have accomplished this. But the Confederate officers told us that we must first have a trial to show that we were soldiers. Andrews had proposed to send a flag of truce through to get from our officers a statement of our true character; but they refused, saying that they credited our own story, and did not need to go to so much trouble.

Finally we fixed the night when we would again test our fortune by a bold effort to escape. But that very day an order was given to send twelve of us to Knoxville for trial. For some reason the court had been carried to that place, or the trial ordered before a new court—probably the former. Geo. D. Wilson was sick, and down in the yard when the order came. The officer of the guard spoke to him about it, saying that it would no doubt be a mere formal trial preliminary to the exchange, and intended to make sure that we really were soldiers as we claimed. The officer told Wilson that he might select twelve to go, as no names were mentioned. Wilson accepted the offer, choosing all his own regiment, the 2nd Ohio first, and afterward his special friends from the other regiments. The officer had suggested that it might be well to select the ablest men, who would do us the most credit. Wilson had no fears about that; but he did think that he would be doing those chosen a favor by making their exchange the more probable and speedy. But it was a deadly favor; for though he knew it not, there was every probability that it was their death warrant!

The reason for thus dividing the party has never been clearly made known. One theory is that the enemy simply wanted by dividing us, to make the work of guarding easier; in that case, after the first lot had been tried, the other would also have been sent for. But I think it more likely that in view of the large amount of sympathy that had been expressed for us, it was thought best to spare part, and only condemn a round dozen—surely enough for example and vengeance! If so, they would naturally wish to get the most prominent members of the party; and knowing how anxious and hopeful Wilson was in regard to an exchange, the plan was hit on of making him unconsciously select the men to die. It is sure in any case that those thus selected were put into the forefront of battle.

This separation at once broke up the plan of escape, or rather forced its modification, for though feasible with twenty-two, it was impracticable with ten. I have little doubt that some of us would have escaped if Wollam had once got the door open.

The news of the departure for Knoxville roused a great deal of excitement in the prison. We were always glad to move; but the thought of parting was painful, for we knew not when we might meet again. For six weeks we had been companions in danger and privation; and in spite of fair words, we knew that we were still in the hands of those enemies who had filled our country with blood, and whose deadliest vengeance hung suspended over our heads by a single hair. A deep sadness fell upon us which was abundantly justified.

With Andrews the parting was peculiarly affecting. He was our leader, and we had been accustomed to look up to him in all emergencies. He was specially marked for vengeance; officers who had encouraged us had uttered no word of hope for him. He bore this like a hero as he was, and continued mild and cheerful as ever; so kind, tender, and helpful, in the prison, so ready to sink his own sorrow in comforting others, or to yield his own preferences,—there was such a touch of sadness in his low, calm, thrilling voice, that we could not help loving him with an affection which the lapse of years has not dimmed. It would not have been hard for us to die for him; but he seemed more than willing to reverse this, and give his life if it could have availed for our safety. On his trial he had not uttered a word which could shake credit in the story upon which we had risked our hopes, though it provoked the question: " How could you be so cruel as to lead men into these deadly perils without giving them fair warning of the consequence?" He had been tried. It was probable that the decision was already rendered, and that it was death. Separating the party rendered us less able to strike the blow that the enemy might anticipate from our known affection, in case of a fatal sentence. We had never heard Andrews utter a word of repining. He had played a fearful game and lost; he was ready if need be to pay the extreme penalty. One evening after we had ceased to sing, and had been silent for a time, he said: " Boys, I have often thought I would like to see what is on the other side of Jordan." We were not anxious, or at least not in haste to that view, and the subject was pursued no fur
curred to the same word again. When we h

for the last time, the twelve came to bid him farewell. I will never
forget his parting words. He pressed our hands, one by one, before we
were taken out of our room for the Knoxville journey, and with a tear in
his eye, and a low clear voice that had no tremor, but unutterable tenderness and earnestness, he said, " Boys, if I never see you here again, try

"Meet me on the other side of Jordan."

to meet me on the other side of Jordan." It was our last earthly meeting! The parting from our nine comrades was only less affecting, and
then we turned our faces toward Knoxville, certain that " bonds and af-
fliction awaited us " there. The memory of my beautiful dream did some-
how cheer me a little, though nothing that followed could make it seem
like a preternatural intimation of good.

CHAPTER XXI.

A STRUGGLE AGAINST DESTINY.

A N evening dark with coming storm is the emblem of the events that soon followed our departure. The ten left behind had much more room in their narrow prison, but sadly missed us. Their singing was less full in volume and the voice of Ross left a blank that no one could fill. But prison life passed on nearly as before until the last day of May. On a warm afternoon when the prisoners were in the yard, resting in the shadow of the jail, and wondering what was happening to us at Knoxville, an unknown officer entered the gate and went quickly up to Andrews; without a word he handed him a paper contained in a large envelope, and turning away walked rapidly out of the yard. Andrews broke the seal and glanced at it, turning pale as marble. It was his *death warrant!* All who saw him felt that some tidings of evil had arrived, and the officer of the guard spoke very gently as he told them that it was now time to retire to their room. When there, Andrews gave his comrades the paper, and they read the fearful intelligence. In one week—June 7th —he was to be executed by hanging! This was Saturday, and on the Saturday following he was to die!

No time was lost in useless regrets. All of our comrades resolved to carry out the plan of escape which was their leader's only chance of life —probably their own also. Andrews was separated from them the same evening and put down into "the hole." There was a knife in the party and they at once began work. On Saturday night they cut into the plank overhead, as this could be the more easily concealed. It was fearfully difficult. One man stood on the shoulders of two others who leaned against the walls, for there was no other means of reaching the ceiling, and carved at the heavy oak plank till weary, when another relay of three would take their place. The cutting was not very noisy, but a little shuffling about, talking, and especially singing, effectually drowned it. A piece was thus worked out during the night large enough to admit the passage of a man's body and the work suspended till the morning and afternoon visits of the jailer had been passed.

After they had returned from their daily airing on Sunday they went to work with new vigor. Now they did not need to conceal the evidences of their work, for before the jailer came again on Monday morning they meant to be free. They knew that dangers were ahead, but the thought of liberty, and their leader's life, was enough to inspire them. They worked hard and sang long that Sunday evening. Swims afterward said that he ought to have known that something was the matter by their singing so mournfully! They hoped to finish all that was to be done by midnight; but they had miscalculated their task.

They had to cut the lock out of the trap-door in order to bring Andrews up from below; then to pick their way through the end brick wall above the ceiling, slowly and carefully, so as not to alarm the guard outside. Their garments had to be twisted into ropes to lift Andrews from below and the last of themselves up to the ceiling, as well as to make a longer rope for the perilous descent from the gabled end of the jail to the ground outside.

When all was done day was just beginning to break faintly in the east. No time was to be lost. In half an hour it would be so light as to render their escape impossible. They were all in the loft and Andrews was given the first chance. Of course all fetters had been removed. The rope was passed out, and Andrews crawled through, and in a moment was swinging outside; but in getting out he happened to push off a loose brick, which fell to the ground and gave the alarm. The nearest guard raised his musket and fired at the man hanging on the rope, but missed his aim. Andrews had his boots in his hand, but, in the excitement, let them fall and could not stop to pick them up. He afterwards sorely needed them. But in his stocking feet, he flung himself over the fence, and through the guard line, repeatedly fired at but unhurt. John Wollam followed, and while he was in the air he was fired at by other guards, but succeeded in getting out of the yard unhurt. Those who had failed to get out, crawled down and put on their irons again, and it was a great mystery how the two men alone had been able to effect their escape.

Wilson, who was present, very graphically describes the excitement of the escape :[1]

"When everything was in readiness. Andrews, who was to go first, went up in the loft. The work of making a hole out through the brick wall under the roof was a much more difficult job than we had expected, and proved to be slow work with our case-knife. It had to be done too without noise. We at last succeeded in getting out brick enough to allow a man to pass out, just as the gray streaks of dawn began to show. If I remember correctly, each man had his boots or shoes off, so that we could avoid making a noise. We could see the dim, gray form of the sentry, and hear his tread as he paced back and forth. It was an anxious moment of suspense, when at last, in a whisper, word was passed from one to the other in the dark prison, that all was ready.

"Andrews crept out and swung down, but in some manner a loose brick or piece of mortar fell to the ground and attracted the notice of the sentry, and almost instantly we heard the report of a gun. John Wollam, who was next behind Andrews, paid no heed to

[1] *Adventures of Alf. Wilson*, page 122. Toledo, 1880.

the shot, but lunged out head over heels.   Bang , oang ! went the muskets, and there was loud shouting—

"'Corporal of the guard !   Post number—Captain—Captain of the guard !   Halt !   Halt !'

"Dorsey, who was following Wollam through the hole, halted between two opinions, whether he had better jump down while the rebel sentry stood beneath holding a cocked

Escape of Andrews and Wollam.

gun with fixed bayonet on him, or crawl back into the old prison cock-loft and bear the ills he was certain of.   He crawled back and told us ' it was all up with us.'   We were crowded in the loft waiting for our turn to go out, and listening to the racket on the outside. Within a very few moments, almost no time at all, the yard was filled with troops, and by their loud, excited talk we learned, to our unspeakable joy, of the escape of Andrews and Wollam.

"The rebels, of course, did not at that moment know wno or how many of their prisoners were out, but we in the loft already knew that the excited sentries had fired wildly.   At all events, neither Andrews nor Wollam were to be seen anywhere, either dead or alive. While we felt the keenest disappointment at our failure to get out, yet we felt a thousand times repaid for our effort that even Andrews had escaped. A heavy load had been lifted from our minds. We took new hope. We knew that Andrews would put forth superhuman efforts to gain the Federal lines, and, if he succeeded, we felt that Chattanooga would, in all human probability, get a visit very shortly from General Mitchel. We thought that if either of the escaped men reached the lines and told our comrades of our desperate situation, that they would at once demand to be led to our rescue.

"The musket firing and the news of the jail-break and escape of prisoners spread through camp and town like the wind, and soon the whole population was in a fever of excitement, and all the available man-hunting force, dogs included, joined in the pursuit.

"It is hardly necessary for me to tell the reader that those of us who failed to make good our escape were now put down in the hole. This would follow as a matter of course."

In our Knoxville prisons we were thrilled beyond measure by reading the following item from the *Knoxville Register* of June 4th, 1862. The paper was kindly slipped into our cage by a prisoner who was less strictly confined. It is produced here as showing the impression of the escape at that moment.

"THE ESCAPED TRAIN-STEALERS!

"Below we give the copy of the despatch that was sent from this city yesterday to the Provost-Marshal at Chattanooga by Mr. Wm. A. Fuller, authorizing him to offer a reward of $100 for the re-capture of the train-stealer, Andrews, Mr. Fuller is the conductor of the stolen train, who made such a heroic pursuit of the thieves, starting with a handcar, and eventually succeeding in their capture. He is in this city, attending the court-martial in session here as a witness. He is naturally indignant that the rascals he made so much exertion to capture should have been permitted so easily to escape.

"June 3d, KNOXVILLE.

"Col. HENRY L. CLAIBORNE.

"Is it possible that the infamous Andrews escaped? Is he pursued? If not, offer in my name $100 reward for his recapture and reincarceration.

"WM. A. FULLER.

"Andrews, we learn, is tall in stature, weighing about 180 lbs., and is about 35 years old. He has short black hair, and a heavy black beard all over his face. In ordinary conversation his voice is fine and effeminate, and his general address is good. We trust that this description may lead to his recognition and arrest."

Andrews and Wollam separated as soon as they left the prison. The former ran a short distance beyond the skirts of the town, after having taken precautions to throw the dogs off his track, and finding it too light to travel further in safety, climbed into a tree with dense foliage, which stood in plain view of the railroad. All day long he watched the running of the trains so close that he could have tossed a pebble on them, and once heard a party in pursuit talking about his mysterious disappearance. The search was patient and complete, but they did not think of looking over their heads!

He descended at nightfall and swam the deep and rapid river, feeling that his best course was to get into the loyal mountainous country through which he would only need to journey a short distance to reach the Union lines. His prospect now would have been good but for the loss of boots and hat in the first rush, and the additional loss of his coat in swimming the river. His course was in the main down the river, but he could not make rapid progress. The sharp stones in the darkness soon cut away his stockings and left the bleeding feet unprotected. He bound them up with portions of his garments as well as he could, and continued on his desperate and painful way. But he was a little too long in finding a hiding-place, and was observed in the morning twilight just as he was crossing an open field in which he intended to take shelter, as he had done the day before. Instantly the alarm was given and pursuit made by men and dogs. With boots and other clothing he might have escaped, for he could probably have made such use of the streams as to elude them. As it was he put forth every effort. Dashing through the woods he regained the river bank much lower down than he had crossed the night before. Believing that he was now unobserved, he swam a narrow channel to a small island, and carefully concealed himself among some drift-wood at its upper end.

But the hunters were determined to leave no spot unsearched. A party with bloodhounds now crossed over from the mainland and explored the whole island. He was soon found, but broke away from them and ran around the lower end of the island, wading in the shallow water to throw the hounds off the track; then he plunged into a dense thicket with which the island was covered, and again ascended a tree. For a long time he found secure concealment here, his foes being frequently under the very tree. They finally concluded that he must have got back over the strip of water to the mainland, and slowly returned to seek him there. Two little boys who had only followed for curiosity were all that lingered behind.

One of the boys happened to look up and said to the other that he saw a great bunch on a tree. The second looked to see what it was—shifted his position—looked again, and exclaimed that it was a man! They cried out in alarm and thus announced their discovery to their friends on shore. The latter instantly returned, and Andrews seeing himself discovered—the story is almost too pitiful to be told!—dropped from the tree, ran to the lower end of the island, seized a small, dead log, and with a limb for a paddle, pushed into the stream, hoping to reach the opposite shore before he could be overtaken. So far as the island pursuers were concerned he might have succeeded; but there was another party with a skiff, lower down the stream, who shoved out to meet him. The helpless man could do no more, and was taken.

The struggle had been one of almost hopeless agony. He had eaten nothing since Sunday afternoon and it was now two o'clock on Tuesday. His back was blistered by exposure, unprotected, to the sun; and his feet were covered with bleeding gashes. He said that he felt so wretched and miserable that the thought of certain death, to which he then resigned himself, had no further terror.

Wollam's atempt to escape was for a time more fortunate and skillful than that of Andrews. He broke through the guards, and ran the gauntlet of hasty shots without injury. Soon he reached the river bank and not wishing to attempt the passage in the growing light, hit upon the happy expedient of making the enemy believe that he was across. To this end he threw off his coat and vest, dropping them on the river bank, and then waded a little way in the water to throw the hounds off the scent; then quietly slipping back, hid himself in a dense thicket of canes and rushes. He soon heard the hounds and men who were pursuing, on the bank above and all about him. He could hear the words they uttered, they were so close. At length they found the clothing and concluded that he had taken to the river. They crossed over and searched with their hounds along the water's edge on the other side for the place he had come out. As might be expected, the dogs failed to find the exit, and after due consultation, they concluded that he was drowned, which being a satisfactory termination, they returned

But Wollam was not drowned. He spent the day in much anxiety and suspense, and when night came he cautiously left his hiding-place and worked his way along the river on the very front of Chattanooga, till he came across a canoe, which he borrowed for the occasion—without seeing the owner—and rowed down stream all night. This was a swift mode of progression. As soon as he saw a sign of dawn he sought a retired place, sunk the canoe, and hid in the woods till night allowed him to proceed. This he did daily for a week. Twice he was saved if he had but known it. General Mitchel had constructed an extemporized gunboat with which to patrol the river, and twice Wollam passed within hail of it. But he had heard nothing of any such Union craft being on the river, and imagined it to be some rebel boat, perhaps searching for him. In the dark it was not easy to see any indications of its character. So the poor boy crept cautiously by in the shadow of the shore without being discovered!

But at last he made the mistake that Wood and Wilson had made long before. He imagined that he was safe and went boldly forward in the day-time! One more night's journey by boat, or half that time put in on foot directly northward would have carried him safely beyond the border. But as he was going forward, congratulating himself on having succeeded so well, a band of rebel cavalry who were making a raid into Mitchel's territory, saw him, and procuring a boat with several pairs of oars, came out to meet him. Wollam saw his danger, and there was a hot chase, but the advantage was all on their side. If Mitchel's gunboat had but appeared on the scene then! He was retaken, and as usual tried to deceive them as to his character; but a Lieutenant Edwards, who had been with the party who captured him the first time, identified him, and he was reunited with his comrades in Atlanta.

When Andrews was brought back to Chattanooga a scene of much apparent barbarity followed. His escape had excited great rage, and pro-

Riveting Chains in the Dungeon.

duced most terrible consequences at Knoxville, which will be narrated hereafter. But they were now determined to give him no further opportunity of snatching their cherished vengeance from their hands. He was put down in the hole with the other prisoners, and all access to the yard was denied. Of course no other visitors could see them. The guard was stimulated to renewed diligence. But as chains and handcuffs had proved ineffectual, something more secure was devised. From the shop of William Lewis, the colored blacksmith before mentioned, a man was brought over and taken down into the dungeon, who riveted a pair of heavy iron fetters around his ankles. Dorsey and Wilson, who were present, describe the scene as ominous and terrible, the dimness of the dungeon, the poor, death-sentenced man, half reclining with his feet across the blacksmith's anvil, the blows of the heavy hammer, as the work of riveting went on! A strong chain, only eighteen inches long, united the two heavy fetters, so that only half a step could be taken at once. The feet were thus fastened, in the same manner as hands are by handcuffs, and the latter were also replaced. When all these arrangements were completed, he was once more left to himself.

Andrews had now but four days more of dungeon life between himself and eternity. Escape was impossible unless there should be a rapid advance of the Federal forces—a possibility which did come very near being made a fact. He applied himself to the great business of preparing to die. Most unexpectedly a letter written at this time and in some way carried through the lines has come to hand, and throws great light upon his character and thoughts at this period. He managed in some unknown manner to get writing material and wrote two or three letters. One, no doubt, was written to his betrothed in Flemingsburg, but never received. Another was written to his mother in Missouri. The contents of the latter can only be given as they are remembered after an interval of many years by one who read the letter. He told his mother that he was to die, and that all he regretted was that he had been able to do so little for his country; that many other sons had left their bones bleaching on Southern battle-fields; that he had tried to do his duty, and was now seeking the pardon and favor of God. There were many other half-remembered expressions similar to those which are given in the letter below.

The following communication addressed to a trusted friend in Flemingsburg, Kentucky, and which from some references to property it contains, has been called "The Will of Andrews," needs a word of explanation. The gift bestowed upon Miss Layton was of trifling value, though most pathetic—a mere empty trunk! But the full significance of this was, no doubt, given, with probably more substantial bequests, in one or other of the missing letters. This letter, which reached Flemingsburg, Kentucky, in August, two months after it was written, being mailed at Louisville, is recorded in the Flemingsburg book of wills, while the original is most carefully preserved. Andrews had directed his friend to draw out his money in the Flemingsburg bank—some $2000, with gold premium and interest—in case he never returned, giving him a check for that purpose; and to lend it on good security, paying the interest as a perpetual bequest to the town poor. The friend was faithful to his trust; and though the money was afterwards squandered in a pitiful way, and gave rise to vexatious law-suits, yet this secured the careful preservation of the letter.

In all probability Andrews wrote first to his betrothed, giving those
sad remembrances and bequests which would not be repeated in a letter
to another, and followed with this more general and business-like communication. The original is terribly misspelled, far beyond the ordinary
misspelling of ignorant persons. This is probably intentional, as a few lines
at the first have no errors. The letter also makes references to our being
detailed soldiers, and the manner of this leads me to think that Andrews so
wrote it that our pretensions might not be contradicted if the enemy
should read it.

'CHATTANOOGA, TENN., June 5th, 1862.

"D. S. McGAVIC, Esq., Flemingsburg, Ky.

"DEAR SIR:—You will be doubtless surprised to hear from me from this place, and
still more surprised to hear that I am to be executed on the 7th inst. for attempting to capture and run a train of cars from the Western and Atlantic Railroad to Huntsville for the
use of Gen. Mitchel. I had a party of twenty-one detailed men from the 2d, 21st and 33d Ohio
Regiments with me. We succeeded in getting possession of the train and travelled with
it some eighty or eighty-five miles, when, on account of an extra train being on the road, we
were compelled to abandon the train, the party scattering and trying to make our way back
on foot. The whole party, however, were captured. I was taken on the 14th of April.
I am satisfied I could very easily have got away had they not put a pack of dogs on my
trail. It was impossible to elude them. I was tried by court-martial and received my sentence on the last day of May, just one week from the time set for my execution. On Monday morning, the 2d of June, I made an escape. I succeeded in getting out of the prison
and run by the guard, they shooting at me but not hitting me. The whole country was
immediately swarmed with soldiers. I succeeded in eluding them till on Tuesday, about
2 o'clock, when I was recaptured and will be executed on Saturday. The sentence seems
a hard one for the crime proven, but I suppose the court that tried me thought otherwise.
I have now calmly submitted to my fate, and have been earnestly engaged in preparing to
meet my God in peace. And I have found that peace of mind and tranquility of soul that
even surprises myself. I never supposed it possible that a man could feel so complete a
change under similar circumstances. How I would like to have one hour's chat with you;
but this I shall never have in this world, but hope and pray that we may meet in heaven,
where the troubles and trials of this life never enter. What the fate of the balance of the
party will be I am unable to say, but I hope they will not share the fate of their leader.
If they return, some two or three of them will call on you and the rest of the friends, and
I hope you will receive them kindly. They are noble fellows, and will give you a full history of the affair. Please acquaint my friends with my fate. I will try to write to some
two or three more before my execution. Tell J. B. Jackson, should there be any little
claims that I neglected to settle, to pay them, and keep the horse. I don't think there are
any, but there may be. In regard to other matters, do exactly as *instructed before I left.*
I wrote several letters, but never received any. Please read this letter to Mrs. Eckles, and
tell her that I have thought of her kindness many times, and that I hope we may meet in
heaven, where we shall enjoy the presence of the Lord forever. Give my kindest regards
to Mr. Eckles also. According to the course of nature it will not be long till we shall meet
in that happy country. Blessed thought! Remember me also to the young ladies of Flemingsburg, especially to Miss Kate Wallingford and Miss Nannie Baxter. Hoping we
may meet in that bet· · country, I bid you a long and last farewell.

"J. J. ANDREWS."

The following was added on the same sheet:

"CHATTANOOGA, TENN., June 5th, 1862.

"D. S. McGAVIC, Esq., J. B. JACKSON, MRS. SARAH ECKLES,

"Flemingsburg, Fleming Co., Ky.:

"You will find one trunk and one black valise: the valise has my name in red letters on the end, the other had my name on a paper pasted on the end: these are at the City Hotel at Nashville, in care of the old porter on the third floor. These, with contents, I present to you. Mr. Hawkins, you will find at the Louisville hotel, a large, lady's trunk, no mark on it, and is entirely empty. Please take it to Mr. Lindsey's, near Mill Creek Church, on the Maysville and Flemingsburg Turnpike, and request him to present it to Miss Elizabeth Layton for me, and oblige,

"J. J. ANDREWS."

(This was proved and recorded as a will, at Flemingsburg, on the 3d and 19th of January, 1863. The money referred to in the clause "do exactly as instructed before I left," was duly drawn from bank and loaned for the benefit of the poor.')

After writing these letters Andrews had but two days to live. He watched for opportunities to send them by faithful hand through to the Federal lines. It was in vain to ask permission of the Confederate authorities, as they had apparently tried to keep everything relating to us from the Federal forces.

The erection of the scaffold began at Chattanooga, but on the next day the movements of the Federals had become so threatening as to produce quite a panic at Knoxville, suspending the Court-martial there and leading to the removal of everything which could be spared, further South. On the 6th of June, the day before that fixed for the execution, General E. Kirby Smith wrote no less than thirteen dispatches² from Knoxville in different directions, the general purport of which was that the enemy was advancing with overwhelming forces, and that Chattanooga would fall and East Tennessee have to be abandoned, and giving directions for lines of retreat and for removing the stores. Of course, to arrange for an execution on the 7th, in the face of an advancing enemy, might have led to a very sudden pardon; and accordingly Andrews and his companions were ordered to Atlanta once more, on the early morning train. There was again the excited crowds, an invariable accompaniment of our frequent transits over this road; but in addition the fact that Andrews was to die, was published, and he was taunted frequently with references to his approaching doom. These he bore with his usual calm, sad patience.

An instance in connection with these persecutions is especially pathetic. Whiteman came on the cars, and, advancing to where Andrews was,

¹ For further particulars see Chap. XXXVI.
² War Records, Series I., Vol. X., Part 2, pp. 592-597.

accosted his former partner. Parrott, who gives the account, was sitting on the seat behind, and could not help overhearing all the conversation.

The merchant said, "What can you do, Mr. Andrews, about that $10,000 I let you have for the purchase of quinine and other things."

Andrews replied, "Mr. Whiteman, this is no time to talk about money. If you had done as I wished you to do in Chattanooga, you would have had all that back, and twice as much more." (Parrott understood Andrews to refer to some proposition that Andrews had made to Mr. Whiteman on his trial, and the failure to accept which was the greatest disappointment that Andrews had then experienced.)

Whiteman continued, "Is that all you have to say, Mr. Andrews?"

"Yes, sir, that is all," responded the doomed man. With a gesture of deep disappointment, Whiteman turned on his heel and walked rapidly away.

The death procession reached Atlanta a little after noon, and the prisoners were conducted by their guard to a room used as barracks, two squares from the depot. Here they were kept under close guard awaiting the completion of the arrangements for the military murder. The foot chains had not been removed from Andrews, and as he walked up into his room with the short, halting step that they required, the clanking was horrible. Not very much was said in these few sad moments. Andrews did speak in his quiet way of the better life, and his wish to meet all his comrades in heaven. His words could not fail of making a deep impression, though hope of vengeance for the coming deed would have been sweeter to the poor boys than almost any kind of a prospect beyond the grave. But soon a body of strange soldiers came up to the building. Their commander entered and asked Andrews in a very respectful tone if he was ready now. The latter answered in the affirmative, and then bade "Good-bye" to the comrades who had passed through so many dangers with him. They were affected beyond the power of words, and could only vow vengeance—a vow made good on many a subsequent battle-field! They heard the clank, clank, of his chains as he walked slowly down the stairs. These chains, so far as we could learn, were never removed!

The procession moved out Peachtree street, the most fashionable and beautiful street of Atlanta, and continued for about two miles from the depot. On the way, the Provost Marshal asked Rev. W. J. Scott, a Methodist clergyman, to accompany them and act as Chaplain. He almost refused, but Andrews spoke in his winning, courteous manner, saying, "I would be glad to have you go, sir." Such an appeal Scott could not resist, and attended him to the last, writing many years after his recollection of the affair.[1] A great crowd, in addition to the strong guard, went along, but there appears to have been no unseemly taunts or disorder. To

---

[1] An Episode of the War: an essay in a volume entitled *From Lincoln to Cleveland*, by Rev. W. J. Scott, Atlanta, 1886.

Mr. Scott Andrews gave substantially the same account of the enterprise that has already been given, colored a little by the fact that he did not wish even in death to say one word that might in any way injure those comrades who had been so true to him.

No element of pathos in the terrible scene was lacking. A few scores of yards from the road, in a little valley, a scaffold was erected. There were thin woods around, and night was coming on. A rope circle fenced off the spectators to a respectful distance. Mr. Scott spoke the words that he judged fitting; Rev. Mr. Conyers [1] led in prayer; Rev. Mr. Connor administered some religious counsel to the patient prisoner, who probably thought that all the sins of which he repented were less than the sin of rebellion of which they were guilty. No coffin was provided, but a few hundred feet away the grave was already open. The signal was given, and the not uncommon bungling of an execution added new horrors. The cotton rope stretched so that the shackled feet reached the ground. "From motives of humanity" [2] the ground was shoveled away, and the soul liberated.

The pathos of this death is indescribable. The drop falls and the merely physical agony is soon over. The body, weakened by the last terrible struggle for life—made not so much for self as for the loving heart in far-away Flemingsburg—cannot long resist. Then the corpse is taken down; the horror-bound spectators still linger. The poor remains of a man of superb beauty and princely endowments are carried to the shallow grave on a little hill crest, and there, near a large stone, "which may mark the grave, if any friend ever wants to know where it is," as a spectator charitably said, he is laid to rest. There is no shroud. The only grave-clothes are the tattered garments left from the last sad race for life. Can the reader conceive anything more pitiful than the view presented just before the damp earth is thrown on the cold, upturned face. The busy brain from which came daring enterprises and cool action is quiet forever. The limbs that toiled so far for patriotism, fame, perhaps for vengeance, and at last for life, labor no more. The heart so true to country and comrades, so faithful under forms of falsehood, is stilled. The utmost depths of adversity have been sounded, and the enemies around can touch him no more. Even the welded shackles which seem to bind in the grave, have lost their power. It is well that man has one refuge from every earthly misfortune: and as evening gathered its shadows over the little heap of freshly turned earth in the wood—a spot long unrecognized—was he not better off than the comrades from whom he had just parted, or those more distant, whose fate was trembling in the balance at Knoxville!

---

[1] *Atlanta Southern Confederacy*, June 8th, 1862.

[2] Captain William A. Fuller's account in the *Sunny South*, 1877.

It is difficult rightly to estimate the character of this remarkable man. That he failed, and brought fearful suffering to others as well as himself, should not blind us to his real greatness; for success is often a happy accident. The manner of his death should not be permitted to cast backward too dark a shadow over his enterprise. In the mind of his countrymen, North and South, there will always remain one blot upon his memory: he sought his results by fraud rather than force. This was not felt so much while the passions of war were raging as afterward; even the enemy who put him to death would have been ready to accept with acclamation an advantage to their own side obtained in the same manner. And standing by the desolate grave, that man must have a cold heart who can utter only words of criticism and condemnation. Certainly the present writer, who spent so many days and nights with him in the same dark cell, has no reproach to speak.

Andrews was noble, brave, refined, courteous and true. The latter epithet may seem strange as applied to one whose trade was in falsehood, and whose perfection in the art of deceit was a constant source of wonder. But deceit was exercised in one direction only. There was no one of our party who had not the most absolute trust in him. If our lives had depended upon his fidelity, we could have gone to sleep without a single tremor. In this our creed was a little like that of the American Indian, who bounds his virtues by his own tribe and thinks they have no application outside. We would have accepted a statement of Andrews, *given in the absence of the enemy*, as conclusive on any point; if the enemy were present we would have given it no value, till we had seen whether he had any motive for deceiving. This is well illustrated by the price Andrews was said to be promised for the work he tried to do. It pleased him to represent it to the enemy as a purely financial transaction, and as of little military importance. He never made any such representation to us *when alone*. Had he told us the same things privately, it would have ended all controversy.

Some days or weeks after the completion of this mournful tragedy,[*] a man came to the old depot at Stevenson, Alabama, which was then used as a store-house by the Federals. He seemed to be a stranger, and went cautiously up to Sergt. Wm. Hunter Myres, of Co. K., 33rd Ohio, and asked to speak with him alone. Myres at once assented, and took him into the room. The man looked to see that no one was near the door or windows, and then said, "I have papers in my possession, which would cost me my life if the rebels should discover them on me. I want to get clear of them." Myres took the papers, and glanced over them, finding the letter of Andrews to his mother, and his "will," already quoted. He was perfectly familiar with our expedition, belonging to the same Company as Parrott, and indeed was spoken of for that place himself. This made it

---

[*] This account was received directly from Mr. Myres by the writer.

easy for him to recognize the great interest of the papers, for up to this time only scattered and partial information had been brought through the lines. On inquiry the man said he was a fireman on the Georgia State Railroad, and that he had been employed for several years in that capacity. His native place, however, was Hagerstown, Maryland, and he had stood the ways of the rebels as long as he could, and was now anxious to get back home. Myres wanted to know how he came in possession of the papers, but he declared that he dared not tell. Finding that he had nothing more to say, he was sent under guard to Huntsville, from which place it was easy for him to reach his old home; and the papers also, after considerable detention, arrived at their destination.

The account of the escape and recapture of Andrews was published in the Cincinnati *Commercial*, about the tenth of June, and reached the sister of Miss Layton, with whom that lady then made her home. As she was already in deep distress because of Andrews's long delay without any message, they did not dare to tell her the perilous situation in which her lover was placed. But near the end of June, the full account of his execution was copied in the same paper from the *Southern Confederacy*. As the end of all her hopes had come, (less than a week before the intended wedding-day) her brother and sister judged it best not to keep her longer in suspense, and the paper was handed to her. Her eyes rested on the following paragraphs:

" Yesterday evening's train brought from Chattanooga to this place to be executed, Andrews, the leader of the engine thieves, under sentence of death, convicted by court-martial of being a spy. He was carried out Peach Tree St. road, accompanied by three clergymen, and escorted by a guard. A considerable crowd followed to witness the execution.

" He was a native of Hancock Co., Va., born in 1829, brought up by pious Presbyterian parents, who now reside in Southwestern Missouri. A good portion of his life had been spent in Fleming Co., Ky. He had no family, but was engaged to be married during the present month."

She did not shriek or cry out, but read it through to the end and went silently to her room, from which she did not emerge for hours; and when she did rejoin the family her face was drawn and pale, and the light had gone out of her eyes. From this time forward she took little interest in anything until the letter to Mr. McGavic, printed above, arrived. Many months after, the empty trunk, so pathetic an emblem of her blasted hopes and the great tragedy that had fallen on her life, was recovered. In the absence of any explanation, for the letter to her was never received, it seemed like a cruel mockery ! Not long after she died, thus rejoining the man she had loved so faithfully through such hopeless sorrow. No brave man perishes that some tender woman's heart is not crushed !

' *Southern Confederacy,* June 8th, 1862.

CHAPTER XXII.

KNOXVILLE.

THE twelve of us who had been separated from the others at Chattanooga were escorted by Col. Claiborne to the cars, ironed as
usual, and committed by him to the care of a band of Morgan's
celebrated guerillas, with the charitable injunction: "These are men,
like other men, and gentlemen too, and I want them treated as such.'
We parted from him regretfully, for his kindness was rare and precious.

Morgan's men were well dressed in citizens' clothes, (for they were not
always uniformed, even in the enemy's country,) and treated us kindly.
They were equally liable with ourselves to be held as spies, and probably
did not feel that our having been captured in citizens' dress was such a
deadly offense! To see us in irons made them very indignant, and they
vigorously denounced their own government for such an outrage, but in
the face of their orders could not remove them. We had started as usual
without rations, on the calculation that we could starve through, for the
time should not have exceeded ten or twelve hours. But the trains in the
South then ran slowly and were often delayed. It was not at all surpris-
ing that we were nearly a day and night on the way. But our guerillas
would not permit us to suffer, buying us pies and all accessible luxuries,
telling us that they had plenty of money, and that when it was gone they
could easily get more from "the Yankees." We hoped that we might
always have Morgan's men as an escort when moved from one prison to
another.

A little after noon the next day we arrived in Knoxville and were
lodged in the old jail. It was a square and massive building, far stronger
than any jail we had occupied. At this time it was used as a military
prison and was filled from top to bottom with dirty ragged prisoners.
Most of these were Union men, but some were deserters from their own
rebel ranks. These constituted the lower class of prisoners, and were
permitted to range over most of the building, which was completely encircled by a strong guard

The higher (or more dangerous) class were shut up in cages like that
which I had found at Lafayette. There were five of these. No doubt
was entertained as to our classification, and two of the cages were at once
emptied for us. One was about seven by nine, and held four, the other
was possibly ten by twelve, and eight filled it quite full, though there was

no such crowding and suffocation here as under the dominion of Leadbetter and Swims at Chattanooga. In this larger cage the noted Parson Brownlow had been confined for a time.

It was now May and the weather outside was warm, but inside this large building it was cool—indeed some nights were too chilly for comfort. We did not receive any new clothes or blankets, and were little prepared for even moderate changes of temperature. But suffering from this source was not serious and the time passed not unpleasantly. We could talk with outside prisoners through the bars and get many an interesting story. Although newspapers were forbidden, good friends would often bring them to us, and we were kept fairly well informed. This place was a great improvement on any which we had endured, and we spent the days in comparative pleasure, and in a great degree of hope.

In looking back over this period such hopefulness seems wonderful. I had but little fear of the result. The chain of reasoning by which I had demonstrated that it would be unprofitable for the rebels to hang us was very plain—possibly because I wished it to appear so! We were visited by Confederate officers who took a great deal of pains to confirm our hopes. Whether this was because as they grew to know us they did not wish us to die, or whether they wanted to keep us in good heart that we might not make desperate efforts to escape, I can not positively determine. Even the Judge Advocate seemed to agree with all our hopefulness. So far as he and the officers of the court who visited us are concerned, I cannot acquit them of a deliberate intention to deceive, of which very strong documentary evidence will be presented shortly. In fact there is one phase of the work of this court-martial upon which no Confederate even can look without sorrow and indignation. But more of this anon.

We here formed the acquaintance of a few Tennesseeans who remained with us during our stay in the South. Peter Pierce was a remarkable man, some sixty years old, who had received a stroke with a gun-barrel, right down the middle of his forehead, which, even after healing, had left a gash more than an inch deep. From this he was familiarly called "Gunbarrel," "Forked-head," etc. He was both very religious and very profane. At one moment he would be singing hymns and the next cursing the Confederacy in no measured terms. He was very generous to Union soldiers, whom he almost adored.

Here it was that we first learned to know one of the noblest men in the world, though as yet we could not see him. Captain David Fry was in solitary confinement—that is, he was kept in a cage by himself—and we frequently wrote little notes to him on the margin of newspapers, and were sure of courteous replies. Afterward he came to be virtually one of our number, to which position we were the readier to admit him, as he also had been a bridge-burner, and far more successful than ourselves.

I also became deeply interested in an old man in an adjoining cage, who was awaiting sentence of death. He had been known as a Union man, and one night a band of three Secessionists came to rob him. He resisted, and they attacked him with pistols and bowie-knives. They chased him for some time till, in dodging around some barrels in an out-building, he got hold of a pitchfork and plunged it into his foremost assailant, and then escaped. The robber died, and the old man was arrested on a charge of murder. I never heard the final result, but think it probable that he perished.

We received with great pleasure a paper containing an account of the escape of Andrews. The next day we reached equal depths of regret on the news of his recapture, though we did not hear of his sentence. But on a day not long after, a paper was handed to us by a man who turned away with a face so sad, and action so significant, that we knew it must contain heavy news. A few moments justified the fear. It was the full account of the execution of Andrews! We had been engaged in all kinds of games and story-telling, for we were always merry, and never cherished gloomy forebodings. But this news hushed all noise and merriment, and we passed the whole day in the most heartfelt mourning for our leader. Yet we did not give up hope for ourselves, for we had always understood that his case was more serious than our own.

Judge O. P. Temple. From a photograph.

The amount of provision we received here was very small, and we suffered much from hunger. A little alleviation was provided by the kindness of outside friends, who gave us a small amount of money with which several loaves were purchased—though the price was high. Ross, who was a Freemason, was able to get some help in that manner which he was very ready to share with his comrades. I also met with a great piece of good fortune. Before leaving Chattanooga I had asked the Captain of the guard if he could not borrow a law-book from some lawyer for me that I might have it to read during the terribly long days; he promised to ask, and did so. Very much to my surprise, he brought the book. When we were moved it had to be returned, and I thought that my prison law-studies were ended, as the same thing could hardly occur a second time. But from some of the Union men of Knoxville I heard that Judge Temple was a most kindly and liberal man of Union sentiments. I put the matter to a practical test by sending him a note asking for a copy of "Greenleaf on Evidence." It came promptly, and soon after I had a visit from the Judge himself. The opportunity for studying law was a grand one. I could make long hours and corresponding progress! At first this new pursuit afforded no small amusement to the prisoners outside, to whom the sight of one of the most "desperate prisoners" shut up in a cage, yet wearing spectacles and reading law from a huge volume by the hour—sometimes aloud, when any body would listen—was exceedingly ludicrous. But I had a double profit in gaining knowledge, and passing the terrible hours pleasantly.

We were soon visited by Captain Leander V. Crook, Judge Advocate, and probably other members of the court-martial, and notified of the trial. He told us that we could employ counsel if we could find any one to serve. I at once thought of my friend Temple who had loaned me " Greenleaf, and although it was not a very good return for his kindness, to put upon him a laborious and possibly dangerous task, yet I could not forbear naming him. Temple was promptly

JUDGE BAXTER. From a photograph.

sent for, and I think it was on his suggestion that we also sent for Col. Baxter. Both these men have since attained to high positions on the bench, and were of first-class ability. We could not have done better had we been able to pay a magnificent fee. The safety of our attorneys demanded that the whole matter be put on a purely professional basis, and they accordingly asked us to sign a note for one hundred and fifty dollars each. The prospect for obtaining the money was not good; but it was never intended that the notes should be collected.

When we learned that it was the intention of trying us one by one we protested that this was useless, as our cases were precisely alike. If one was guilty, so were all; if one was a prisoner of war simply, the same was true of all the others. But it would probably have looked too absurd to

put twelve men on trial as spies at once, and our request was refused.
We then asked that one be tried and the result in his case be accepted for
all. We also offered to tell just what we did, thus saving them all trouble
about proof, for we did not wish to deny being on the train, or engaged
in its capture. But they gave us a clear intimation that they knew their

The old Court House in Knoxville where the Court-Martial was held.

own business best, and we were obliged to take it in the manner of their
choosing

The nature of the charge against us gave some uneasiness. It was of
being spies, and of lurking about Confederate camps as spies, and hinged
only on our going South without anything being said of the capture of
the train, or our return. This was suspicious; but we were led by their
explanations to think that possibly they only wanted to get it into such a
form that we could consistently say, "Not Guilty," without denying what

we really did. But afterward we found the intent far more serious. The charge was of violating a certain section of their Articles and Rules of War; and the specifications were two—first, coming to Chattanooga, and lurking as a spy about the Confederate camp there; and second, going through Dalton and Camp McDonald to Marietta, and lurking as a spy in those places. There was not actually any lurking about any of these camps or posts. We only sought to get through as fast as possible.

But military law is very stern and summary. When men are making it their duty and ordinary employment to kill each other, it is not to be expected that they will stand long upon the dispatching of an enemy in their power if fully convinced that it is on the whole to their advantage to do so. What we have good reason to complain of in this case is not that some of our number were put to death—that was very probable from the first. But nothing could justify the atrocious rigor of our long imprisonment, the manner in which false hopes were encouraged, or the awful suddenness by which sentence was executed. No apologist for these things has been found *on the Southern Side*.

In three things military law differs widely from civil. There is no challenge of jurors; the laws of evidence are very loose, and a two-thirds majority convicts. These things go far to explain the deadly character of courts-martial, which are in their very nature "organized to convict." But we did not understand this so well then as afterward.

The trials began. One of our number was taken out, the charges and specifications read, a few witnesses heard, and then he was returned to us. The next day another comrade was treated in the same manner. In no case did the proceedings occupy any great period of time, and as each one was a mere repetition of those that went before, the members of the court soon became very inattentive. The only real question was one of interpretation, as there was no dispute about facts. Did the mere fact of our having come into the territory under Confederate authority, in ordinary dress, render us worthy of death? And if so, would it be to the interest of the Confederacy to inflict the extreme penalty, and in what time and manner? A tabular view may perhaps put the whole case as it was presented to the court more vividly before the reader. These items only are given which were brought to the knowledge of the court, as no others could have had an influence on their verdict.

Against us.	*For us.*
1. Though Federal soldiers, we were without uniform.	1. We were detailed, without our knowledge or consent, for what we believed to be a purely military expedition.
2. We had passed Chattanooga, Dalton, and Big Shanty, where were rebel troops; and reached Marietta, where there was a rebel military school.	2. We passed no rebel pickets and did not enter any camps.
	3. We had a military object, which we ex-

ecuted as promptly and quickly as possible

4. We dressed like citizens to avoid
alarming the citizens of the country we
passed through

5. When taken before regular Confederate authority, we claimed to be soldiers,
giving our regiments correctly

6. That many Confederates who were
captured within Federal lines when not in
uniform, were yet held by the Federals, and
that it was unwise to provoke retaliation.

7. That many orders had been issued by
Confederate commanders, calling upon citizens to burn bridges within the enemy's
lines. Some who had obeyed were now in
Federal hands.

8. That the Confederate Government
had expressly authorized their citizens to do
as we had done.

A great part of the testimony above in our favor depended upon our
own admissions, but as this was adduced by the enemy and not contradicted, they could not know how much of it was overdrawn—especially in
reference to our ignorance, and the involuntary character of our service.
Judges Baxter and Temple were far abler men than any on the court, and
therefore managed to arrange the testimony in the best possible manner.
They declared to us that all the plans of the prosecution had been deranged by our course and confidently anticipated a verdict of acquittal.
But alas! they were more familiar with civil than military procedure. The
same thing was intimated by officers of the guard and by members of the
court itself.[1] It is no wonder that we were confident and believed that a
protracted imprisonment—perhaps still in chains—was the worst of the
evils we had to fear.

No one of the prisoners was allowed to hear the pleading of counsel
for or against him, and each case was completely concluded the same day
it was begun; but the decision was reserved that all might be rendered
and approved together. I have since been told by leading Confederate
officers that they grew very tired of the trials, feeling a deep compassion
for us, and were rejoiced to have them interrupted.

George D. Wilson related an incident that occurred while he was on
trial, which showed how hard they were put to their wits to secure effective
evidence. A young lieutenant volunteered to testify as to at least one
place where we had passed a Confederate picket line. When put on the

---

[1] Possibly those who talked in this manner were sincere, but were outvoted in rendering the verdict.

stand, he declared that we passed the guard at the river ferry at Chattanooga on the evening of our first arrival. This was very good, but immediately the president of the court arose, and said that the young gentleman was badly mistaken, as he, himself, had commanded the guard at
Chattanooga that day, and no sentinels were stationed at the ferry. This
raised quite a laugh at the expense of the lieutenant, who did not volunteer any more testimony !

The request to hear the pleading of counsel was made by us and de-

The Court-Martial.

nied. This was the more strange because Andrews had been allowed that
privilege; but in this, as in several other particulars, the soldiers were
treated more harshly than their leader. After several of the trials had
passed, Judge Temple visited us in prison and read the plea which had
been composed by Judge Baxter and himself, and read on each trial. It
was an able paper and worthy of their subsequent fame. They contended
that the whole case against us consisted in our being dressed in citizen's
clothes instead of our regular uniform; that this was nothing more than

what Confederates frequently did, sometimes from necessity and sometimes for their own advantage; that they had many regiments in service that were not yet in uniform, and that they had expressly encouraged guerilla bands to raid Federal communications, who by every interpretation of law should fare worse than we. And they cited the instance of Gen. Morgan having dressed his men in Federal uniform and passed them off as part of the 8th Pennsylvania Cavalry, by which means he succeeded in reaching and destroying a railroad. Some of these men had been captured by the Federal Government and treated as prisoners of war. They stated further that we had freely and plainly told the object of our expedition, which was purely military, for the destruction of communications, and as such authorized by the usages of warfare. Judge Advocate General Holt pronounced this "a just and unanswerable presentation of the case." But we never could tell how the Judge Advocate answered this; probably he did not wish his paper to be brought into comparison with that of Judges Baxter and Temple !

Each day began and finished the trial of a man. I do not remember the order in which they came. There seemed to be no rule followed, and probably they were taken just as they happened to come on the prison list. The table in the court room was covered with newspapers, bottles and novels. For a mere formal trial this was well enough; and I do not suppose that it was intended to be much more, or that the members had any doubt as to the result from the beginning.

But the Confederates were not destined to carry out their intention of bringing us all to trial, one at a time. Indeed, it proved most fortunate for some of us that they would not accede to our request that all be tried at once. General Mitchel had sent Gen. Negley with a considerable force toward Cnattanooga, and thus produced the panic of the 6th of June to which reference has already been made, and which caused the scaffold of Andrews to be transferred to Atlanta. In such an emergency, the officers composing the court-martial hurried to their regiments. Kirby Smith expected the immediate fall of Chattanooga, and of the rebel dominion in Tennessee. Had Mitchel and not Buell been in the chief command, he would not have been disappointed; but as it was, this advance, which Buell strongly disapproved, saved some lives. Probably it was expected that as soon as the danger had been repelled, or matters had been settled in a new department, the trials would be resumed.

But when the Confederate forces, with Smith himself, were at Chattanooga to repel the danger there, another as formidable arose nearer home. The Federal Gen. Morgan,[1] who had long been besieging Cumberland Gap, took advantage of this diversion, to turn that position by a difficult flank movement through the mountains, and to threaten Knoxville itself.

---

[1] War Records, Series I., Vol. X., Part 1, p. 57.

The first movement stopped the trials; the second rendered our speedy removal necessary.

A strong guard came in one morning, accompanied by two or three men bearing large bundles of ropes. These were suggestive and somewhat alarming. Some supposed that we were to be taken out for immediate execution. But we soon learned that it only meant another change of place. The guard were kind enough to explain that our irons were used in sending some prisoners to Richmond, so they were obliged to use ropes for us. They tried, however, to make up for any defect in the material, by a little extra liberality in amount. They first tied our hands very tightly together; then fixing our arms securely in the loops of long ropes, bound them down firmly to our sides, after which we were coupled two and two. We had never been accustomed to so much solicitude in our own country, and found it always a novelty at each change. I should have mentioned that when placed in the cages our irons were removed for the first time by the Confederates—one indication of hopefulness that was made the most of by the more sanguine members of the band

There was now a considerable difference of opinion in our party as to our future prospects. Wilson was hopeful and sure that we could not be convicted. The greater number of us sided with him. There was one notable exception. Marion Ross had not been sanguine from his first entrance on the raid. But now he was quite gloomy, without being able to assign any reason for it. The only cause I can conjecture for his not sharing our hopes is the following, and it is pure surmise. Ross was our only Freemason, and he was tolerably well advanced in the order. The Judge Advocate belonged to the same fraternity, and also many members of the court and other officers that visited us. It may have been that some of these gave him an intimation "on the square;" he could hardly have failed to ask them—and I understand that they would be bound to answer, and even if they did not, their silence would be significant—as to whether we were likely to be acquitted. If he did receive a hint in this way which he was not permitted to share with his comrades, I can scarcely conceive a position more pitiable. Often when we were most boisterous in innocent sport during those last weeks, he would look around on us with a sad air, as if he could tell a story that would overturn all the hopes we were resting upon

While the guards were arranging the abundant cotton rope they had bestowed upon us, I had an amusing passage-at-words with the adjutant who was superintending the operation. I said to him as politely as I could,

" I suppose, sir, our destination is not known?'

" It is not known to you at any rate, sir," was the somewhat gruff rejoinder.

There was a little laugh, and I felt rather beaten; but a moment later

came my chance for revenge. He turned again to me and said, in a dicta-
torial manner:

" Who was it that run your engine through ? "

I bowed and returned in the blandest manner, but loud enough to be
heard by all in the room: " That is not known to you at any rate, sir."

There was a hearty laugh, and the Adjutant, reddening to the eyes,
turned away, muttering that he believed I was the engineer myself !

A number of East Tennesseeans were removed with us. Among them
was Captain David Fry, and I now had the opportunity of learning something of his most eventful history. Early in the fall of 1861 he had gathered
a company of his Union neighbors, and under his guidance they had run
the gauntlet of guarded roads, and rebel scouts, till they reached the
Federal camp in Kentucky. He was here elected Captain, duly commissioned, and served for some time at the head of Co. F. 2nd. Regt. East
Tennessee Vols. In October of the same year he was sent by Gen. Carter
back into East Tennessee, for the purpose of burning the bridges on
the great East Tennessee Railroad, preparatory to a general Federal advance for the deliverance of that section of the country. The government had then determined on such an advance through Cumberland Gap,
and as our own Gen. Mitchel, who was in command at Cincinnati, had urged
the scheme, with the offer to raise all the necessary troops in Ohio and
personally lead them, there seemed, to those who knew his energetic
character, no reasonable doubt of success. The force at the disposal of
the Confederates was not large. But the fear that Mitchel might by
marching across the department of Kentucky, then under other command,
offend delicate military sensibilities, caused the permission which Mitchel
had received for such an advance to be withdrawn, and for nearly two
years nothing effectual was done !

But Fry, with all assurances of support, was far on his perilous way,
and knew nothing of this change of plan. He soon had collected enough
of his neighbors, who had perfect confidence in him, to do his work
effectually. The great bridge at Strawberry Plains was burned with many
others, and all reinforcements by rail were cut off. He held a considerable mountain district in Tennessee and North Carolina for many weeks;
but at length superior rebel forces were accumulated, and forced him to
retreat. Some of Fry's comrades then returned to the Federal force that
lay not far from Cumberland Gap; possibly he might have done the same,
but this would be to leave those who had assisted him in the work of destroying the bridges to be hunted down by the rebels and to abandon their
wives and children to the tender mercies of their enemies. He, therefore, believing that the time would be short, resolved to remain, and hold
the country in spite of adverse fortune till the Union army came. All
this time the state was open to Federal invasion and was defended almost

solely by Federal military etiquette! The struggle of Fry was heroic. A number of sharp skirmishes were fought, but the rebels sent reinforcements as fast as their severed communications would allow. During the whole winter Fry kept the field, on one occasion seriously threatening Knoxville. In March, however, he despaired and determined to abandon the unequal struggle. With some six hundred men he tried to fight his way through to the Federal forces. At the foot of Cumberland Mountains, about ten miles from Jonesboro, he was attacked by an overwhelming rebel force, and defeated, his men scattered, and himself badly wounded. Soon after he was captured and taken to Knoxville, where he

would have been hanged had it not been for a very vigorous letter¹ with threats of retaliation from Gens. Morgan and Carter. He had thrown a citizen's coat over his uniform in his last effort to escape and this was held to constitute him a spy. He was kept in the most rigorous confinement until removed with us from Knoxville, and for months afterwards he shared our fortunes. The hardships endured in this whole adventure were indescribable. His hair and beard were black when he left Kentucky camp, (as shown in the accompanying photograph,) but when I knew him a few months later both were almost white.

CAPT. DAVID FRY. From a photograph.

On this long journey from Knoxville directly to Atlanta we carried no rations, and as we now had no guerilla friends to supply the neglect of their officers, we were obliged to *fast* through. A comparatively weak guard was sent along, as the pressing emergencies of the Confederate service which had compelled our removal also made it difficult to spare heavy guards for the prisoners being removed in different directions. Our attendants had barely provisions enough for themselves, and did not, therefore, feel that they could share with us. On the way the populace— it was usually the worst elements of the towns that would thus gather

¹ E. Kirby Smith to Gen. Carter, April 19th, 1862. War Records, Series I., Vol. XVI., Part 2.

around the stations—taunted us with Andrews's death, and charitably hoped that we would share the same fate. But officers talked with us in a different manner, assuring us that we would not be hurt. This threw us off our guard, dividing our counsels, and causing us to lose a most favorable opportunity to escape,—the very effect, I am confident, that it was designed to produce. The Captain who now commanded the guard was a simple, honest-seeming man, who talked freely and assured us that we were in danger of nothing worse than being kept close prisoners until the close of the war. He had heard, he declared, high officers at Knoxville, and also members of the court-martial say that it was required by public policy to put to death Andrews, who had deceived them so terribly, but that not one of the enlisted soldiers would suffer. I do not think that he was insincere, but that he had been told these things that he might impress them upon us. Does the reader think that I am uncharitable? Let him closely read the sentence of the court, printed on a subsequent page, especially the clause requiring death to follow immediately on publication— a clause approved, with all the rest, by E. Kirby Smith, and then judge for himself!

Ross and myself favored the attempt to escape, but for different reasons. I was deceived by the assurances of the enemy as completely as Wilson; but I dreaded the continued imprisonment, and was willing to take a good deal of risk to escape that; Ross would say nothing, except, "Let us go if we can;" Brown and Knight, always bold and fearless, were ready for the effort. But Wilson, who had great ascendancy over us, had not recovered from an injury which would have rendered his own escape difficult; he thought some of us would be sure to die in the attempt (which was likely): while if we did nothing, we would soon be released by the collapse of the rebellion, which he thought very near, as he still had faith in McClellan (which I had lost). We were all night on the cars, and toward morning the guards were nearly all asleep, so that by watching our chance, and giving a word here and another there, I managed to know the sentiments of nearly all our party. Had the word been given, even those who opposed the scheme would have responded as well as the others. Though so well tied, it was with new cotton rope, which stretches greatly, and by shifting around and "settling" in it, the most of us had our fastenings in such a condition that we could have been practically loose whenever we chose. The guards, with their muskets, were in the same seats with us, and we would have had the muskets in an instant, and they would have been at our mercy. I never felt more anxiety to undertake an enterprise. Capt. Fry also favored it, but the opposition of Wilson was too strong. Alas! he was throwing away his last chance of life, and knew it not! All these discussions and proposals took place north of Chattanooga and in the night, where we could easily have uncoupled our car, and, with our muskets, have struck at once into the mountains of East Tennessee.

When morning came and we passed south of Chattanooga we felt that the opportunity was gone.

Finally we reached Atlanta. Here Andrews had perished but a few days before, and it was to be the last earthly abode of some of our number; and from it none of us were to escape without great perils, and long and wearisome imprisonment. Before we left the cars a man calling himself the Mayor of Atlanta came up, and began to insult Captain Fry, telling him that he knew him well, that he was a great rascal, and that he hoped soon to have the pleasure of hanging him. Then turning to us, he boasted that he had put the rope around Andrews's neck, and was waiting and anxious to do the same for us! I was afterwards told that this man's name was Col. O. H. Jones.

From the depot we were marched quite a long distance. The afternoon was delightful. The insulting mob did not follow us, and as we looked upon the beautiful residences which we passed, everything seemed so calm and peaceful that it was difficult to realize our perilous position.

The city jail to which we were taken was a large square brick edifice, smaller than that at Knoxville, but still of considerable size. The lower story was occupied by the jailer and his family, while the four rooms in the upper story were devoted to prisoners. Captain Fry with our party occupied one of these; the Tennesseeans, brought at the same time, were put in the room just across the entry from us. The eight comrades who had been separated from us at Chattanooga were already in another of these rooms; and the fourth had various tenants, sometimes being occupied by negroes who had been allured from their lawful masters by the North Star.

At first we were comparatively well fed here. Mr. Turner, the jailer, was a Union man at heart and sympathized with us; but he was soon suspected of being too favorable, and an odious old man named Thoer was put with him in the double character of assistant and spy. The change was at once manifest. The food came down to the starvation point, so that there was reason to fear that we might actually perish with hunger; and all those little attentions which prisoners put such a pitiful value upon were withdrawn. This made our condition worse than it had been in Knoxville; but we had been trained in a thorough school as far as hardships were concerned, and made no complaint. Indeed the first week we spent here was marked by comparative quietness and hope. There was no word of any further court-martialing, and those of our number who believed our lives would be spared made converts seemingly of all the rest —all but poor Ross, who looked so mournful and hopeless, while refusing to argue on the subject, that we sometimes thought that our terrible hardships had possibly somewhat affected his mind.   If he had received Masonic intimations of danger, which he was bound not to impart to us, his gloom and depression are easily explained.   But the overthrow of all our hopes was at hand.

CHAPTER XXIII.

A DAY OF BLOOD.

AFTER the lapse of twenty-five years my hand still trembles as I endeavor to copy the account of the most fearful tragedy of the civil war. The young men who died on this day of horror were from twenty to thirty years of age. They were guilty of no crime; for their offense in its real essence was only that of being willing to meet danger for their country's sake. Had they been selected to lead a forlorn hope in storming the enemy's entrenchments they would have felt that duty forbade them to decline the perilous honor. They were selected for work of great peril, and as good soldiers they did not shrink from it. In the *manner* of performing the duty upon which they had been sent, there was nothing to which the enemy himself could find the slightest objection. Indeed, Confederate officers had frequently gone out of their way to express admiration for the dauntless bravery of the men whom they now sought to overwhelm with dishonor in death. To have died in the execution of their task would have been sorrowful, but only an incident of war. Death in the moment of passion and vengeance following capture would have seemed far less terrible; but to first keep them so long, to encourage hopes that were only doomed to disappointment, and to cause sufferings of such fearful character in imprisonment, has not been justified.

Yet it must be distinctly understood that the people of Atlanta were in no degree responsible directly or indirectly for the tragedy. So manifest were their sympathies on the other side that no account was allowed to be published. Yet the whole city was filled with the fame of the deed and with emphatic condemnation of it. This feeling was so general that Provost Marshal Lee referred, in an official report [1] five months after, to "the great number of sympathizers." But we will hasten through the fearful narrative.

The 18th of June was a bright summer day. Our party in the jail were making merry with games and songs, utterly unsuspicious of evil. But one of our number, looking out of the window, saw a squadron of

---

[1] War Records, Series I., Vol. X., Part 1, p. 639.

cavalry approaching and called attention to it. There was nothing unusual about this, for we often noticed bands of troops on the streets; but they now halted at our gate and surrounded our prison. This was unusual and startling.

The doors downstairs opened. We heard the shuffle of feet in the hall, and the clink of officers' sabres as they ascended the stairway. We held our breath in painful attention, while they paused at our door, unlocked and threw it open, and then one of the number, stepping before the others, read the names of our seven comrades who had been tried at Knoxville. They were ordered to respond and stand in a line before him, which they did. Robinson was sick with fever, but a guard assisted him to rise, and he stood with the rest. Then they were all told to follow over into the opposite room, while the Tennesseeans there were brought in return to us.

With throbbing hearts we asked one another the meaning of these strange proceedings. Some supposed our comrades were about to receive their acquittal; others, still more sanguine, that they were to be parolled, preparatory to an exchange. But we had no confidence in these suggestions even while we made them. It would not have been necessary to surround the prison for such purposes; and the faces of the officers who had entered our room were solemn and stern.

I was sick, too, having suffered a good deal recently with malarial fever, but rose to my feet oppressed with unutterable fear—the most deadly I ever remember feeling. A half-witted fellow who had been put in with the Tennesseeans, came to me and wanted to play a game of cards! I had been fond of the game, but never played it after this day! Now I struck the greasy pack from his hand and bade him leave me.

From over the way we heard the sound of voices, muffled and indistinct because of the two iron doors between; then the opening and shutting of doors, the passage of several persons up and down the stairway, and last the sound as of solemn reading.

A little while after—I cannot judge of the length of time spent in such fearful agony—the ministers in the other room think it must have been more than an hour—the door opened, and our comrades came back, one by one, but the change in them was fearful! My own friend, Geo. D. Wilson, was leading, his step firm, and his form erect, but his hands firmly tied, and his face pale as death. "What is it?" some one asked in a whisper, for his appearance silenced every one.

"*We are to be executed immediately,*" was the appalling reply, given in a low tone, but with thrilling distinctness. The others followed him into the room, all tied ready for the scaffold. The officers were standing in the door, and barely granted them the privilege of taking us once more by the hand before death. Then came the farewells, hopeless in this world! It was a moment that seemed an age of measureless, heart-breaking sorrow.

What had occurred in the other room while we were separated? The narrative[1] of the ministers will make that plain.

Rev. W. J. Scott was requested by Col. G. J. Foreacre, then Provost Marshal in Atlanta, to visit some Federal prisoners at the city jail who were about to die. On his way, Mr. Scott called on Rev. Geo. G. N. MacDonell and asked him to go along. At the jail they were taken into the room where our comrades were. Scott says:

"They impressed me at once as a body of remarkably fine-looking young men. I could but notice also their cheerfulness under such painful environments." He told them that he was the bearer of unwelcome tidings. This arrested their attention, but they were still unprepared for the blow that followed. Then Scott with the brevity which was the best kindness, with a few questions answered, gave the full truth, every word being like an added stab; telling them that they had been found guilty at Knoxville—of being spies—that they were to die—to die by hanging—and at once! Their natural and indignant protests were waved aside as something with which the ministers had nothing to do; their only business was to help the doomed men by prayer and counsel to

Rev. W. J. Scott. From a photograph.

prepare for death, and the hour was at hand. Anxiety and even horror was in an instant depicted on every countenance. When they asked "How soon?" he answered, "In less than two hours." This was probably a merciful over-statement. The hearts of the preachers upon whom had been rolled the fearful task of first communicating this terrible intelligence, were very heavy. Scott adds—"They were gallant men, who would have stood unshaken in the imminent deadly breach. They were picked men, chosen for their soldierly qualities; yet in a moment every cheek blanched to the lily's whiteness. In another moment, however, they rallied and appeared firm and unflinching." Scott and MacDonell

---

[1] "An Episode of the War"—a chapter in a book published by Rev. W. J. Scott, in Atlanta, 1886, entitled *From Lincoln to Cleveland.*

then gave them such counsel as the dying need, recited to them appropriate Scripture passages, and prayed with them.

What followed is so extraordinary that it is fully given in Mr. Scott's own language with only two remarks. The "few hours' notice" was virtually no notice at all, as according to Mr. Scott's own words, all the time was taken up with clerical and official preparation. From the moment the awful news was communicated there was no pause save for the prayer of the minister, the reading of the sentence, the binding for the scaffold, and the clasp of hands with friends. This was all.

We had often said to each other that, no matter who else might perish, Ross in some way would escape by reason of his high standing as a Mason. Probably the following narrative shows better than anything else the fearful resolution with which this deed of blood was carried through to the utmost extreme. Mr. Scott continues:

"As we rose from our knees one of them—I am not sure at this late day whether Ross or Campbell—gave me a Masonic signal which craftsmen are only permitted to use in seasons of supreme peril. I recognized it instantly and took him aside and satisfied myself that he was a 'son of light.' No one who has never been raised from a dead level to a living perpendicular can appreciate my feelings. I said with a faltering voice:

"' My brother, I will do what I can for you consistently with my obligations to the government to which I owe allegiance.'

"He replied: 'I ask for nothing more. We are about to be executed with only a few hours' notice. We had no intimation of it until you informed us. Now, can you not prevail on the military authorities to respite us one or two days?'

"I replied: 'I will make an honest effort.'

"The other prisoners must have heard a portion of the conversation, for they seemed quite elated.

"I knew that I must act promptly, so leaving brother MacDonell to talk with them, I left the cell and went down into the front prison-yard where a squadron of cavalry were already drawn up. They had, I found, been waiting quietly for our appearance. Colonel W. J. Lawton had on that day assumed command of the post. He was an old and highly esteemed personal friend. I told him what had transpired in the cell, and urged him to respite them at least until the next day; that to execute them on such short notice would be utterly indefensible; that he could easily cut off all possibility of escape. He was a man of generous impulses, and I saw he was greatly troubled and perplexed. He replied:

"' I agree to all you say. I would most gladly afford them relief, but,' he continued, 'my orders are peremptory. I am required to execute them to-day and have not the slightest discretion. If I disobey my orders I am liable to be cashiered and disgraced.'

"He proposed to show me his orders, but I told him his statement was sufficient.

"I was compelled to return and announce my failure. I was then asked if I would transmit some messages to their friends. I said certainly, if the military authorities would allow it. They then dictated their messages, brother MacDonell writing three in his memorandum-book and I writing four in mine. There were but slight verbal differences in their messages, and the following may be taken as a sample of the whole:

"' I am to suffer death this afternoon for my loyalty. I am true to the old flag and trust in God's mercy for salvation.'

"The name of the party and number of his regiment was attached.

" The messages were not sent because of some technical objections at the War Department." [1]

Immediately after this failure to get the least respite in the inexorable orders, the officers read the sentence of the court-martial. It is here appended, and will explain the feverish hurry of the executioners:

SENTENCE OF THE COURT-MARTIAL.

" HEADQUARTERS DEPARTMENT EAST TENNESSEE,
" KNOXVILLE, June 14, 1862.

" *General Orders, No.* 54. *VII.*

" At a general court-martial held at Knoxville by virtue of General Orders Nos. 21 and 34 (Department Headquarters, April 15 and May 10, 1862), whereof Lieutenant-Colonel J. B. Bibb, of the Twenty-third Regiment Alabama Volunteers, was president, was tried George D. Wilson, private Company ' B,' Second Ohio Regiment, on the following charge and specifications, to wit:

" *Charge.*—Violation of Section 2d of the 101st Article of the Rules and Articles of War.

" *Specification 1st.*—In this, that the said George D. Wilson, private Company ' B,' Second Ohio Regiment, not owing allegiance to the Confederate States of America, and being in the service and army of the United States, then and now at war with the Confederate States of America, did, on or about the 7th day of April, 1862, leave the Army of the United States, then lying near Shelbyville, Tennessee, and with a company of about twenty other soldiers of the United States army, all dressed in citizens' clothes, repair to Chattanooga, Tennessee, entering covertly within the lines of the Confederate forces at that post, and did thus, on or about the 11th day of April, 1862, lurk as a spy in and about the encampment of said forces, representing himself as a citizen of Kentucky going to join the Southern army.

" *Specification 2d.*—And the said George D. Wilson, private Company ' B,' Second Ohio Regiment, U. S. A., thus dressed in citizen's clothes, and representing himself as a citizen of Kentucky going to join the Southern army, did proceed by railroad to Marietta, Georgia,—thus covertly pass through the lines of the Confederate forces stationed at Chattanooga, Dalton, and Camp McDonald, and did thus, on or about the 11th day of April, 1862, lurk as a spy in and about the said encampment of the Confederate forces at the places stated aforesaid.

" To which charge and specifications the prisoner pleads ' Not Guilty.'

" The court, after mature deliberation, find the accused as follows: Of the 1st specification of the charge, ' Guilty.' Of the 2d specification of the charge, ' Guilty,' and ' Guilty of the Charge.' And the court do therefore sentence the accused, the said George D. Wilson, private Company ' B,' Second Ohio Regiment (two-thirds of the members concurring therein), as soon as this order shall be made public, ' to be hung by the neck until he is dead.'

APPROVAL AND ORDER FOR EXECUTION.

" The proceedings in the foregoing case of George D. Wilson, private Company ' B,' Second Ohio Regiment, are approved.

" The sentence of the court will be carried into effect between the 15th and 22d days

---

[1] *From Lincoln to Cleveland* by Rev. W. J. Scott, pp. 148–162. Atlanta, 1886.

of June, inst., at such time and place as may be designated by the commanding officer at Atlanta, Georgia, who is charged with the arrangements for the proper execution thereof.

" By command of

" Major-General E. KIRBY SMITH.

" J. F. BRETON, A.A.A.G.

" To Commanding Officer of Post at Atlanta, Ga."

This sentence allows no respite for a moment, but *as soon as it shall be made public* the accused is to be hung by the neck until he is dead. The words, " made public," have only one interpretation. They mean, when the accused himself is notified. They cannot refer to newspaper publication, for this was never intended, and did not take place till the Confederacy, with all its blood and iniquity, had passed from the realm of existing things—until its archives had been captured at Richmond and the seal of secresy removed by loyal arms. This clause was placed deliberately by the court in the sentence and signed by the commanding general. For what purpose? Two conjectures may be offered with some plausibility: we were deceived with hope to make us the quieter in prison, and it was probably intended when the awakening came to give us no time for the desperate efforts to escape that might be anticipated. When this clause was first inserted, it was probably thought that all the band would be convicted in the same manner, and then in one terrible hour all would be swept away with no opportunity to leave any word behind ! The refusal to send a harmless message to friends—a privilege that would not be denied to the most infamous criminal—agrees with this view.

It was the manner of death rather than death itself which seemed so horrible to our comrades as they took their last leave of us. Most of them were also without any clear hope beyond the grave. A day, even, to have sought Divine favor would have been a priceless boon. Wilson was a professed unbeliever, and many a time had argued the truth of the Christian religion with me half a day at a time; but he said, " Pittenger, I believe you are right now ! try to be better prepared when you come to die than I am." I could scarcely release his hand as he muttered, " God bless you," and turned away.

Shadrack was careless, generous, and merry, though often excitable, and sometimes profane. Now he turned to us with a forced calmness of voice which was more affecting than a wail of agony, as he said,

" Boys, I am not prepared to meet my Jesus."

When asked by some of us, whose tears were flowing fast, to think of heavenly mercy, he answered, still in tones of thrilling calmness, " I'll try, I'll try, but I know I am not prepared.

Slavens who was a man of immense strength and iron resolution, turned to his friend Buffum and could only articulate, " wife—children—tell "— when utterance failed.

John Scott was well educated, and had left a very pleasant home in Findlay, Ohio. Father, and mother, brothers and sisters have always been among the most respected of the citizens there. He had been married but three days before enlisting, and now the thought of his young and sorrowing wife nearly drove him to despair. He could only clasp his hands in silent agony.

Campbell had a half smile on his strong face, but it was terribly unreal, with no light in it as he pressed our hands, and even muttered an unconscious oath, saying, Yes, boys, this is — hard."

But Ross was a marvel and wonder to us all. The cloud that had rested upon him so long seemed to have completely rolled away. All foreboding and fear were gone in the presence of the reality. Others were bitterly and terribly disappointed; he was not. The gaunt spectre he had so long faced now came out of the shadow, and lo! it was disrobed of all terror! He was perfectly erect, with easy grace; there was not a sign of dread, while his eye beamed and his whole face became radiant with the martyr's joy. "Tell them at home," he said, in a clear vibrating tone, "if any of you escape, that I died for my country and did not regret it."

Brown, Knight, Buffum, Mason and myself—all that were left of the Knoxville party—were even more affected than our comrades, for we had not the awful excitement of coming death to sustain us. Had there been a gleam of hope of success how gladly would we have thrown ourselves on the guards, and fought for the lives of our brothers! But the officers and the guard filled the door and the entry, while the jail-yard was also full of enemies. The sense of our absolute helplessness was most agonizing.

All this transpired in a very few minutes, and even then the Marshal and others with him in the door showed signs of impatience, and urged that their time was short. I cannot help believing, for the sake of our common humanity, that they wished to hasten only because the scene was becoming too painful for them to bear.

Very brief leave-taking was permitted with the eight who were in the other room. Robinson, who could scarcely stand, was hurried off with the rest. We heard the dreadful procession descend the stairs, and then from the window saw them enter the death-cart and drive away. It was surrounded by cavalry, and thus passed out of sight. In about an hour the procession returned. The cart was empty!

On leaving us the procession had taken a course which soon carried them out of sight over the summit of an adjoining hill, and continued in an easterly direction till it reached the Atlanta city cemetery—a distance of probably two miles. What thoughts crowded through the hearts of the doomed men we know not; but it is to be hoped that in this last hour of life, they realized that God was more merciful than man, and found that pardon which is never denied to those who sincerely seek.

The cemetery is beautifully located and finely kept. The scaffold had been built in a little wood at the south-eastern side of the yard, then outside, but since included in its boundaries. A monument to the Confederate dead has since been erected in this cemetery, and a large portion of land deeded by the cemetery association for their burial, and it was at the edge of this plot that the great tragedy took place. No element of melancholy horror was omitted. A shallow trench had been already dug within a few feet of the long and hideous scaffold, so that the men as they drove up could look upon their own open grave. The scaffold which had just been completed, consisted of a single long beam extending from one tree to another, to which the ropes were attached, and a narrow platform of loose plank extending under this, so arranged that the knocking out of props would cause it to fall. A considerable number of spectators were present, but not nearly so many as attended the execution of Andrews— no general gathering of the citizens being permitted—indeed the preparations had been carried on as secretly as possible.

Captain Fuller, who had chased the men on the cars and attended the trial at Knoxville, was here also to see the end. He had been moved to come by a promise which he as a Mason had made to Ross, that he would mark the spot of his burial, and notify his father, in Ohio. He was faithful to his promise, though the notification, owing to the policy of the Confederate War Department, could not be made until the close of the war.

Our comrades mounted the scaffold by means of steps from behind, and then stood, all seven, side by side, with the ropes dangling beside them. At the foot of the steps, Fuller shook hands with Ross, for whom he declares he had come to feel a deep friendship. The clergymen, with their souls in indignant protest against the manner of death, had not accompanied the procession. There was no help, and in a few moments death in its most awful form was to come.

Yet the bravery of the seven was such as to command the admiration even of their foes. Captain Fuller had attended many military executions during the war, for such things were fearfully frequent on the Confederate side, yet he says that he never saw men die as bravely as these. With uncovered faces they looked steadily and serenely on the surrounding foe. But they were not to die without a word of testimony that should be long remembered, and which to some hearts then present seemed the death-knell of the Confederacy.

Wilson was their spokesman. He asked permission to say a word before death, and it was freely accorded. Possibly the surrounding hundreds expected to hear some word of pleading or confession—some solution of what still seemed mysterious in the great raid. But if so they were mistaken. I have received an account of this address from more than a score of persons who were present—soldiers, citizens, and negroes,—and it made the same impression on all. Wilson was a born orator, and he now spoke with marvelous skill and persuasive eloquence. He had conquered fear and banished all resentment; and his calm and dispassionate earnestness was such as became a man on the threshhold of

The Speech of Wilson on the Scaffold.

another world. He began by telling them that though he was condemned to death as a spy, he was no spy, but simply a soldier in the performance of duty; he said that he did not regret dying for his country, for that was a soldier's duty, but only the manner of death, which was unbecoming to a soldier. Even those who condemned them well knew that they were not spies; then leaving the personal question, he declared that he had no hard feelings toward the south or her people, with whom he had long been well acquainted; that they were generous and brave; he knew they were fighting for what they believed to be right, but they were terribly deceived. Their leaders had not permitted them to know the facts in the case, and they were bringing blood and destruction upon their section of the nation for a mere delusion. He declared that the people of the North loved the whole nation and the flag, and were fighting to uphold them, not to do any injury to the South, and that when victory came the South would reap the benefit as well as the North. The guilt of the war would rest upon those who had misled the Southern people, and induced them to engage in a causeless and hopeless rebellion. He told them that all whose lives were spared for but a short time would regret the part they had taken in this rebellion, and that the old Union would yet be restored, and the flag of our common country wave over the very ground occupied by this scaffold.

There were tears coursing rapidly down the cheeks of many Confederate soldiers; the emotion of a number of negroes who were a long way off, yet in easy hearing of the trumpet-like voice, was almost uncontrollable. One of them said to Captain Sarratt two years after, " Massa, if that man had only spoke a few minutes longer, they could never have hung him in the world." A rebel officer was heard to mutter, " Why don't they stop him? What do they allow such talk for?" But it was not so easy to stop a dying man whose words were so kind and persuasive, and whose eloquence was of that highest type which throws a spell over friend and enemy.

So the tide of truthful speech flowed on till many of the poor men in the rebel ranks heard for the first time the full arraignment of their own guilty government with a clearness which carried conviction, and then with the bold prophecy of coming triumph for the glorious cause—a prospect which seemed to lift the speaker above all fear of his own death—the hero closed, giving the sign for the deed of shame—dying with this glorious prediction on his lips !

Five corpses only remained dangling in the air—a dreadful spectacle ! But still more terrible to witness was the accident which added repulsive horror to the scene of sublime martyrdom. Campbell and Slavens were very heavy, and broke their ropes, falling to the ground insensible. In a little time they revived, and called for water. When this was given they asked for an hour in which to pray before entering the future world that lay so near and dark before them. But the Confederate authorities did not wish to protract the spectacle. Possibly from one point of view this was humane, since there was no prospect of pardon; yet the wickedness of denying all notice is made more apparent by this incident. As soon as the ropes and the platform could be adjusted, they were again led up the steps, once more faced the expectant throng, and were again hung. The awful work was at last complete !

No coffins had been provided. As soon as life was pronounced extinct, first in the five and afterward in the two, the bodies were laid in the shallow trench just wide enough for their length, and long enough for all the seven to lie close together—a brotherhood in death as they had been in life. Here the earth was filled in, and they remained till, at the close of the war, the national government removed their bodies to an honored spot in the beautiful national cemetery at Chattanooga. A monument should mark both this spot and that in Atlanta, where heroism in death shone so brightly.

The thirteen who remained in prison—Wollam had not yet been returned to us—suffered scarcely less than their comrades. The bitterness of death was upon us also. We did not think that vengeance would stop with those who had fallen. The hope we had so long cherished was overturned at a blow. In Knoxville we had urged that all should be tried together, or that the sentence of one should stand for all. There was no reason for giving any preference to one over another, and no indication that such preference was to be given. But even if we had not believed that only a few days or hours of prison life lay between us and the scaffold, the parting from our loved friends whose voices were yet lingering in our ears while they themselves had passed beyond the gates of death, was enough to break the stoutest heart. There were tears then in eyes that would not have quivered in the presence of any danger

But I could not shed a tear. A cloud of burning heat rushed to my head, and fever seemed to scorch through every vein. For hours I scarcely could realize where I was, or the loss that had been suffered. Every glance around the room, revealing the vacant places of friends, would bring our sorrow freshly upon us again. Grief for our comrades and apprehension for ourselves were inseparably blended. The suddenness of the shock by which we were separated seemed to reveal a spirit that forbade us to hope, while it was a terrible aggravation of the pain of parting. Thus the afternoon hours slowly drifted by under a shadow too dark for words. No one ventured as yet to speak of hope.

The first distraction in this terrible hour we owed to our friendly jailer. He asked us if we would like to be all put in one room.

We were eager for this privilege, and he brought over the eight who were in the front room and placed them with us. We were now fourteen, including Captain Fry, who still remained. There would have been much to talk about in our separate experiences in Knoxville and Chattanooga at any other time, but now the thought of the lost swallowed up everything else.

At length some voice suggested—rather faintly at first, for only a few hours before it would have met keen ridicule—that it would be well for us to pray. The thought was warmly welcomed. Not the slightest objection was offered by any one, and we at once all knelt. One member of the party has lately told me that while he knelt with the rest, and was careful to say nothing to discourage us, yet he never led in prayer, or said anything to indicate that he had changed his life-purpose. I did not notice the exception at the time, as every head was bowed and every face covered. Captain Fry was first requested to lead us, which was peculiarly appropriate, as he had always maintained a consistent religious life, and now seemed to feel our great sorrow as if it were his own. He prayed with deep earnestness, strong sobs mingling with his fervent petitions. Then others led, and we continued until all but the one already alluded to had prayed in turn; then those who had prayed before began again. There seemed to be some help in simply telling our trouble. On my own part, I do not think that there was a great deal of faith, at least so far as temporal deliverance was concerned, but there came a calmness and a passing away of bitterness that was restful to our tired hearts. We besought God mainly that He would prepare us for the fate that seemed inevitable, and that as he had led us into great trials, he would in some manner sustain us there. We kept on praying with but short intervals till the sun went down. As twilight deepened into darkness—the emblem of our own lives—so our petitions grew more solemn. God seemed nearer than ever before. In the darkness it appeared easier to behold the heavenly light. We began to ask for deliverance in this world as well as in the hour of death, and to have a hope, very faint and trembling, that it might be granted. Then little by little we began to profess our purpose to live religious lives while we were spared, whether the time was long or short. I do not know that there was anything clear and definite in the way of conversion or sudden change on the part of any; but when it is remembered that in the forenoon we had amused ourselves by all kind of games, that profane words and jests were not uncommon, and that we would have been ashamed to speak of prayer or of religion in any way except as a mere theory, it will be seen that there was no slight alteration in us already. From that hour I date the birth of an immortal hope and a new purpose in life. And in this experience I am not solitary.

It is an interesting fact, which the rationalist may explain as he will, that from the time of that long prison prayer-meeting—from early afternoon to midnight—the fortunes of our party began to improve. There were fearful trials still before us, not much inferior to any that we had passed; we long held our lives by the frailest thread; yet till the close of the war, though many perished around us, death did not claim another victim from our midst. We committed ourselves to the Lord, not expecting deliverance in this world; and in His boundless mercy He bestowed upon us all we asked, and far more than we had dared to hope.

CHAPTER XXIV.

AFTER THE TRAGEDY.

OUR arrangements for sleeping in this room, thanks to the humanity of Mr. Turner, were much better than we had hitherto enjoyed under Confederate rule. We had two or three cotton mattresses and several blankets, having in this respect little to complain of. When we lay down to rest on the night of the execution, our prayers and the very violence of our emotions caused us to sleep well. But few things in our whole prison experience were more fearful than awakening the next morning. The chill light of a new day, the dispelling of dreams that may have been very pleasant, and have brought home vividly before us—always made the morning hour the most dreary of the day. But on this occasion, we looked around and saw the places of our friends vacant, and all the great sorrow of our bereavement again rolled over us like the incoming of the sea.

But we wished to *do* something. A small Bible was borrowed from Mr. Turner when he came to bring our scanty breakfast—Mr. Thoer, who was always with him to see that he gave us no undue indulgence, did not object—and then we had reading, singing, and prayer—nearly every one praying, so that it might rather be called a morning prayer-meeting than "family worship," though the latter was the title used. We now resolved to continue this practice as long as our prison life lasted. But this was not yet enough. What could we do for our own safety? The vigilance of the guards was such that any effort to escape seemed hopeless. Oh! how bitterly did we rue not making the attempt on the road from Knoxville!

Some one suggested that we write a letter to Jefferson Davis himself, placing all the facts in our case fairly before him and asking his mercy. It was urged that perhaps all this bloody persecution was the work of subordinates, and that the chief would disapprove it if brought directly to his own notice. This led to a hot discussion. None of us in our deadly peril had any pride as to sending such a petition if there was a prospect of doing any good. But it might call attention once more to us, and only lead to a swifter sentence. Against this it was contended that our case could not possibly be made worse, as it was now hopeless; and that such a petition or letter in respectful terms might reach the heart of the man who had the power, if he would use it, to disapprove the sentence of a court-martial even after it had been rendered.

I did not myself like the idea. I had built such high hopes on the representations we had made to the court, and the assurances received in return, that I thought it was simply trouble for nothing, that probably our communication would never reach him, and if it did, I thought him no better, indeed rather worse, than other Confederates, and that the chance of any mercy was exceedingly slight. But Mason spoke of his wife and children, and hoped for their sake that nothing possible would be left undone. This appeal could not be resisted, and I agreed to write if the others would furnish the substance of the letter. We induced Mr. Turner to get us paper, pen, and ink, and I composed myself to write my first and last letter to the President of the Confederacy. A photograph of the letter is given herewith, including our autograph signatures, and some interesting endorsements. I now distinctly remember the circumstances of the composition, though until its discovery in the archives of the war department, a few months ago, it was completely forgotten, or rather confounded in memory with another document of similar character and later date.

Obviously the first thing to do was to state our case so that the President might know who we were, and be as favorably impressed as possible. I wrote our statement in the same manner it had been presented to the court at Knoxville, referring to its records in perfect confidence that nothing would be found there to contradict us.

The first paragraph was read and approved as far as it went. But Mason said, "It's too cold. You only argue the case as if we had no great interest in it." Brown agreed with Mason: "Yes, Pittenger," he said, "put it on a good deal stronger. Get right down and beg to him." I did not like this idea very well (probably as much because I thought begging would not be likely to help us as for any other reason): but I added the following sentences, "O! it is hard to die a disgraceful and ignominious death; to leave our wives, our children, our brothers, and sisters, and parents, without any consolation. Give this matter your most kind and merciful consideration. Give us the mercy you yourself hope to receive from the Judge of all."

This was pronounced very good, but some one suggested that we ought to make the offer not to serve against the Confederacy. This was superfluous, because we were now in Confederate power; and they could make their own conditions before exchanging us. But I put it in also. Buffum said, "Tell him, Pittenger, 'that all we want is to be let alone.'" This was understood to be a quotation from one of Davis' own speeches, and raised quite a laugh; but it did not go into the letter. Brown again said,

To His Excellency, Jefferson Davis, President
of the Confederate States of America;

Atlanta, June 15th 1862.   Sir;

We are the survivors of the party that took the
engine at Big Shanty, on the 12th of April last.
Our commander, Andrews, and 7 of our comrades
have been executed. We all (with the exception
of Andrews) were regularly detailed from our regiment
in perfect ignorance of where we were going, and what
we were to do. We were ordered to obey Andrews, and
every thing we did was done by his order; he only telling
his plans, when he wished us to execute them. In this
we are no more to blame than any northern soldier; for
any one of them in our circumstances would have been obliged
to just as we did. For fuller details we refer to the evidence in the cases that have been tried. No real harm
was done, and as far as thought and intention is concerned we are perfectly innocent.

O! it is hard to die a disgraceful and ignominious
death; to leave our wives, our children, our brothers
and sisters, and our parents without any consolation.
Give this matter your most kind and merciful consideration. Give us that mercy you, yourself hope
to receive from the judge of all. We will all take an
oath not to report or to any thing against the Confederacy. If this cannot be done, at least spare
our lives until the war is closed, if we have to
remain in prison until that time

                    Wilson Brown 21st Ohio
                                    Co H

            Wilson Brown 21st Ohio Co H
            William Bensinger   "   "   Co G
            Elihu H. Mason   "   "   Co B
            John A. Wilson   "   "   Co C
            John R. Porter   "   Co F
            Mark Wood   "   Co C
            Robert Buffum   "   Co H
            William Knight   "   Co E
            Wm. Pittenger, 2nd Ohio   Co G
            Daniel A. Dorsey 33d O.V.   Co H
            Jacob Parrot   "   K
            Wm. Reddick   "   "   Co B
            M.J. Hawkins   "   "   Co A 33d

that I ought to tell him that we wanted our lives spared until the close of the war, even if we had to stay in the prison till that time. This excited quite a discussion, as to whether that would be really desirable, but Brown said that McClellan would take Richmond in a short time—a week or two—and then the whole Confederacy would vanish like smoke, and we could take back all that we said. So that petition also went into the letter, and it was signed first by Brown and then by all the rest. It was handed to the Provost Marshal on his visit to us the same day, and sent with his endorsement and that of the Commander of the Post to the Secretary of War. What its ultimate fate was we cannot tell, except that it was captured in the rebel archives at Richmond. It remained for a later letter to another person to show us beyond all doubt the true disposition of the rebel President toward us.

From this time forward we had religious exercises morning and evening, and found them a great consolation and support. They began and closed the day aright, and thus added sweetness to all its hours, supplying a subject of thought not bearing directly upon our own gloomy prospects, and thus enabling us to maintain better mental health. We always sung a hymn or two on these occasions. Indeed there was nearly as much singing as at Chattanooga, but of a far different and more inspiring character. Instead of " Nettie More," " Carrier Dove," and such harmless sentimentality, we sang " Rock of Ages," " Jesus, Lover of my Soul," and others of a pronounced spiritual cast This greatly astonished the guards. They were given strict charge to watch us closely, with the statement that we were the most desperate characters in the whole United States; then to hear us singing " Methodist Hymns," and to know that we had prayers, morning and evening, was a contradiction they found it hard to reconcile. Soon the story of the heroic death of our comrades and our own religious bearing was noised about Atlanta, and no doubt there were many expressions which gave some ground for the bitter complaint of " sympathy " made afterward by the Provost Marshal in his report.[1] But we cared comparatively little for this, of which, indeed, we then knew nothing. We had never expected to receive much help from the people outside, and would not have dared, for fear of treachery, to accept it if offered. But we wished to find that peace in believing that we had heard of Christians possessing. What would we not now have given for the counsels and assistance of a minister we could fully trust !

It is a delicate matter to speak of the beginning of one's own religious life—to say neither too much nor too little; but in the hope of guiding some other who is feeling after the truth, I will venture, using the light that twenty-five years have thrown back on those early days.

[1] War Records, Series I , Vol. X., Part 1, p. 639.

After the terrible 18th of June I am not conscious of any experience of a religious character for several days, except a profound and burning conviction that it is folly to wait for death before trying to be right with God. I might be sinful or wicked again, but the idea that the great business of life may safely be left to the last could influence me no more! Just *how* to be religious was a puzzle. I knew if I had a command to execute from an army officer I would do it, if in my power, no matter how difficult or dangerous; and I wished intensely that it was just as easy to be religious as to be a soldier. But there was the question of right feelings and right motives that did not seem to come into play very much in the army; for if a soldier did his duty, he was not apt to be asked how he felt about it; I had the belief that I must have joy and rapture in thinking of death, a readiness to shout God's praises, which I did not feel; and for a time it seemed as if I could never reach a genuine conversion. I diligently read the Bible which we had borrowed, but while I enjoyed many things in it, little direct guidance for me was found.

I asked counsel of Captain Fry, for whom I had the greatest esteem and respect. But it was so easy for him to believe, that I thought his case must be very unlike my own. I also spoke to J. R. Porter, the only one of our number who had a clear religious faith, and seemed to be happy in it. His first answer was very striking. I asked how he felt about death. He thought that I referred to our worldly prospect, and answered that probably we would soon all be put to death. "But what is your feeling about death itself?" I continued. He said:

"I am not afraid to die, if it is God's will: I trust Him now, and I expect to trust Him to the last." He took my hand, and there was a steady light in his eye that made me believe every word he said. But when I asked him *how he got* such a faith he could only tell me that he went to a Methodist "mourners' bench" two years before and sought till he found it. This did me no good, for there was no such place accessible here.

In sore perplexity I read the Bible from day to day and prayed, taking my turn in praying aloud and reading with the others. At length I thought I began to see that trusting Christ meant something like taking his words and teachings for my guide, trying to do all that he commands, and leaving the result, while I did this, with him. This was not that sudden transformation that I had hoped, but I soon found that it opened up a good many things that I had never dreamed of. One of these seemed especially strange under the circumstances. I had yet but a slender hope of ever escaping from the prison except by the way of the scaffold. But in spite of that dark prospect the question came as an absolute test of my obedience, "Will you, if satisfied that it is God's will, be ready to give up the profession of law if you ever get home, and go into the ministry?"

The first and spontaneous reply was, "No!" I had studied law and meant to practice it if I ever got where law reigned. But at once the self-response was clear, "What kind of obedience is this?" I saw that I was not sincere in proposing to enlist under Christ as my Captain, unless I would really obey him. It would be a poor allegiance that stopped short with the things I wanted to do. For a long time I could not pass this point. The difficulty when communicated to my prison companions seemed utterly absurd. "Try to serve God in the prison, where you are," they said with a rough plausibility, "and don't bother about preaching, being a lawyer, or anything else, when you get out, for you never will get out." This seemed good advice, but it would not bring a serene mind or the victory over the fear of death which I so much desired. One after another of those in the prison found the comfort I lacked; and it was not till wearied and worn out with the struggle that I vowed if God would only give me His peace, I would serve Him as sincerely in the prison or out of it as I had tried to serve my country, and in any way that He might direct. Oh! that this vow had been always more faithfully kept!

From this time I did have a steady conviction that I was on the Lord's side, and that I had a right to commit myself and my life to His keeping. The prison did not prove a palace; its discomforts were still felt keenly, and the prospect of death by the gallows did not appear more inviting. I would not have been the less ready to make any desperate venture for escape; but I had a hope which went beyond the prison and the scaffold— beyond any contingency of earthly fortune, while it did not take away any real earthly good.

It was now July, and the heats of midsummer were upon us. The days were long and weary, and the heat fearfully oppressive. The natural consequence was that a good many of us suffered in health. The Englishman, Wood, was prostrated with fever for nearly a month, and at this time his life was despaired of. His comrades would have greatly mourned his death, yet they administered comfort to him in a style worthy of the best of Job's friends. They said, "Now, Wood, if you get well, you will only be hung; you might as well let yourself die, when you are so nearly dead, and you will thus outwit them." But Wood did not relish the counsel, and recovered, "just for spite," as he often declared afterwards.

We soon had great friends in the negro waiters of the prison. They gave us every help in their power, and it was much that they could do. Finding that newspapers were almost more than meat or drink to us, they taxed their ingenuity to get them. They could neither read, buy, nor borrow papers; but would watch till the jailer or some of the guard finished reading one, and slyly purloin it. Then it might pass through a dozen black hands, and when meal-time came it would be put in the bottom of the pan in which our food was brought, and thus be handed in to us. Usually the paper was returned in the same way to avoid suspicion. The guard and officers would talk with us, and finding us about as well posted on war news as themselves—though all newspapers were strictly forbidden—they came to think that we had an instinct for news such as the bee has of cell-forms.

Having found the negroes so intelligent and useful, I questioned them about other matters and learned that they were better informed than I had imagined. They could not be misled by their rebel masters, for they had adopted the simple rule of disbelieving every thing told, even while professing unbounded credulity. In some way they got news of their own which was often wildly erroneous, but colored by their preferences. They continued to insist that McClellan had captured Richmond, for months after he had been repulsed from the town.

They believed that all northern soldiers were unselfish, fighting only for the rights of all men, and considered it a privilege to help us in any possible way. Many of them had heard that President Lincoln was a negro or a mulatto; and as it had only been talked about among the poorer class of whites without being told directly to them, a few believed it; but the greater number had so little faith in anything the whites said that they disbelieved this also. I never talked with a negro yet who seemed to have the slightest doubt of the victory of the Union troops, and in their own freedom as the result of the war. Their instinct in this direction excelled in truth and penetration the reasoning of many able men. I had extensive opportunities of observing the negroes, as the front room on the same side of the entry as our own was appropriated to them, and there was a secret mode of communication. I never found one without an ardent longing for freedom and a belief that soon the war would give it to him.

One morning Turner came to our room, and said, "Do you know John Wollam?" We hesitated to answer either way, being very anxious for any news of him, as no one of us had heard anything since the day he escaped from the Chattanooga jail; but we feared to compromise him. While we were trying to recollect whether we had ever heard of such a person, John put an end to all perplexity by striding up and saying with a laugh and in his broad, hearty way, "How are you, boys?" We were glad and sorry in a breath to see him. He joined in our religious exercises with much good will. Now all the survivors of our party were together again.

We all remember with deepest gratitude the Rev. Geo. G. N. Mac-Donell, one of those who attended our comrades on the day of death. We did not see either of the clergymen then, or know who they were, or indeed that there was such attendance till long after. Whether Rev. Mr. Scott did visit us or not I am unable to state with positiveness. A minister came, and I was afterward told that his name was Scott, but he may have been another person, as I think this one was not a resident of the city. The interview in this latter case was unpleasant. The preacher had been brought in by our old jailer on the very natural presumption that persons who prayed and sung so much would like to meet a clergyman. He promised the officer of the guard that he would talk only about religion. But his first question built up an impassable barrier between us. He asked how we could be so wicked as to come down there, and fight against the South, and try to overturn their government? We had been trying to repent of our sins, but had not got so far as that particular one yet, and answered a little tartly by asking how he and his friends could be so wicked as to rebel against a good government. He answered by a reference to the North trying to overthrow slavery, and I asked him, if it was possible that he, a minister, was an apologist for slavery! It happened that he was a zealous defender of the institution and very sensitive on this point; and so much noise was soon made in the discussion that the guard removed him. He did not come again.

But our interview and subsequent acquaintance with Rev. Mr. Mac-Donell was of a very different character, though it also opened unpromisingly. In his first prayer he petitioned very earnestly that our lives might be spared *if consistent with the good of the Confederacy!* This offended some of us, but the better opinion was that if sincere in his loyalty to the rebel authority, he could hardly have prayed differently. So kind was he to us afterward that some thought he might possibly be a Union man in his real sentiments; but he has since assured me that he was not in the slightest degree, and that all he did for us was at the dictates of humanity and religion. We had a very pleasant interview. He gave us valuable counsel, and I felt it a great privilege to talk over religious questions with one so intelligent and sympathetic. When he left, he promised to send us some books, and did not forget to promptly forward them. These we took good care of, read thoroughly to all in the room, and then returned, asking for more. These he generously gave, and we thus continued till we had read nearly his whole library. Those only who know what a dreadful weariness it is to pass days without any definite employment can realize the great boon these good books bestowed on us. It made the prison room a veritable school; and in view of our religious efforts the character of the books was just what we would most have desired. I did not care, as in Knoxville, for law-books; but the fact that many, though not all, of the minister's books were of a theological and religious cast only made them the more welcome. This Atlanta jail was my Theological Seminary!

Our food here grew less in quantity till we reached the verge of starvation. For weeks together all we received was a little corn bread, the meal ground with all the bran in it, and half baked without any salt, and a little pork that was mostly spoiled. Even this disgusting food could not overcome the inexhaustible good humor of some members of the party. Knight would arrange the small particles of *fresh meat* in rows or circles on the old spoiled piece, and say, holding it up to view, "Now you see it;—now you don't," as it disappeared down his throat! But there was not enough even of this miserable fare to satisfy the slenderest appetite! Neither was such starvation unavoidable on the part of our enemies at this stage of the contest, whatever it may have become later; for Atlanta was the store-house of the South, where had been accumulated the spoil of Kentucky and Tennessee in immense quantities. But this was to be reserved for future emergencies, and not wasted on prisoners! We believed that Turner would have fed us better on his own responsibility if it had not been for the odious Thoer.

To fight against despondency, and to provide such employments as would prevent the awful dreariness of prison days from eating away heart and soul, taxed our energies to the utmost. The fate of our companions hung over our heads continually, and their memory was never long absent from our minds, but we strove to provide that regular occupation for mind and body which is the necessary condition of health and sanity.

On the terrible 18th of June we threw our cards out of the window, and resolved to engage no more in that game. Aside from its gambling associations, it was inseparably connected for us with the terrible shock we then received, for we were at cards when we discovered the guard approaching. But we carved a checker board on the floor with a nail, and it was occupied at all leisure times by eager players. We also formed ourselves into a debating society, and spent a specified number of hours each day in discussing questions of all kinds. No one was excused, and some became quite expert speakers, and frequently the interest would be intense. Yet at the close of the fixed time, which was determined by the relief of the guard outside, or guessed at from the lengthening shadows of the jail by the person appointed for that purpose, we would change the order of exercises, suspending the question till the next day. The advantage gained and the actual information imparted in this way were not small. What one knew became the property of all. The time and labor expended were no loss, for time was abundant, and the hardest of all labor is to do nothing.

But there is no employment upon which I look back with more pleasure than that for which the Minister's books furnished us the material. With fifteen persons in a room not more than eighteen feet square it was needful to preserve quiet if any reading was to be done. We therefore appointed regular reading hours—two in the forenoon and the same in the afternoon. During this time no one was permitted to speak above a low whisper, and all noise and running about were also forbidden. The rules were sometimes broken, and penalties had to be applied, but usually the order was excellent. Those who did not wish to read might sleep. Sometimes the books were read silently, but for a part of the time in nearly every period a volume of general interest would be selected and read aloud. These books would often furnish subjects and arguments for discussion in the debating periods that followed. We gained a great deal of knowledge in our novel school which has been of life-long value. Books of travel, adventures, history, biography, and theology—no fiction —were freely read, and brought the freshness of the outside world into our dreary captivity. MacDonell, who was greatly interested in the use we were making of our time and his books, asked if he could also send us some papers. The Commander said that he might if they were old and religious. He accordingly sent us a number of journals of that character which we greatly enjoyed; but we examined them with even more interest when we found by experience that a late number of the Atlanta dailies was not unlikely to be found in nearly every bundle. Mr. Mac-Donell had no sympathy with the idea of trying to punish men by cutting them off from all knowledge of the living world.

We had also our times, less firmly fixed but still coming in each day, for physical exercise. Those who had least interest in the reading and debating were foremost here. It may be thought that our little room would allow of no great use of the muscles. But this is a mistake. We were especially anxious to keep our strength up to the maximum, not only for health but for the critical use which we might—and did—find for it before long. The two large windows of the jail, although there were bars across them, afforded us light and air; and if our food had been abundant and nourishing, we might have remained nearly as strong as when we left our own camps. Marks were made on the floor for jumping, and we could tell from to-day what progress we were making in this art. One of the mattrasses was placed in the middle of the room for handsprings; and as three of the number were very expert to begin with—Brown, Knight, and Hawkins—the latter having been for a time in a circus—we were soon all taught a great variety of feats. Wrestling and boxing excited even more lively interest, and we were scientifically trained in both. In fact, whatever one knew in any one department, he felt it a duty and pleasure to communicate as far as possible to all the others. The result was a feeling of comradeship and of brotherhood that was of the greatest value. I doubt whether—notwithstanding the great sorrow behind us for our leader and companions, and with our own future most lowering, half starved, with only the remnant of clothing we had brought from camp, and in a prison swarming with vermin—we were not after all happier than any equal number of Union prisoners then in Confederate hands!

In the morning when it was light enough to see well and all had fully aroused themselves, the signal was given for prayers. At first this was before breakfast, but we changed afterward, for the reason that after our breakfast, little as it was, we could be in a more thankful spirit than when the fierce morning hunger was entirely unsatisfied. After this change had been made we simply talked and yawned and grumbled around till the jailer came with the anxiously expected food. The doors were then opened and the negro waiters would hand in a pan with our allowance. This was shared out carefully, and soon eaten. Then the pans were taken by the jailer, and when alone, the one appointed for that purpose would read a chapter in the Bible, sometimes more, and we would sing, and then in due turn, one or more would lead in prayer. After a short interval we would have our two hours for reading. Often this was a period of great interest. Those who had read most in former days were often called upon to explain what was read to the others, and our rules would be relaxed by vote far enough to permit this. Next came debate. Sometimes this was dull, at other times full of fire; then the muscular exercises would follow, after which every one did what he pleased till Turner again opened the iron doors—there were two, one of iron slats, and the other of solid iron—and handed in our dinner, the only other meal we received, which usually came between three and four o'clock. When this was eaten, we again had a reading period, after which was no more prescribed employment, until in the twilight we had worship once more, and a period of singing. We did not sing nearly so well as in Chattanooga, for Andrews and Ross were not with us, but we improved through practice. Then followed what often seemed the sweetest part of the day. We talked as we pleased. All our past lives were brought into review; the books we had read were recalled, and the conversation went on sometimes in little groups, sometimes with the whole room listening to a single speaker. We never had any light, and could do nothing but talk. The darkness seemed to make us converse more freely; and often it could not have been far from the middle of the night when we ceased. .

Buffum was very fond of stories and I had read a good many, which he would induce me to tell. Once he got me started on Bulwer's "Strange Story," which I had read before leaving camp, and he was so much interested in the beginning, that he persuaded others to listen. I had to begin again, and the whole company were attentive. Where memory failed, it was easy to bridge the chasm with a little invention, and in the darkness the weird story was very fascinating to one hearer at least. Soon after Buffum's liberation he hastened to buy the story that had thrilled him so much, but declared himself greatly disappointed in it! No doubt the difference in surroundings was mainly responsible for the result.

But there were times when all our resources would fail, and the awful tediousness of those terribly long summer days be fully felt. Brown, who was one of the most restless of mortals, would amuse himself as long as he could endure it, and then suddenly break off from whatever we were engaged in, and commence pacing the floor like a caged bear; when he could stand this no longer he would catch hold of the latticed door, and shake it with all his immense strength till it rattled, and then say.in the most piteous tones, which we only could hear, "Please, kind sir, let me out. I want to go home!" A good many of us felt the same way, though we did not think it worth while to say so! Mason was the most despondent member of the party. The thought of his wife and children sometimes overwhelmed him, and he utterly despaired. He would sit, as long as we would allow, with his back against the wall, and his knees drawn up to a level with his face, while he would say, "Darker and darker, boys; saltpetre won't save us!" then Buffum and Brown, forgetting their own burdens, would turn in and tease him until by making him half angry or diverting his attention, they would rouse him out of his despondent mood.

There have been many battles fought and fierce controversies waged over the Scriptures. The most serious strife that I remember to have seen in our prison had the same origin. Porter was reading the jailer's Bible one morning, and Alf. Wilson wished to refer to the book to settle some point raised in controversy. Porter did not heed the request for it, either because he did not hear it or was too busy himself to be disturbed at that moment. Wilson asked again a little more peremptorily, and the book was not yet forthcoming. Then Wilson who was standing up, while Porter was sitting, gave him a kick, not very hard, but sufficient to show a little temper. Porter did not resent this, only closing the Bible and handing it up, saying, "Take it, if you want it." At once Knight took up the quarrel for his friend, and words flew fast and furious. A fight was out of the question, for we would not permit anything of that kind, but the wordy war could not be quelled so easily, and at intervals during the whole day it would break out afresh. But like all our prison quarrels, it soon passed away.

Never before did we realize the full meaning of confinement as we did here, in the great length of time, and the heat and languor of the long summer days. The changes and even the horrors of former experience, and the frequency of our removals, prevented the blank monotony from settling down on us then as it now did, after the first few weeks had rolled by and no intimation of our fate been given. It was like the stillness and death that brood over the Dead Sea.

We would sit by the windows in the sultry heat and watch the free birds as they flew past, so merry and full of joyous life, and foolishly wish that we were birds, that we might also fly away and be free !

Turner here gained for us a great indulgence. Two of us were permitted to go down in the yard, at long intervals, to do a little washing. None of the clothes with which we left camp had been washed or replaced, neither had we been permitted the slightest opportunity of washing. To say that our shirts and ourselves were dirty would be a mild statement! but this was the first opportunity to remedy the matter, and it was eagerly accepted. The two who went down could take the shirts of the remainder —it was no great hardship to do without them in a July or August noon in Atlanta !—and give them a very slight cleansing. It would not do to be at all rough in the washing, for they were very frail, and we did not flatter ourselves with the prospect of obtaining any more.

One day it came my turn; it was then three months since I had stepped out of the room, and the unobscured vision of the blue vault above made it seem like a new world. I looked up at the beautiful snowy clouds, my eyes dazzled by the unusual light, and thinking of the yet brighter world beyond, where were neither wars nor prisons. With the thought came also the fear that, once out of danger and prison, I might forget my prison-made vows and thus lose my claim to the world of light. Such a sense of weakness and helplessness came over me, that it was almost a relief when called back to the room behind the bars and the iron doors !

Our dreams were at once a sweetness and an aggravation of our suffering. Perhaps they made, in part, the difference between the cheerfulness of the evenings and the gloom of the mornings. In day-time, friends and happier days seemed separated from us by an impassible gulf, but at night and in sleep there were no barriers; then we were at home enjoying love and freedom.

Often have I seen in dreams the narrow street and the familiar buildings of my childhood town rise before me, and have felt a thrilling pleasure as I wended my way toward the old home. But waking from these incursions into a forbidden paradise was sad beyond measure, and the cold bare walls of the prison never looked half so dreary as when seen in contrast with such visions of delight.

An anecdote here may fitly illustrate the affection for what we called, to the provocation of guards and citizens, "God's country." During the study period, I had read aloud Bishop Bascom's sermon, contained in one of MacDonell's books, on "The Joys of Heaven." All listened with profound attention to the magnificent description, and when it was finished, Brown rather startled us, by saying in his most matter-of-fact way, "Well, boys, that is very good; but I would like to know how many of this party, would rather be there now, safe from all harm, or back in Cincinnati?" The question was a good one for discussion, and it was debated with great animation; but the majority maintained, no doubt sincerely, that they would rather be in Cincinnati—for a while at least!

We found amusement and ultimate profit in opening secret communication with every room in the prison. These are trivial details, but there is an interest in seeing how with the slenderest resources things almost impossible are accomplished. We could shoot a stick with a string attached to it under the door to those on the opposite side of the entry, and written messages could thus be sent. If the stick did not reach its destination, it was pulled back for another effort. There was a chimney between our room and the one on the same side of the house, and by pulling out the pipes which led into this chimney, messages could be passed from one hole to another. In preparing for our final effort to escape, these communications were almost indispensable.

One morning a number of prisoners were put in this room, and as soon as the guard were out of the way, we resorted to our usual mode of "telegraphing," and found to our surprise and pleasure that two of them were from the 10th Wisconsin regiment, in our own brigade. They told us that we had long since been given up for dead, and that our comrades were vowing vengeance for our murder. They were greatly surprised to find so many of us alive. I have little doubt that the Confederates wished the impression to prevail that we had all perished, and therefore kept us so close, and would not let even the dying messages of the seven pass the lines. The other two were soldiers of the regular army who had been captured on the coast of Florida. The four remained with us until we were taken to Richmond. From them we learned much as to the position of affairs outside, which the rebel papers had not given. The disappointment we felt in learning that McClellan's repulse from Richmond was really as bad as we had heard was indescribable. We had hoped that the war was near its end, but now hope seemed extinct.

As each month dragged its slow length along, we were startled by the thought that we were still alive—that the bolt had not yet descended—and we wondered how much longer it could be delayed. Could it be that even the years would pass and find us still enduring the dreadful monotony of the Atlanta jail? The long reprieve we were enjoying seemed incomprehensible. Why did they not put us to death if they meant to do it; if not, why were we not placed on the footing of other prisoners of war? There seemed no reason for any middle course. I have no doubt now that we were forgotten in the rush and hurry of great events, but we did not dream of that, then.

Most of our number were tobacco-chewers and were driven to numberless expedients to obtain what many of them declared they valued more than their daily bread. They begged of the guards, of the jailer, and of the negroes; and as one tobacco-chewer sympathizes with another, they did not always beg in vain. The supplies thus obtained were economized to the utmost. Not only was the chewing continued till the last particle of taste was extracted, but "the remains" were then carefully dried, and smoked in cob pipes !

A few articles that could be spared, such as handkerchiefs and vests, were sold to the guards, as also a coat which Andrews had left with Hawkins at his death, and the proceeds invested in tobacco, apples, and onions. As I did not use the first or last I was generously accorded a double portion of the second !

But I wanted books more than all else, for the generosity of the minister did not supply all wants in that direction. Accordingly I sold my vest, the only article of clothing I could spare (and that I needed bitterly afterward), and a pocket-book, which was left when its contents were rifled; and with the price Turner bought for me three books—all gems—"Paradise Lost," "Pilgrim's Progress," and "Pollok's Course of Time." The first of these I began to commit to memory, and made considerable progress in that direction. I also used it for noting on the margin important dates in this history. I brought it, with the "Course of Time," through all my wanderings and have them yet. The second, after reading thoroughly, I presented with an appropriate inscription to Rev. Mr. Mac Donell, as the only testimonial in my power of gratitude for the great service he had rendered us. These books, with the use of Mac Donell's library, very much lightened and shortened those almost interminable four months.

20

CHAPTER XXV.

A DARING ESCAPE.

FOR a long time there was little to interfere with the routine of our prison life. We appeared to be completely separated from the great outward world as well as from the terrible struggle that was convulsing the nation. In regard to attempting an escape, Andrews and G. D. Wilson were no longer with us to preach caution; but I had now changed sides, feeling that circumstances were so different as to make the effort very nearly hopeless, while failure would not only lead to more rigorous confinement, but awaken attention and cause a new trial. Our strongest men were dead; two or three others were sick; and I did not think we had the chance of success that had inspired us earlier. These views prevailed for the time; but we all agreed that if there should be any prospect of another court-martial, we would strike at once, even if death was sure; for if we were again summoned it could only mean death.

Besides the reduction of our force there were many other unfavorable elements in our situation. Atlanta was virtually the centre of the Confederacy. Our forces, under the command of Gen. Buell, had been driven back across Tennessee, and a journey of three hundred miles would be necessary before we could reach the shelter of the old flag. We were in the edge of the city, and would be obliged to run at least a mile—probably more—before gaining the nearest wood. There was a heavy guard constantly on duty, and such vigilance was used that the first start would be difficult if not impossible.

But none of us had any objection to talking about escape, as it gave variety to conversation and imparted the knowledge of each to all. The final result of this was a most extraordinary improvement in travelling power. All our former attempts were discussed and the particular causes that led to the capture of each person were carefully gone over, and vows made not to fall into the same mistakes the second time if we ever had the good fortune to get in the woods at night again. These discussions awakened confidence, and were in some measure a substitute for experience. How to approach a house in search of food or information, how to travel by the stars at night, for in the fall months there was but little fear of the persistent clouds which had been so serious a hindrance before; how to throw dogs off the trail, which was more necessary than formerly, for we would probably not be armed; the divergent courses which would most likely divide pursuit, and bring us to the Union border at different points;—all these things were fully considered, and in due time the fruit was seen.

But if these topics were of absorbing interest when we had resolved not to attempt to escape except in certain contingencies, they were wonderfully freshened when those contingencies, in worse form than we had dreamed of, were actually upon us. A series of events now began which was to convince us all that we had no hope save such as we could wrest from our enemies by our own hands. The same events also secured the official preservation of most interesting rebel documents regarding us.

In the first agony of that day of death at Atlanta we had written a letter to Davis, the Autocrat of the South. We did not expect a direct answer; but so far as we ever knew, it produced no result at all. It is of course possible that this paper, which put our position in a strong and favorable light, procured our long respite, so unlooked for; but probably we were simply forgotten in the confusion of Confederate affairs. Now it was proposed to write again, but to address a humbler person, who might be readier to respond. Gen. Bragg at this time commanded the department —it probably was fortunate for us that Kirby Smith was otherwise engaged,—and we naturally wrote to the former. There were strong objections to writing at all, which suspended the proceeding for several days. It was urged that if we were forgotten, such a letter might bring the unwelcome fact of our existence to mind and lead to our death. But the great majority did not believe that we could be forgotten so soon, and some argued that it was better to find out our enemy's designs even if it did incite them to renewed acts of hostility, for at the worst death was better than such imprisonment continued indefinitely. It was August now, and the prospect of a speedy ending of the war, which had seemed so bright in June, was greatly darkened. If we were kept much longer it meant a slow and lingering death; while if they attempted to try us, we had all agreed to make the utmost possible efforts for escape. Such reasoning prevailed, and we secured a narrow slip of paper and wrote the following letter. The substance was the contribution of the whole party, while as usual I furnished the words. The document with all its endorsements is printed in the War Records,[1] and the original carefully preserved at the War Department.

---

[1] Series I., Vol. X., Part I, p. 635.

LETTER TO GEN. BRAGG.

" Petition from the survivors of Andrews's party, who took the engine on the Georgia State Railroad in April last, to Major-General Bragg, commanding Department No. 2."

"ATLANTA JAIL, August 17, 1862.

"RESPECTED SIR :—We are United States soldiers regularly detailed from our command to obey the orders of Andrews. He was a stranger to us, and we ignorant of his design, but, of course, we obeyed our officers. You are no doubt familiar with all we did, or can find it recorded in the trials of our comrades. Since then, Andrews himself and seven of us have been executed, and fourteen survive. Is this not enough for vengeance and for a warning to others? Would mercy in our case be misplaced? We have already been closely confined for more than four months. Will you not, sir. display a noble generosity by putting us on the same footing as prisoners of war, and permitting us to be exchanged, and thus show that in this terrible war the South still feels the claim of mercy and humanity?

" If you will be so good as to grant this request, we will ever be grateful to you.

" Please inform us of your decision as soon as convenient."

" W. W. BROWN,	WM. PITTENGER,
WM. KNIGHT,	2nd Ohio Reg't,
ELIHU MASON,	WM. H. REDDICK,
JNO. R. PORTER,	JNO. WOLLAM,
WM. BENSINGER,	D. A. DORSEY,
ROBT. BUFFUM,	M. J. HAWKINS,
MARK WOOD,	JACOB PARROTT,
ALFRED WILSON,	33rd Ohio Reg't,

" 21st Ohio Reg't,

"All of Sill's Brigade, Buell's Division.

"Respectfully forwarded to GEN. SLAUGHTER,

" G. W. LEE,

" Commanding Post."

Photo-lithographs of Endorsements by Gen. Bragg and others.

Endorsements on the Petition of the Andrews Raiders by Jefferson Davis and others. Photo-lithographed.

On these endorsements we observe that Lee, Commanding Post, promised to forward the paper, which was all we asked of him. Bragg is entirely non-committal, not even using the opprobrious names to which we were so well accustomed of " engine-thieves " and " spies." But he feels that the case is large enough for the attention of the authorities at Richmond. Randolph, the Secretary of War, has the same indisposition to deal with the matter on his own authority, being in this very unlike our own Stanton, but makes a recommendation that was perhaps as reasonable and humane, as we could have hoped. To spare our lives and by holding us as hostages, to make us contribute to saving the lives of some unfortunate rebels in Union hands, would have been a policy which might have been pursued at first with good results, and which could have been boldly avowed and justified.

But the nature of Davis did not incline toward mercy. He seemed to prefer exacting the pound of flesh. With his own hand he wrote the order for an inquiry to be set on foot as to why we had not been hung with our comrades—for this is the plain English of it. He does not say that if nothing is found to justify a discrimination in our favor we must yet be dealt with in the same manner, but this is the fair inference, and it was so understood when the inquiries were set on foot. But the answers show that it was not easy to gain the desired information.

"HEADQUARTERS, ATLANTA, GA., Sept. 16, 1862.

"HON. G. W. RANDOLPH, Secretary of War, Richmond, Va.

"SIR:—Your communication of the 11th inst. is duly to hand. In reply, I have respectfully to say that the arrest, incarceration, trial and execution of the prisoners you refer to occurred before I took charge of this post by your order. I found a number of prisoners on my arrival, and among them the men named in the petition transmitted.

"Inclosed I transmit the papers handed over to me by my predecessor. Since the reception of your letter, I have endeavored to find Captain Foreacre, and ascertain something more, explaining what I was not conversant with in the transaction, but as his business takes him away from the city, I have not as yet had an interview with him. I will still seek occasion to find him, and give you all the information learned from him. You will please find inclosed the names of the engine-stealers and bridge-burners who are confined in the jail of this city. It is entirely out of my power to answer you as to 'why fourteen of the engine thieves were respited while the others were executed, and whether or not there is anything to justify a discrimination in their favor?' as I am not informed in relation to the proceedings of the court-martial that tried the men.

"I am, sir,

"Respectfully your obt. servt.,

"G. W. LEE,

"Commanding Post, and Provost-Marshal."

"*List of Prisoners sent to Atlanta, Ga., June 13, 1862, from Knoxville, Tenn., by command of Major-General E. Kirby Smith:*

NAMES.	RESIDENCE.		CHARGES. SPIES.
1. Wilson Brown	Ohio.	Court-martialled and sentenced	Engine-stealing.
2. Marion Ross	"	"	" " A "
3. W. H. Campbell	"	"	" " "
4. John Scott	"	"	" " "
5. Perry G. Shadrach	"	"	" " "
6. G. D. Wilson	"	"	" " "
7. Samuel Slavens	"	"	" " "
8. S. Robinson	"	"	" " "
9. E. H. Mason	"	"	" "
10. Wm. Knight	"	"	" "
11. Robt. Buffins	"	"	" "
12. Wm. Pettinger	"	"	" "
13. Captain David Fry	Green Co., Tenn.		Bridge-burning and recruiting for Federal army.
14. G. W. Barlow	Washington Co., Tenn.		Obstructing railroad track.
15. Thos. McCoy	Morgan Co.,	"	
Peter Pierce	Campbell Co.,	"	Prisoners of war.—Federal soldiers.
John Barker	Estill Co.,	"	
Bennet Powers	Lincoln Co.,	"	

NAMES.	RESIDENCE.		CHARGES. POLITICAL PRISONERS.
Ransom White............	Morgan Co.,	Tenn.	Citizens aiding the enemy.
John Walls...............	Blount Co.,	"	Trying to go to Kentucky.
John Green...............	Union Co.,	"	Rebellion
John Tompkins...........	Washington Co.,	"	"
Henry Miller.............	Sullivan Co.,	"	...Suspected as a spy.
William Thompson........	Arrested at Bristol......	"	" " "

" Respectfully submitted by order.

" WM. M. CHURCHWELL,

" Colonel, and Provost-Marshal."

List of Prisoners in Atlanta City Jail, September 16, 1862.

ENGINE-STEALERS.

M. J. Hawkins,	W. Knight,	M. Wood,
J. Parrott,	W. Pettinger,	W. W. Brown,
W. Bensinger,	W. Reddick,	R. Bufman,
A. Wilson,	D. A. Dorsey,	David Fry,
E. H. Mason,	J. K. Porter,	J. J. Barker.

BRIDGE-BURNERS.

T. McCoy,	H. Mills,	Jno. Walls,
B. Powers,	G. D. Barlow,	R. White,
Jno. Green,	P. Pierce,	J. Tompkins,
	Jno. Wollam.	

The seven death sentences were also enclosed, which were precisely alike, and none of which gave the slightest information regarding those who were spared. These lists also were equally indefinite. The mere fact that we were charged with "engine-stealing" would hardly justify the death penalty, even with the word "spies" interlined. But it might require a court-martial, and this would have been to us the signal of death. The reader must bear in mind, however, that we as yet knew nothing of this save indirectly in a way to be stated hereafter. But Lee made one more attempt to gain information and forwarded the result to Richmond the next day.

"HEADQUARTERS, ATLANTA, GA., September 17, 1862.

"HON. G. W. RANDOLPH, Secretary of War, Richmond, Va.

"SIR,—I respectfully forward to you hereby all that I have been enabled to obtain from my predecessor, Captain Foreacre.

"The documents relating to the cases, so far as I know anything about them, were forwarded to you on yesterday.

"I am, sir, very respectfully,

"Your obedient servant,

"G. W. LEE,

"Commanding Post, and Provost-Marshal."

"ATLANTA, GA., September 16, 1862.

"HON. G. W. RANDOLPH, Secretary of War, Richmond, Va.

"DEAR SIR,—Your letter of September 11, 1862, to Major Lee, provost-marshal, has been shown me by him, and, as far as I am acquainted with the matter, General Smith only sent from Knoxville instructions and orders to have seven of them hung, which was promptly attended to by myself.

"The remaining fourteen were reported to this office only for safe-keeping,—some having been tried, but not sentenced, and others not tried. The only office which can properly answer your inquiry is that of Major-General E. K. Smith.

"I have the honor to remain,
"Your obedient servant,
"G. I. FOREACRE."

The above letter from Capt. Foreacre is in error in saying that some of us had been tried. It also contains one of the most heartless sentences that ever came from rebel pen. "Gen. Smith *only* sent from Knoxville instructions and orders to have seven of them hung, *which was promptly attended to by myself.*" (Italics mine).

While this correspondence was pending Capt. Lee visited us in search of information. We took the liberty of telling how our seven comrades had been put to death without warning. He seemed shocked, and said that he heartily disapproved such a proceeding, for which he was in no way responsible, as he was not then in authority. He also told us that he had been asked from Richmond to state the reasons for our being dealt with more leniently than the others. Some of our numbers would have gladly made up a new story, but the time was short and the case too difficult. So they simply said that it was the judgment of the court that we were not so much to blame as those who perished. The Marshal listened, but did not seem satisfied, and it will be noticed that he makes no reference to the statement of the prisoners, in his letter to the Secretary of War. He promised to let us know what conclusion was reached at Richmond, but never did so. Indeed, it was scarcely to be expected; and the idea of an officer holding such a consultation with the men who were at his mercy seems absurd. The next time we saw him was under widely different circumstances. Whether Davis ever did obtain the information he desired as to why our lives were prolonged is doubtful. The court-martial to which he was referred had long since dispersed, and possibly many of its members had fallen by Union bullets, while no record of its proceedings have, so far, been discovered. But we received a few days after sufficiently alarming information.

The visit of Lee convinced some of our party that it was time to act, since if the rebel authorities had turned their attention again to us, we had nothing to hope. But I was of another opinion. The difficulties of an escape were so enormous that I judged only the feeling that we could

be saved in no other way would enable us to attack the guards with sufficient desperation to secure success; and I was also convinced that no mode
of escape was feasible which did not involve the conquering of the guard.
That I was right was proved by the fact that in the actual attempt the
Tennesseeans and the other soldiers in the jail completely failed, though
they had been anxious for the attempt before the trial came.

Several circumstances conspired to end all hesitation. The jailer
looked at us with a great deal of compassion, and said to the soldiers in
the front room that he was very sorry for those Ohio boys, for he feared
they would all be hung. This was duly "telegraphed" to us. Then the
guard was strengthened and there was an appearance of unusual vigilance.
A solitary horseman was one day seen riding over the hill toward the jail.

We could not from our room see the gate,
and we waited with some interest to learn
whether he would ride past, but he did
not; after a little while he rode slowly back
the way he came with his head down as if
in deep study. We "telegraphed" to the
front room to see if they had noticed him.
It was some minutes before the answer
came, but when it did, it was like the knell
of doom. Geo. W. Walton, the taller of
the regulars confined there, wrote us a note
saying that the man came up to the gate,
and calling Turner out, had charged him
to keep a very sharp look-out on those
"engine-thieves," for orders had been received from Richmond to have them all
court-martialed and hung. Walton and

GEORGE W. WALTON. From a photograph.

his companions advised us to escape if we could, and promised to d
what they could to help us. This was great news indeed, and all contro
versy as to escaping was ended. But how?

As usual there were two classes of proposals—the one for secret work,
the other for open force. The first wished to try the same plan in substance that had been employed at Chattanooga; to saw off a few bars from
one of our windows with a knife which Wilson had, wait for a dark night,
and, making ropes of our blankets, descend, one by one, slip past the
guard, climb the high fence, and then hunt the way to the borders of the
Confederacy. My objections to this plan were obvious. We were liable
to be discovered in the cutting of the bars, which would be a slow process.
It might be some time before a dark night was found, for it was now in
October, and the weather was very fine; and in the meantime the court
might begin its deadly work, or irons be welded upon us as had been done

with Andrews. If none of these obstacles arose, and we were ready to start, it was almost certain that not more than one or two could get away by this means. We had but two windows and they were closely watched. There were seven guards on duty all the time, who were relieved every two hours, and their vigilance directed toward us. It was by no means likely that many persons could climb down a rope right from windows thus watched, alight, and scale a fence in plain view, before the alarm was given. I would have had no hope of success.

But in the other plan, we could not fail if every one did his exact duty, and we were now so well acquainted that we had perfect confidence in each other. It was simply to attack our foes in broad daylight. When our food was brought in the afternoon, and the door opened we could rush out, seizing and holding perfectly quiet the jailer and his assistant, threatening them with death if they moved, unlocking all the doors so that we might have the assistance of all the prisoners, and then charge upon the seven soldiers below, dispossessing them of their muskets in the first rush; and if this was done without noise or alarm, march them up into our room, and gag them there. It was not likely, however, that we would be able to keep everything quiet enough for this; in which case we were to run as soon as an alarm was raised, for we knew that there was a strong reserve close by, and did not feel able to reckon with any more than the seven rebels on hand. In our own room there were fifteen men, of whom two, Wood and Dorsey, had been sick so long that, though full of enthusiasm, they could not promise to be efficient in a fray. There were four prisoners of war and a rebel deserter in the front room, and some ten or twelve other prisoners in the remaining rooms, so that our force was formidable in numbers, at least. But we did not base our plans on any help outside of our own room, and the issue proved this to be wise.

In such an attack, the element of time and exact planning of every man's work so that there is no confusion and hesitation, are of vital importance. We arranged with the utmost nicety. Captain Fry was to begin the movement, for he was the oldest, and we gave him the post of honor; I was to stand by and help him with the jailer and the watchman Thoer, if the latter was on hand as he usually was; probably I was given this place from the correct view that with my poor eyes I would be of more service in a scuffle in the hall than in the glaring light outside. Then Buffum, who was as agile as a cat, was to snatch the keys, and, waiting for nothing else, to open all the doors above. There were three, and the fitting of keys from the bunch under such excitement was likely to make this take some time. I think no one of us felt that Buffum had a desirable office. But it was advisable to have all the prisoners released if only to distract the pursuit. All the others were arranged into two bands with leaders, to slip down the stairway at the proper time and break out on the guards at the front and rear doors simultaneously. Then quickness, courage, and desperation were to be pitted against loaded muskets and bayonets, and the issue left to the God of battles.

We had also chosen our comrades and routes. We were to travel in pairs and in every direction. Captain Fry was to be my partner, and all the rest considered that I was fortunate, for he would be at home in the Cumberland mountains, toward which we were to journey. The intended course was marked out for each couple and everything done to forward the movement on which we believed depended our last chance of escaping the gallows. We did not forget to make most earnest supplications in prayer, and to vow, in the old-time manner, that we would render faithful service to the Lord of Hosts if he would aid us in this great emergency.

It was afternoon when we received the intelligence which determined our action and we could not very well be ready to start that day. So the work was set for the following afternoon. We patched our shoes as well as we could, and made cloth moccasins to protect our feet, for many shoes were worn out. We gave messages to each other beginning with the form, " If you get out and I do not— " for we could not tell who would be the fortunate ones in the effort, or how many might fall. We had a strong conviction of success, but whether seven guards would allow their muskets to be taken without using bullet or bayonet against some of their assailants with fatal effect, seemed more than doubtful. I have made ready for battle more than once, but never had so deep and solemn a realization of the uncertainty of the issue as on this occasion.

The last night that we ever spent together was a very quiet one. We sung but little,—only the usual number of hymns in our worship. But we talked late and thoughtfully. We were never all to meet again in this world, and the shadow of the separation was already upon us.

The next day was long and tedious. All our usual exercises had lost their zest and we could do little but discuss in low tones the coming effort. We would not risk an attack in the morning, for that would have given our enemies all day to search for us. It was far better to wait for evening, even if the day seemed well-nigh endless. Slowly the sun rose up, reached the meridian and disappeared behind the jail. We watched the shadow slowly moving up the hill opposite our window till it had well-nigh reached the line on the summit that usually marked our supper-time. The hour was come ! We shook hands with a strong, lingering clasp, for we knew not how many of us might be cold in death before the stars came out. Captain Fry, who was tender-hearted as a child, wept at the parting. He had two coats and as he had immediate use for one only, he loaned the other to me. It was a wonderful boon, for I was nearly destitute of clothing. Everything that we felt ought to be taken we secured about us, so as not to be in the way in the coming struggle. We still had on hand a lot of books belonging to the preacher who had sent them in not long before. We had not dared to return them for fear of arousing suspicion, and I carefully piled them in the corner, and wrote him a note thanking him for the use of them. I also wrote a few words of presentation and gratitude in the " Pilgrim's Progress," one of the three books I had purchased in the prison, as the other two were fully as many as I wished to take with me in the desperate race for life. There was nothing more to do, and we sat around in rather uneasy suspense waiting for the coming of the jailer with the servants to bring our food.

At length the noise of shuffling feet and the voices of the colored women who carried in our provision was heard in the hall—a sound always welcome, for we were hungry enough to make the coming of our miserable dinner a great event. The door was unlocked—only one was kept fastened in the day-time now—our food was handed in and the door locked as usual, while the company moved on to give rations to the other prisoners. We did not wish to make the attack until the slender morsel of food was eaten or secured, as it was too valuable to be neglected. With great satisfaction we had noticed that the cross old watchman Thoer was absent. It so happened that he had been called away that day, and his place had not been supplied. So far as I remember this was the only time that he failed to be at his post, and Marshal Lee blames Turner severely for coming to us without some helper. It was not less fortunate for Thoer than for us that he was not present! We ate a few bites of the food and secured the rest: we were ready.

Again we heard the shuffling feet in the hall as the waiters returned. For a moment I felt a sharp knife-like pang shoot to my heart. So keen was it, that I thought for a moment that my physical strength was about to fail in this time of sore need—something that had never happened, or even been feared before, for always in the presence of danger I had possessed more than usual power. But in an instant it passed away, and I looked about to see if all were at their post. A glance was enough to show that there would be no flinching. The men looked pale but their teeth were firmly set, and they were leaning slightly forward like a horse straining on the bridle. If there was any fear it was that they would strike even too soon. As for Captain Fry, whom I had seen weeping a few moments before, he was perfectly calm and his face wore a pleasant smile.

As the jailer unlocked and opened the door for the bread-pans to be passed out, Fry stepped forward in such a manner that it could not be closed, and said very quietly as if it were the most natural thing in the world:

" A pleasant evening, Mr. Turner."

We had no thought of hurting the old man, if it could possibly be avoided, and hoped to frighten him into surrendering and giving up the keys without any alarm.

"Yes, rather pleasant," responded he in a dazed and bewildered manner. He could not understand what Fry had come out there for.

The action of the next few moments was so quick and under such a fever of excitement that accounts of both words and deeds vary widely. I have reconciled them as far as I can, not always following my own recollections when the preponderance of evidence is strongly against me.

"We are going to take a little walk this evening,—we are going out of here," continued Fry standing close to him, and looking in his eye to see the first symptom of a motion. I was by his side equally watchful. There was no fear of an alarm being given by the colored women. They were frightened nearly white, but were our friends, and had enough of their wits about them to remain silent.

Turner s e e m e d undecided. "How about the guards?" he said in a feeble tone. We were nearly all in the entry now, for there had been a slow, almost unconscious edging forward, and half a dozen low, quick voices answered, "We'll attend to the guard, Mr. Turner."

"Well, you can go then," he said trembling, while his face seemed to grow even whiter, for our looks were not pleasant.

ROBERT BUFFUM. From a war-time photograph.

"Well, give us the keys, then, and you'll not be hurt, said Fry, while Buffum reached out his hand to take them.

The action seemed to rouse Turner like an electric spark. "You can't do that," he said, and then sprang back, and opened his mouth in the cry, "Guar—" when my hand closed over his mouth, and stifled the incipient alarm. It was scarcely fair, but in a moment the three of us were upon him; Fry had clasped him round the body and arms in no gentle embrace, Buffum had wrested away his keys, and was off like a shot and unlocking the doors as if his life depended upon it, while my hand had effectually stopped all noise. He bit my finger with all his might, but the teeth were not sharp enough to do any real injury, and the other bands were gliding down stairs.

Buffum unlocked all the doors easily till he came to the last one, in

which were the four Union soldiers and the rebel deserter. Not one in the other rooms dared come forth! But the deserter was intensely anxious. He encouraged Buffum saying "Don't hurry, it will come in a moment." Finally it yielded, and the poor fellow flung himself out like a shot from a cannon, and was soon in the front of the flight; but the four others remained inside.

One of the points about which there is a little difference of opinion among us is whether the attack on the guard below was not a few seconds too quick—a most natural error, under the circumstances. If all the doors could have been first unlocked, and all who were willing to go have been ready, the jailor bound and gagged, and the whole number, led by those who had been appointed for the work, had burst upon the guard together, it is possible that the victory would have been more decisive, and the number of escapes larger. But Knight thought, as he was gliding cautiously down the stairway, that there was a movement in the group of guards by the front gate indicating alarm. In this case the attack could not be delayed, and it was made with surpassing boldness and success.

Porter and Bensinger led at the back door. The former grabbed the gun of a guard that stood near, and jerked so hard that the guard suddenly letting go, Porter fell flat, but was on his feet in a moment. Bensinger caught the sentinel who was disarmed and held him perfectly quiet. Another enemy was in the corner of the yard, and, seeing the rush, brought his gun up, but before he could take aim, Porter's musket had covered him, and he was emphatically and briefly warned that any movement would forfeit his life. The third guard in the back yard was at once overpowered and knocked down, though not seriously injured, and in a moment the victory was completely won in this quarter. It was time, for matters had not gone as well in the front yard.

Knight and Brown went down the stairway as softly as cats, but the passage behind them was filled for a little time with those who were to make the attack at the back door, which left them without support, two men against four, and the latter armed. But they did not for a moment

WM. BENSINGER. From a war-time photograph.

Seizing the Guard.

hesitate though the guards were, unfortunately, not near the door. Knight darted on the nearest, who was by the fence, and as he was bringing the gun down to a charge, Knight seized it with his left hand and struck its owner so powerfully with his right that the gun was instantly released. Brown had dealt with another in an equally effective manner and reinforcements for the prisoners were now coming; but two other guards who were close to the gate, instead of standing their ground, ran out and raised a great outcry. Wilson, Dorsey, and others, threw some loose bricks which happened to be handy, after them, and prepared to charge out through the gate, when Knight heard the running of the reserve guard up the road, and flinging away his musket gave the word, " Boys, we've got to get out of this;" then hurrying through the hall and down to the lower corner of the yard, was in a moment to the top of the fence, being the first over, but was closely followed by the others. The fence was nine or ten feet high and was no slight obstacle, but it was soon passed and then followed a most desperate and exciting chase.

All of this took but a moment. The negro waiters had kept perfectly quiet, looking on the proceedings with the greatest interest, and only beginning to scream when the noise outside convinced them that they might as well contribute their share. Buffum had just succeeded in opening the last door, and flinging it wide with an impatient, "There now," when the thrilling outcry from below warned him that his own departure must be no longer delayed. Fry and myself had been engaged in securing the jailer, who though old, was powerful, and fought vigorously, but had not finished, when we were warned by the uproar that all thought of a quiet departure was at an end, and that there was no longer a motive in holding on to Turner. We all rushed down stairs as best we could, well knowing that we would now be last in the flight, which was *not* the post of safety. The deserter passed us all like a tiger on the leap—I never saw such speed in a narrow place—and getting to the back door found two guards awaiting him with bayonets at the charge. He seized one in each hand, cutting himself severely, but flinging them aside so forcibly that the men were very nearly overthrown, and then with the same swiftness continued over the fence, and on to the woods, soon being in advance of all the fugitives. I learned that he escaped to Washington, but months after returning south secretly to visit his family, was captured, recognized, and hanged.

Buffum followed after him and got over the fence without difficulty, but though a very brave man and a hard fighter, he was a poor runner. One rebel who was quite swift-footed kept right after him, gaining continually, and threatening to shoot him if he did not stop. To this Buffum paid no attention, for a running man has an unsteady hand and an uncertain aim, but he soon stopped because of utter exhaustion, just as the

man overtook him. Now Buffum thought it the right time to try his
"Yankee wit;" so he threw himself down and said, "I am so done out
that I can go no further; you run on and catch that fellow," pointing to
a fugitive running a short distance ahead. But the man saw the design
and with a great oath declared that he had *him* now and meant to keep
him. Poor Buffum was allowed a very brief time to rest, and then was
marched back again to the prison. I have scarcely a doubt that if he
had not waited to unlock the doors he would have made good his own
escape.

Captain Fry and myself were close together in going down the stairs,
he being a little in advance. At a glance he saw there was no chance in
the front yard—the way we had intended to take—and at once turned to
the back door, which was left open by the passage of the rebel deserter.
He got over the wall with little effort, but finding himself chased as Buffum
had been, he used a little strategy. A good many shots were being fired
in all directions, and he suddenly threw up his hands and fell flat. Those
who were following him passed on after unwounded game, and when the
way was clear he arose and resumed his course. He was seen again and
had a most desperate chase, but reached the shelter of the friendly woods.
Wilson and Wood in their account written from Pensacola pronounced
him killed, and for a long time his death was firmly believed in. But he
succeeded in reaching Kentucky in safety, although his trip took more
than two months,—longer than any of the others. He journeyed north-east
through Georgia into North Carolina, then into East Tennessee, and finally
across a corner of Virginia into the Union lines in Kentucky. During
most of this time he travelled by night and hid in the woods and mountains by day. For twenty-one days at a time he was not in a house.
These hardships almost completely wrecked him physically, but he recovered, reëntered to the Union Army, became a colonel, and rendered most
efficient service to the close of the war.

At no time in all my southern experience did I find defective vision to
be such a dreadful misfortune as just now. My eyes were easily dazzled
by a sudden increase of light, and as I came out of the obscurity into the
broad light, for some seconds I could scarcely see at all. In this interval
I was parted from Fry by running to the front door according to our
original plan. There were two frightened guards in the gate tossing their
guns about and seeming not to know what they should do. These were
not dangerous looking and I ran up to them—for now the power of seeing
had come back; but just as I was about darting out of the gate, I saw
the stream of guards outside. They called on me to surrender, but I was
not ready for that and hurried back into the yard. A sentinel tried to
shoot me at point-blank range, but fortunately, his gun failed to go off.
I got back into the jail and now started out the back way—the course I

should have taken at first. There were a number of guards in the back yard by this time, but in the confusion I got through them, and to the top of the fence. What was my dismay to see a considerable number of self-possessed soldiers outside waiting with lifted guns to shoot any one whose head might appear above the fence. I jumped very quickly down on *the inside*. One hope yet remained. I ran into the building and out at the front door, thinking that now the front gate might not be guarded, and that in this least likely way I might slip through. But it was vain; a large number of soldiers were on the ground and they were being carefully posted. I saw that the first panic and all the advantages of surprise were over. I ran back into the jail to try the back door once more, but a sentinel was now standing at it and several soldiers followed me into the building. I did not care whether they fired or not, for I now utterly despaired. I went up the stairway, the guards not molesting me, and looked out at the chase which was continuing over the adjacent hills.

It was a wild and exciting spectacle. Company after company of soldiers came up. The bells of the city were ringing, and shots were being fired rapidly, while loud commands and screams were mingled. I feared that many of our number were or soon would be killed. Then I left the window and went to the front room where the prisoners of war were, and to my inexpressible surprise found that they had not gone out of their room at all! They said that there was too much risk in it—that it would not be possible for any one to get off.

Parrott and Reddick were captured inside the prison yard and Buffum outside. Bensinger, who had been so gallant in the struggle in the jail-yard, had a fearful experience afterward. After the first race for the woods he was discovered by some men with dogs. For some three hours they pressed him sorely. He could get out of sight of the men, but the hounds clung to his trail, like bloodhounds as they were! When wearied almost to death, he found a stream of water, and by running for a long distance in that, was able at last to get away from them. But he was utterly exhausted in this long and critical chase, and, being alone, was in no fit condition for the terrible journey that lay before him.

The next evening he went to the negro quarters on a plantation and was received by the slaves with the sympathy they were always prepared to extend to fugitives. But the planter also saw his approach, and, coming unawares upon Bensinger, revolver in hand, forced his surrender. A messenger sent to Atlanta brought a company of cavalry very promptly, and we had the melancholy pleasure of welcoming back our comrade.

Mason was brought in shortly after, and the account stands:—six recaptured, and eight of our own party, with two others—Capt. Fry and the deserter—escaped. This was a better result than we had any good right to anticipate, when planning the attempt.

The most lamentable part of the story was the case of Barlow, a young East Tennessee soldier, who alone attempted to go with us. He was only 18, but brave and very amiable. No harm came to him in the fight, but in jumping down from the fence he broke his leg just above the ancle, and was dragged back to his cell in a very rough manner. No attention was paid to him till the next morning, and very little then. He died from the injury after enduring great suffering.

Flight to the Woods.

From the window where I was I had a good view of the proceedings below. In a short time all the force of the place, including a regiment of cavalry, was drawn up in front of the jail. I heard Col. Lee directing the pursuit. He was in a towering passion, and shouted out his orders in a very angry tone. Said he, "Don't take one of the villains alive. Shoot them down and let them lie in the woods." He ordered pickets to be placed at the ferries of the Chattahoochee, along the railroads and at all cross rods. I was glad to hear such arrangements, for these were the very places we had promised to avoid. The following report[1] written a month afterward, gives the Commander's views of his own success in the pursuit. He is as much mistaken as to outside sympathy in any effective form as in regard to the number of killed and wounded. It is marvelous, considering all the firing, that not one of our party was hurt at all! The only life lost was that of poor Barlow in jumping down from the fence.

"ATLANTA, GA., November 18, 1862

"CLIFTON H. SMITH, ASST. ADJ. GEN., CHARLESTON, S. C.

"SIR : I have the honor to enclose the report of prisoners now confined at this post, as requested in your communication of the 15th instant. I take leave respectfully to remark that when I took charge of this post I found the bridge-burners and engine-thieves confined here in the jail of the county, under a contract made by General E. K. Smith. My force being limited, I could not put a very large force at the jail building, but immediately placed a much stronger force than had usually been stationed there. Notwithstanding, they were enabled, as I have every reason to believe, from outside influences, which I was unable to counteract by the force then at (my) control, to make their escape.

"I found out afterwards that the jailer, contrary to my oft-repeated orders, went, alone and unarmed, into the room in which they were confined, and being immediately overpowered, 13 of them succeeded in making their escape. Three of these were, after their escape, killed by my guard, and one or two wounded. One of them was afterward recovered, and reconfined. I immediately made arrangements to have them all removed to suitable barracks, and a much superior building as regards strength, and in a more central part of the city, where I now have them all properly and strongly guarded.

"There is no blame attaching to the guard. The escape was owing in part to the fact that the jailer, as I above remarked, went in improperly, and I think in part to the fact that they had sympathizers outside. I made long and diligent search for these prisoners, but from the unusual facilities afforded and the great number of sympathizers, I was unable to recapture them all. There were no papers turned over to me by my predecessors, with the exception of the proceedings of a general court-martial which sat at Chattanooga, which papers were all forwarded to the Secretary of War by his own orders, said papers referring to those who were executed.

"I am, General, very respectfully, your obedient servant,

"G. W. LEE,

"Commanding Post and Provost-Marshal."

"P.S.—I will simply add that the facts above stated were duly reported through Major General Jones to the War Department."

All night long the guard talked over their adventures. There was a hammock belonging to one of the prisoners in the front room, and this he kindly permitted me to occupy. It was suspended right before the window, and I could hear and see much that was going on. The guards had their reserve around a camp-fire close by, and I could hear their discussion of their parts in the affray. Generally they lauded their own bravery to the skies, telling how they had served the prisoners who had broken out upon them. Occasionally, one who had not been present then, would suggest that it did not show a great deal of bravery to let unarmed men snatch their guns from them, but such hinted slanders were always received with the contempt they deserved and the work of self-praise went on. One wondered at the speed of the Yankees who had been kept in prison so long. Another of a philosophical turn of mind accounted for it by saying that they had received so much practice in running away in all the battles they had fought, that it was no wonder they were fleet of foot! This sally received prodigious applause!

This was a doleful night. As I heard one after another of the guards tell how he had shot one of the prisoners, how another had been wounded and had no doubt crawled off somewhere to die, and speak of the great preparation for the search, it did seem very doubtful whether any would survive. And my own prospect was very disheartening.

"Sadly I thought of the morrow."

The guards were not slow in saying that the two or three who had been captured would be hanged at once. I knew that they did not have any more real knowledge on the subject than I had; but confident opinions are not without their weight. And n this case my own opinion was to the same effect. The startling news which had precipitated our desperate venture was vividly recalled. There was no reason that the feeble remnant of our band should not now receive the full punishment intended, and it was to the last degree unlikely that opportunity for other escapes would be afforded. The death which for six months had stared us in the face, and which had been so strangely averted up to this time, would not be much longer delayed. And such a death! It was inconceivably horrible as looked at in the darkness of the night when the other prisoners, who did not share my danger, had sunk to rest,—for they were only "prisoners of war!" There was no vision of glory now to hide the monster death from view or wreathe him in flowers. There were no eyes of friends to behold the last struggle, and sure to give sympathy if nothing more—only ignominy and darkness, beyond which no loving thought might ever pierce! But the chill horror of the scaffold and the thought of the heartless jeering crowd rose before me only for a moment, when a thrill of sweetness and a strange tender joy came to my heart. I remembered that I had a Friend who knew all about it; that I was not alone, but had given myself for life or death to one who never forsakes. Almost for the first time, some of the deep emotional experiences of religion became mine. I knew that I was not as I had been on other occasions when the thought of death had been brought before me. If I had to walk through the valley of the shadow of death, I would not be alone. I looked out into the clear night and up at the stars, and their beauty seemed warm and loving. I knew I had a home and a life which no malice of the enemy on the morrow could deprive me of.

The next morning the jailer put me back in the room I had occupied with the remainder of my comrades. I was much amused when he told me that a man had put his hand over his mouth and nearly smothered him, but he added with great seeming satisfaction, " I bit his finger terribly, and gave the rascal a mark he will carry to his grave." However, he had not bitten as severely as he supposed, as I had received only a slight scratch that healed in a week or two. I had always been rather a favorite with him, and he had no suspicion that I was the guilty person, as his fright had prevented him from observing anything closely. He spoke in strong complaint of the ingratitude of our companions, saying that he had been kind to them, and this was the return he got for it. While we remained with him he watched more closely, though he supposed he was flattering us by saying that he had no doubt the men who had gone off were much worse than we.

CHAPTER XXVI.

IN CAVE AND MOUNTAIN.

SOME time in the forenoon several officers came to see us in no mild humor, and one roughly demanded the course our boys intended to travel. I had no delicacy whatever about giving the information; I even took a good deal of pleasure in telling him that they had said that Atlanta was in the middle of what was left of the Confederacy, and that they were going to travel toward the outside! The officer was so well satisfied with this information that he asked no more questions!

But it was a herculean task upon which our brave boys had entered, and my statement to the Marshal was literally correct, though it could be of little use to him. I cannot look back upon what they did without thinking that in dangers encountered and obstacles overcome, the proudest exploits of Livingstone or Stanley were not superior. The following sketches show a vast advance on their part in skill as travellers, when compared with the attempted escapes six months before. Then, all sooner or later failed. Our long discussions and planning had borne good fruit.

A whole volume would be required for the adventures in detail of these different parties, but some brief account seems needful for the completeness of our story. The different narratives have been furnished me for publication by the parties themselves, or are condensed and revised from accounts published by them, some very near the time of the return of the fugitives. The story in each case was of such extraordinary character that local newspapers were glad of the privilege of laying it before their readers. These different accounts have been carefully compared, and any paragraphs which have only passing or local interest have been omitted.

The first story is that of our engineers, and is furnished by Knight, with a few additional items by his companion, Brown. They were the foremost of the whole party in the run for the woods, as they had been in attacking the guard. Knight says:

" We started for the woods that were about a mile distant. We ran through gardens or anything else that came in the way, kicking the pickets off several garden fences as the easiest way to get over them. To say that we were tired out when we reached the woods does not half express our condition. The woods was but narrow, and we only stopped there long enough to catch our wind, and then pressed on again. Mason, Dorsey and

Hawkins were now with us. We next entered a big field, in the middle of which was a deep ravine with brush grown up in it. Here we rested and took our bearings, and then travelled on. Mason now began to get sick, but we worked him along till dark ; then through the night we moved along slowly, secreting ourselves a good share of the time.

"During the day we hid, but some men caused us to change our position once or twice to avoid them. The second night we also made but little progress on account of Mason's sickness, and after hiding all day again, we still found him no better. We held a consultation, Mason urging us to leave him to his fate and save ourselves. But we decided not to leave him in the woods, at any rate."

(The fidelity with which one member of the party would stick to another under the most perilous circumstances finds many illustrations. But it was not necessary that Dorsey and Hawkins as well as the other two should share the fate of Mason. Besides, it had been resolved to travel in couples, and this party was too large.)

ELIHU H. MASON. From a war-time photograph.

"We selected a house that was on a little cross-road between two main roads about a mile apart, and with woods near. So Brown and I went to the house and asked for lodging, telling the man that one of our number was sick. He did not wish to keep us, but we told him that we were going to stay, as we could go no further. We thought it best to tell him that we had broke jail in Atlanta, and were part of the so-called 'engine-thieves.' This made him still less willing to keep us, as he said that he was already suspected for being a Union man. But as we insisted, and told him that he might inform any one who called of our character, he yielded and gave us our supper. Mason was put to bed, and we staid up with him part of the night, and then went to bed ourselves and had a good sleep. We had not yet determined what to do, but we ate our breakfast in a back kitchen and then went into the large house to get our hats, which had been left there. Just then three men walked in. They talked a little time about the weather and the war, and we began to edge toward the door. They asked us if we had not broken jail in Atlanta? We told them that we had." (Was this strange candor desperation ? or was it because there were only three questioners?) "Then they advised us to surrender, saying that the ferries and roads were all guarded, and that part of our comrades had been shot down in the woods, and that they had come to take us. Brown said : ' No, we won't ; now you see if we do.' We jumped out of the back door, and made for the neighboring wood, jumping the fence and running like two deer ; they ran out at the front door and round the end of the house, calling, ' Halt, halt,' but we did not halt. We had no choice but to abandon Mason, who was taken back to Atlanta the same day. We undertook to cut across the road to get in behind them, as we feared to cross the open plantation ; but we failed and had to take the risk. The old fellow with whom we had stayed turned his hounds loose and put them on our trail. We had a big hill to go down and then one to go up before we could reach the woods. We put in our best licks, and could hear the hounds coming their best. We got down the hill and across the flat, and were climbing the hill, when we saw that the hounds were about to overtake us, and we prepared for battle by stopping in a stony place and getting a pile of rocks ready. We waited for them to come close up, and took them at short range. We rolled them down the hill ; and then, as the Southerners used to say after a battle, ' We won the victory, but we evacuated the ground :' for by

Victory over the Bloodhounds.

this time we could see our three callers coming around the road near by, on horseback, to get ahead of us. We got to the woods as soon as possible, and when we were out of sight, changed our course so as to get away from the horsemen. They got part of their hounds rallied so that they would follow along and howl on our track, but they could not be made to close up on us any more. After a while we would see those horsemen heading us off again, and then we would cut in another direction, and the hounds would give them our course again. About noon we came to a small stream of water. We plunged into that, and would stoop down and take up our hands full of water, and drink as we ran. We kept in the bed of that stream for a couple of hours, and then the hounds lost track of us. Soon after we reached Stone Mountain, about eighteen miles east of Atlanta, and went on the north side of it and concealed ourselves in the grass till dark, when we picked out the North Star and travelled by it.

" For twenty days we travelled by night and hid by day. Each day, for several days, we could see them after us, sometimes with dogs. For six days after we left Mason we were without a bite to eat save what the woods furnished, such as nuts, bark, buds, &c. On the seventh day we were going along a little stream that had willows on both sides, and which ran through a field we wished to cross, as there were mountains on the other side, and we thought we could get on their slope and be travelling in the day-time. We had great good fortune here, for we found two ears of corn on the bank and a flock of geese in the creek, one of which we captured by means of the corn, and then getting into the mountains, we commenced to pick our goose. If anybody ever picked a goose without scalding, they know what a kind of job we had ! When we got tired, we took each a leg and pulled it in two ! then we went along eating our raw goose, taking first a bite of it, then of corn. That goose lasted several days !

" On the tenth day we reached the Chattahoochee River. There was a rail fence alongside of the woods, and we took two large rails, crossed them near one end, lashed them fast with bark, and putting our clothes on the highest end, we floated at the other till we got across. We took a good sleep in the thick cane-break on the other side.

" The next thing of importance to us was an orchard from which all the good apples had been picked ; but we ate and carried away as many of the poor sour ones as we could. We also got some strippings of tobacco that had been left in a field alongside.

" The same evening we came to a drove of small pigs and began to figure for one. Finally I stood behind a tree with a club ; Brown bit off little pieces of apple and pitched to the pigs ; and soon one little fellow commenced to pick up the pieces ; then Brown kept working backward till he passed the tree where I stood, and when the pig followed up I shot him with my stick ! Previous to this we found an iron strap that had fastened a shovel to its handle, and one part of it had been worn thin. This we rubbed up a little on a stone and it made a very good butcher knife. We split the pig, and with each a half, ran up the side of the mountain, and waited till dark for a wonderful feast ! We could see over the farm where there was a fire burning out in a back field, and we went there and roasted the pig most of the night. Then we had one of those feasts you read about ! That pig lasted us till we struck the Hiawassee River in the corner of North Carolina. Here we had thought we could run a boat down by night and hide in the day-time. But when we saw the river we changed our mind, for it was a swift stream, full of great rocks, where we could not have run a canoe by daylight. But we tried to cross and each picked out a rock that we intended to reach. I plunged in and missed my rock, as the current swept my feet away while I tried to put my hands on it. I looked to see what had become of Brown after I had got on a flat stone much further down the river than I intended. He was sitting on a stone laughing at me, though I did not see where the laugh came in. He asked me to wait and see if he could get that far down ! He started, and landed far below me, and did not feel so much like laughing !

" But the tug of war came when we were on the other side, for now we had to cross the mountains which we had been travelling lengthwise before—mountains ' it took two men and a boy to see to the top of !' It was so rough that we were four days going eight miles.

" In a deep mountain valley beside a river we met two men armed to the teeth. We all stopped as if we had been shot, but quickly moved on again. We simply spoke when we met, and all seemed glad to get by without anything more to do with each other. Soon we came to a small log cabin, the door of which was fast. We waited a long time to see if anybody came, and as none did, we concluded to investigate it. Brown stood guard, and I climbed down the wide outside chimney, and found nothing inside but two ears of corn. I went up the ladder and there was a bed all made up ready for a person to get in. I turned down the covers and found a rifle and all the furniture but powder ; and as there was none of that I did not care to take it. But I carried the two ears of corn out of the chimney with me. There was fire in some of the stumps on the mountains and we roasted the corn, and that was our supper. We went down the river a little further and camped for the night.

"The next morning we continued down the river, making good progress, for now the trail was better; when suddenly, around a sharp bend, we came to quite a large house with two men sitting on the porch. We concluded to go up to them and ask how far it was to Cleveland, and maybe we could get something to eat. When we hailed and asked,

Brown and Knight Capturing a Pig.

one of the men came down to us, and told us it was sixty miles. Then we asked for some dinner, telling him that we were sick soldiers. He said we could have some dinner, but for his part he was opposed to the war. We were, too!

"We got some water and some soap, which improved our appearance a good deal, and finally dinner was ready. There were two ladies, one old and the other young. The old lady was one of the kind that do a good deal of thinking and say just what they think. After we were down at the table, she said she wished the Yankees would get there, so that they could get some Lincoln coffee. Then I said that I wished so, too. Then she accused us directly of being Yankees ourselves, and as we concluded that those two men could not arrest us, anyway, we said that we were Union soldiers, and belonged to the party called 'engine-thieves.' They had all heard of that raid, and now made us welcome indeed ! They invited us to stay a week and rest up. We were willing, if we could be secreted somewhere. They told us that they could hide us where the ' First Great Rebel ' could not find us. I told them that was just where I wanted to go!

" We had a good time, being kept in the back room during the afternoon. They put a large dog out on picket, and we told them army stories and sung songs till dark. Finally the dog barked, and then stopped. They told us not to be uneasy, as the dog's master was coming. They told him whom they had in the back room, and he came in and took us by the hands, and laughed and cried, and told us some hard tales about his being imprisoned because he was for the Union.

" It was now settled that we should stay for a few days. A large basket of grub was

The Fugitives Entering a Cave.

prepared, and their boy, pretending to be going coon-hunting, made ready a large torch. We were to follow a short distance behind, with quilts and provisions. We first went down the river, and then turned up the mountain and went up, up, till I thought we would never get to the top. We turned into another ravine, and again went up, up, till we came to a solid wall across our ravine. It looked as if the top of the mountain had slid down and barred the passage. Our guide turned a little to the left, and among the bushes he got down and showed us a hole big enough to crawl in. He entered with his torch and we followed. There was a good-sized room in the cave, and he said we could have all the fire we wanted, and hallo as loud as we pleased without danger. It did look as if the Evil One would have quite a task to find us here. Then he gave us countersigns and promised to come again, and left us to enjoy our good fortune alone. We began to eat the provisions brought along, and continued till it was nearly gone. We would eat and lie down, but get hungry again before going to sleep, and eat some more. Finally we quit lest our friend should not come back in time ; but he did, and brought plenty of food with him.

"For five days we were fed and rested in this safe retreat ; then our kind friend took us down again to the river and gave us a guide whom we followed over the mountains. After a long time I noticed a light in the woods and that he was making for it. There we found an old house standing alone, and surrounded by the forests. When we got to the door, my guide opened it, and to my surprise it was full of men. They told us to come in, for we were among friends. We had a good hand-shake all around, and then one old man asked us if we had any money. We told him we had not. He said that our looks showed that we had no clothes, and turning to the company, he said, 'We must get them clothes and money, for men cannot travel without them.' We were taken to a barn and kept till the next night, and were then given a suit of clothes, (this was the first they had received since leaving the Union army,) and ten dollars each, and a guide who was to receive thirty dollars for taking us three nights' journey.

"This placed us across the Tennessee river, when we were sent with instructions from one house to another.

"This was comparatively easy travelling, and we passed rapidly and safely on till we reached our own lines. We had spent forty-seven days and nights, passing over some of the roughest country that ever laid out of doors !"

The rough and simple language of this sketch covers a truly heroic achievement. The devotion to their sick comrade, who was, in spite of all, returned to us in prison, and gave an account of their adventures that far, led to their singular battle with the hounds and the still more terrible race in which they escaped from horsemen and dogs. The journey through North-eastern Georgia was exceedingly difficult, especially for night travel; and when they reached the Hiawassee it was too high up in the mountains for navigation. In this they differed from Dorsey and Hawkins, who began with them, followed the same general course, but kept a little more to the west. The difficulties still before them were formidable, and with winter coming on, and the days growing continually colder, it may well be doubted whether it would have been possible for them to get through at all if it had not been for the romantic incident of the cave, and the help given by the loyal mountaineers. It is hard to speak in too high terms of the generosity of these people, whose devotion to the Union was a passion, and who were always ready, though at the risk of their own lives, to give help to all Union men or soldiers. If these Southern Union men had been trusted sooner by this and other parties much suffering would have been avoided. It is curious to notice how the several parties had their peculiarities; Knight and Brown had this experience with the Unionists; Wood and Wilson excelled in river travel, and in taking an unexpected route; Dorsey and Hawkins in using the friendship of negroes; Porter and Wollam combined the experiences of nearly all the others.

CHAPTER XXVII.

THE LOYAL MOUNTAINEERS.

L IKE others, Dorsey and Hawkins had an exceedingly hard race to gain the woods. The subsequent story is told very clearly and vividly by the former. He says :

" Passing a little way into the woods, I found Brown and Knight leaning against a tree, gasping for breath. I leaned against the same tree. None of us could speak. I thought for a moment or two that each breath would be my last. As we recovered a little, one gasped, ' Guess we'd better go, boys.' On we went, but not so fast as before, for none of our pursuers were now in sight. We were soon joined by Hawkins, Mason, and the escaped deserter, so that we were six in all. We lay in an open field that night, judging it to be safer than the woods, and huddled together as a partial protection from the cold. All night long we heard the baying of the hounds and the frequent discharge of fire-arms. The distance from which these sounds came indicated that the pursuers were beyond us, and that our best chance was in hiding and allowing them to pass still farther ahead. The next day we were fortunate enough to discover some luscious wild grapes, which we devoured with the greatest relish. Our mouths afterwards were very sore, and the grapes may possibly have been the cause of the injury. The same day we were surprised by some citizens with shotguns, but outran them and escaped.

D. A. DORSEY. From a war-time photograph.

" Brown, Mason, and Knight left us, the latter being sick. The deserter continued with us a day longer. He then wished to visit a house for food, but we, though very hungry, did not think it advisable, and parted with all good wishes. I have heard that he got safely to Washington, D. C., but, returning to his home in Northern Georgia, was arrested and executed as a deserter from the Confederate army, into which he had been conscripted at first.

" On the fourth day out we met two of our pursuers, who were apparently coming back discouraged, but easily eluded them by hiding under some bushes. We now began to travel more rapidly, hiding by day and continuing on our way by night, directed by the stars, which Hawkins understood very well.

"On the eighth day out we came to the ferry of the Chattahoochee River, far to the northeast of Atlanta. We took rails from a neighboring fence, and began to build a raft, when we observed a lighted torch approaching the opposite side of the river. When it came nearer we saw that the party accompanying it were negroes, two in number, with four dogs. Hawkins, who had spent some years in the South, and understood the disposition of the negroes, felt disposed to trust them. Accordingly, we asked them to ferry us over, which they readily did, we giving them a little tobacco we had, and which we could not use because of our sore mouths. They professed themselves Unionists, and we told them that we were Union soldiers. The fact of belonging to the railroad party we did not disclose until we were within the Union lines. One went for provisions, while the other remained with us, as if to allay any suspicions we might entertain. They told us that we were forty-eight miles northeast of Atlanta, in the region of deserted gold-mines, and proposed to hide us in one of those mines, supply us with quilts and provisions until we were well rested, and then direct us on our northward way. It would probably have been better to have accepted their kind offer, which I think Hawkins wished to do, but I had some fear ; so we declined.

"The one who had gone for provisions returned with a goodly supply of boiled pork and beans, mashed Irish potatoes, sweet potatoes and corn-bread. What a feast ! It was the first food worthy the name we had eaten for six long months ! We did it ample justice, and what was left carried away with us. Our African friends also gave us a piece of a broken butcher-knife, that was of great service. They also gave us invaluable directions, telling us where the rebel troops lay, and where we could find a colored slave, who would ferry us over the Hiwassee (which runs down from North Carolina into the Tennessee), as they had done over the Chattahoochee. We assured them that they would soon be free, and parted with a mutual ' God bless you !'

"With thankful hearts we pressed on, made a good night's journey, and then laid by until evening of the next day. Seeing a house on the edge of the woods, we watched it until assured that only an old man and woman were there, when we went boldly up to it and asked for supper, which was given with some reluctance.

"Early in the evening journey we came to a small stream, and attempting to cross on a fallen tree, I fell into the water, and was thoroughly soaked. From this cause I suffered greatly with cold. Some hours after we came to a barn, the mows of which were filled with corn-blades. We were glad to bury ourselves out of sight in the fodder, where we grew warm, and slept all day. It was comfortable, but we paid for it by a terrible fright. Some cavalrymen came into the stable under the mows, and took out their horses. We could hear their conversation and the jingle of their spurs, and scarcely dared to breathe. But they left us in safety. We stayed a day longer, as the bed was the best we had found since our first capture. But a negro boy came up to hunt eggs, and found us. He was so frightened that we could not pacify him, and, fearing an alarm, we hastened to the woods once more. Some negroes were again met, starting on their favorite amusement,—an opossum hunt. On application they gave us a magnificent treat,—a hatful of apples, a half ' pone,' and two or three pounds of boiled beef on a bone. This supply lasted for several days.

"On the night journey we were much annoyed by the barking of dogs at the houses we passed. Once we were seen, but, pretending to be rebels on the way to our regiments, we succeeded so well in lulling suspicion that an old man sent a message by us, to his son, who was in the rebel army, and added some corn-bread for the messengers.

" I here became more lame than ever, by reason of an unfortunate misstep, and had to walk by leaning part of my weight upon my faithful comrade. We came to a wide river we could not cross, and, going back into the fields, lay on the damp ground till morning. If I ever *tasted* cold, it was then. Hawkins became reckless from suffering, and was disposed to go to a house, even at the risk of capture. But I demurred, and we waited for an opportunity to communicate with the ever-faithful negroes. We found a hut, and, watching it some time, saw none but a black woman there. She readily responded to our appeal for help, gave us a hot breakfast, a fire to warm by, and some parched corn to carry on our journey. She also directed us to a ford. Thanking her from our hearts, we returned to the river, got over, and concealed ourselves in the woods on the other side.

"With the coming of night we once more took up our weary way. Towards morning we saw a large encampment of whites and colored people. All were asleep except one of the latter. We approached, and, in a whisper, asked him who they were. He told us of the retreat of Bragg's army from Kentucky, and that these were a band of fugitives coming South with their slaves to avoid the Union army. To us this was a serious matter. A large army, with all its baggage, and the country full of pickets, directly in our front, was a great addition to our danger. The colored man gave us all the scanty information he possessed about the position of the army. Hawkins, spying a covered skillet near the fire, winked at our friend, got an assenting nod, and reached for it. It had two baked sweet potatoes, which we appropriated, and departed as noiselessly as we had advanced. Twenty-four hours after, we had reached the Hiwassee River. We called lustily for the ferryman, and, to our exceeding delight, were answered by the very slave our colored friends on the Chattahoochee had said would be willing to ferry us over.

"With his counsel, for we trusted him with the secret of our being escaped prisoners, we resolved to go down the

MARTIN J. HAWKINS.   From a war-time photograph.

Hiwassee to its junction with the Tennessee. To do this, however, it would be necessary to pass round the rebel camp at Charleston, a few miles farther down the river. This was Friday, and by waiting until Saturday, the young man could get a pass good until Monday, and could pilot us around Charleston. We resolved to wait. He treated us royally,—shared his scanty allowance of food with us, for he had only a slave's rations, doctored my ankle, kept us in his best bed—*a feather one*—over night, though, for prudential reasons, we hid in the woods during the day, and, on starting, gave us a bottle of molasses and a piece of pork. We floated down near camp in a 'dug-out' canoe, then left the river and *surrounded* the enemy. Our pilot was obliged to leave us before we got back to the river below the town, but he put us on the banks of a small stream, which we had only to follow down to its mouth. This we reached by two P.M., and amused ourselves by cracking walnuts and hickory-nuts in a solitary place until dark, when we hunted up an old dilapidated canoe. It was a miserable boat, and gave us enough to do in bailing as well as paddling it. We soon saw a better craft, with good paddles, tied up, and, as the owner was not there, we 'traded' without difficulty.

"The stars were shining brightly when we again pushed off, and the water was as clear as crystal, though not deep. We dried our wet clothes, and felt very much more comfortable. Save an occasional whisper between us and the soft ripple of the oars, silence was unbroken. This was the most peaceful and satisfactory night's travel we had yet made.

"At daybreak we hid the boat and nestled away in some dry leaves, and after the sun got high enough to warm us, slept by turns till afternoon. Then we noticed an island half a mile farther down the river, and, as we had seen nobody the whole day, and the place appeared perfectly solitary, we resolved to explore it. Nothing was found, but we saw a house on the east bank, which we watched until sundown, and seeing only women about it, resolved to try for supper. We got a good square meal, but judge our dismay at finding a good number of ladies, and soon after a few men also at the place. It was a 'quilting,' and they were to dance that evening. But we told a new story. We had been working at a saw-mill in the mountains, were now out of employment, and were going to Chattanooga to look for a job. They warned us that we would be arrested at Chattanooga, and would have to go to jail or join the army. They seemed to care nothing for the war, and to have no disposition to molest us. We assured them that we would be all right in Chattanooga, as we were *personally acquainted with General Leadbetter.* They looked doubtful, and in parting they said, in a rather insinuating manner, that they wished us a safe journey *to Chattanooga.* Probably they were Unionists, but we dared not risk a discovery. I tried, unsuccessfully, to steal a quilt, which we greatly needed. The night was overcast, the water was very shallow in places, and some tree tops were in the way. We had to get out, pull our boat out of these obstructions and into deeper water, and then, wet to the skin, to re-embark and paddle on.

"About midnight we came to what seemed to be a ferry, where the river was deeper and wider than it had been before. Suddenly two shots were fired at us. We lay down in the bottom of the boat, and, taking in our paddles, let her float down stream, while we did not move a muscle. I suppose it was a picket of the enemy, who, after firing once, concluded that our boat was only a floating log, and took no further trouble. After getting, as we supposed, out of danger, we again seized the paddles, and an hour of vigorous work brought us to the river's mouth, and out on the broader Tennessee.

"We were very reluctant to abandon the river navigation, but it was manifestly dangerous to continue it farther, and useless as well, unless we were prepared to take the risk of running by Chattanooga. So we rowed to the north side of the Tennessee, and turned our trusty craft adrift, while we started across the mountains. The first road we crossed gave evidence of the passage of a large body of troops, and this warned us that we were probably in danger of becoming entangled in the scouts and detachments of Bragg's army, now on its retreat from Kentucky. Two boys we found by a fire in a school-house—they had been out 'coon'-hunting—confirmed this report. Soon we saw their camp-fires, and ascending a mountain, where we supposed we would be safer than in the valleys, waited for morning. When it came, an appalling sight met our view,—a large division of Bragg's army, with its seemingly endless baggage-trains, well guarded by cavalry, was spread out beneath us. All day long we watched their movements from our eyrie with breathless anxiety. We resolved at night to turn to the northeast instead of keeping due north, as we had intended. Before we had gone far, Hawkins whispered in my ear, 'Dorsey, we mustn't crack any corn to-night.' Rebel pickets and scouts were no doubt on every side of us. The mountain-side was steep and covered with loose stones, where travelling, even by day would have been difficult ; at night in the presence of the enemy, it was terrible ! We came to a picket, and were only saved from running right into it by the snuffling of a horse. We slipped away a short distance from the road, and lay down. Soon a squad of cavalry passed up the road, and we crossed it right behind them, anxious

22

to get out of that dangerous neighborhood while the sound of their hoofs drowned any noise we might make. We moved very cautiously, again ascended the mountain-side, and near daybreak came to a halt and went into camp,—that is, hid in the brush.

"When the light came we could see the enemy no more, but heard his wagons rumbling off in the distance. The immediate danger from that source was over. Our stock of provision, which was only a little parched corn, was almost exhausted ; and as the mountain seemed to be uninhabited, we resolved to move forward in the afternoon. We found a negro, who, for a wonder, could not or would not give any provisions or information. Late in the night we rested, tying some bushes together to make a rude shelter, and both sleeping, for nature was almost overcome. Food and water were also very low, but in the morning we pressed on, halting when our waning strength failed, and going on when strength allowed.

"Very impressive were some of the hours spent in watching on the Cumberland Mountains. One of us would sleep in perfect trust, while the other watched and thought. The lofty peaks, the wide landscape, and the rising and setting sun were doubly solemn in the profound silence, and amid the mighty forests of that region. I can never forget the beauty of nature associated with so much of peril.

"But there were other hours of very prosaic toil. Once we had to force our way on hands and knees through a mass of briers a quarter of a mile wide. Several times we hunted persimmons by moonlight,—Hawkins shaking them off, while I crawled on hands and knees *feeling* for them. Many adventures similar to those already narrated were encountered. Near a ford of the Sequatchie River we found a quilted skirt hanging out, which we appropriated, tore in two, and making a hole in the middle of each piece for our heads, found ourselves possessed of passable undershirts, which we needed sorely, as it was now colder than ever. The ford was waded with our clothes taken off and tied on our heads.

"For two days more we travelled and rested alternately in the mountains, hungry, wet with the rain that now began to fall, and as solitary as if we were the only inhabitants of the globe. Near sundown of the second day we heard some wood-choppers far below us. We were so weak that we repeatedly fell as we descended the mountain-side. Hunger was so extreme that we resolved to try for food from them, using the best story we could frame. We told them we were Confederate soldiers, who had been left in a hospital, from which we had run away, and were now trying to get to our regiments ; also that we were without money, and wanted food. They refused to do anything for us ; said that soldiers had already eaten them nearly out. This reception encouraged us. To test them further we talked of *our cause*, its justice, certainty of success, etc. They did not pretend to agree with us, and finally told us that we were in what had been called 'Lincoln District,' because only two votes were cast there for secession. The conversation led them to a flat avowal that they were Union men. We then cautiously revealed the fact that we were soldiers on the same side, and the hospitality which had been denied before was now readily extended.

"The name of these gentlemen was Moyer. Confederate soldiers almost daily passed their house and great care was required to avoid them. We were offered a bed that night, but declined it, and lay in front of the open fire 'upon our arms' (clubs or canes), not even taking off our shoes, lest we should be surprised, for we only half trusted our host, notwithstanding his hospitality. Next morning there was a snow of two or three inches on the ground. I believe we would have perished if we had laid out that night without food or shelter.

"The younger Moyer, son of the elder, piloted us after a good breakfast to another friend where we got an excellent dinner, making *three straight meals in succession*—the first *meals* since the breakfast at the Crutchfield House in Chattanooga, more than seven months before.

After dinner we were given a cordial invitation to remain and attend a whipping that was to take place a few miles distant that night. The war having interrupted the due course of law, stealing was usually punished by a flogging. But we had no taste for such a scene. Declining the invitation, we were conducted by a new guide to the home of ' Red Fox,' which we reached after night.

" Mr. ' Red Fox' was not at home when we arrived, but came in during the night.

" His father-in-law, aged ninety, was an enthusiastic Union man, and declared his intention to use his old rifle if the rebels ever bothered him or his neighbors. The old gentleman literally forced upon us a dollar—the last one he had.

" Early next morning ' Red Fox,' whose true name I knew at the time, but have since forgotten, put on his hunter's gear, took his rifle in hand, and ' went hunting ;' of course we followed, and of course we saw no game, except a few thousand Confederate soldiers. Being quite lame, I was kept just behind our faithful guide, and Hawkins brought up the rear. About noon we crossed a well-travelled road, and a few rods further on we all hid behind a big log to avoid being observed by any chance passer-by, while ' Red Fox,' who was at the end of his beat, sketched out our path and divided his jerked venison with us. We were about to separate, when we were startled by the approach of a brigade of Confederate infantry, who passed along the road we had just crossed.

" To Hawkins and myself the tread of soldiers and the rattle of their tin cups, pans, etc., was very familiar, but not welcome.

" After they had got well by, our friend bade us good-bye. We could only pay him with thanks, with which he was fully satisfied. Few persons outside of these districts have any idea of the services that were rendered and the dangers encountered by those noble mountaineers.

" Hawkins and I resumed our march, but unfortunately lost our way. A negro on a mountain farm, whom we met, was either too ignorant or too mean to tell us anything— a rare exception to their helpfulness. That night was spent in the woods. We made a bed in the leaves, tying the bushes together at the top to form a kind of protection.

" The next day's travel resulted in but little progress, and late at night we came to a cabin and sought shelter with the bachelor occupant, who had a couple of male friends lodging with him. We were able to give a satisfactory account of ourselves to these gentlemen, and after breakfast asked what our bill was, and were answered, ' one dollar.' Had we been unable to pay this, suspicion would no doubt have rested upon us, and our hunt for the North Star might have been suddenly interrupted, for there was a camp of rebel guerrillas not far away. The dollar bill donated by the aged father of ' Red Fox' met the emergency.

" In addition to the two men lodging with our host, there was a young fellow proba- bly about eighteen years old who seemed to eye us pretty closely, and when we departed, offered to show us the main road to ' Jimtown,' for which place we had made inquiry. A little way out of the woods the young man frankly told us that he believed we were Union men trying to get to the Federal lines.

" He also declared himself a Unionist and offered to aid us, told us our host was a bitter rebel, and that a force of Confederate guerrillas were camped a few miles away. He put us on a path that led us to a ' Squire's,' for whose house we had started two days before. The Squire in turn directed us to Jimtown, and gave us the name of a friend in that place, upon whom we called about bed-time, and were by him stowed away in the loft of his cabin. Jimtown was a small mountain village composed of log houses.

" We were in imminent danger here from guerrillas. The fact of our presence got to the ears of a woman whose husband was imprisoned in the South as a Union man. Her rejoicing that we had escaped, and anxiety to get some account of her husband, led her to involuntarily betray us, and we had to fly in charge of a pilot to safer quarters, which were found before morning some distance away in dense woods. Poor woman, we knew nothing of her husband.

"From here we travelled without a guide, going by the directions given from time to time as we met friends to whom we had been directed. Usually the route was marked out on a paper, and we travelled 'bridle-paths,' avoiding main roads. We frequently lost our way and had to travel by night, appealing to our old-time friend, the North Star, for guidance, and managed to come around to the right place notwithstanding we were several times lost between 'stations' for two or three days. Just over the Kentucky line we reached a friend who, after providing us with a good supper, bed, and breakfast, directed us to another friend at Monticello, where we were received in that whole-souled, hospitable manner for which the Kentuckians are so justly noted. Our host (whose name I believe was Phillips) was a well-to-do planter and slave-holder. From him we learned of the divisions and dissensions that had sprung up at the North in regard to the prosecution of the war, the slavery question, etc. I cannot express the feelings I had on hearing of the fact that the people of the North were thus divided. I felt that if they knew the true inwardness of the rebellion as we did, they would be a unit on the question of suppressing it; and I well remember that we travelled with a heavier heart than before, as we wended our way northward. Our kind host wanted to have his negro shoemaker mend our shoes, which we thought were past redemption; but we were still in great danger of being picked up by scouting parties of Confederates, and had not deemed it prudent to trust implicitly any one. We feared to lose sight of our shoes even to have them repaired, and respectfully declined.

"After another good night's rest, and a better breakfast than we had yet obtained, our friend piloted us through the little town. There was a camp of guerrillas close by, and although our host was a Union man, he made friends with all, and kept any officers or soldiers from either side who might call on him. He gave us a note to his son-in-law, Dr. McKinney, at Somerset, whose house we reached the same day, and by whom the same hospitable reception was accorded. The doctor next day secured passage for us on some wagons that were going to Lebanon for salt, and we found our journey on foot suddenly and very happily ended. Two teams, two white men, one negro and ourselves constituted the party from Somerset to Lebanon, a distance of about seventy miles, which was made in two and three-quarter days. We camped out two nights.

I would like to tell how the old Star Spangled Banner looked to me as we saw it floating grandly in the evening breeze at Lebanon on that day—the 18th of November, 1862—but language fails me.

"At that time there was a convalescent camp at Lebanon and a few soldiers on duty as guards, but no regular troops. Some Confederate soldiers on parole were there also, and others were sick and wounded—it was soon after the battle of Perrysburg or Crab Orchard.

"We were not very cordially received by the officers in charge, as we bore a striking resemblance to the parolled Confederates. When we told them who we were and where we had been, they seemed to doubt our story. Then we added that we had spent six months in the prisons of the South, and could stand a little more of the same treatment if they saw fit to lock us up. They then sent us to the barracks and promised to forward us northward in a few days. We here gave our true names for the first time. The fear that a large reward had been offered for any of us had heretofore kept both names and share in the railroad adventure a secret

"At the barracks, almost the first man we met, was George James, of my own Company. As we approached, I called him by name; he looked intently at us for a moment, then, throwing up both hands exclaimed, 'Dorsey! is that you?' He remembered Hawkins, also, and at once called another comrade from his company. It was a happy meeting, but the sad tidings we had to convey of the death of our leader and seven comrades, and uncertainty as to the rest, marred the occasion. From these friends we learned all about our comrades in arms: who had fallen in battle, who had been wounded, who discharged, and something about the friends at home. We borrowed a little money from these comrades and spent it for provisions.

"I think that we were sent to Louisville, Ky., the next day. This was a regular military post and contained quite a number of troops. As Col. Moore of our regiment was at his home in Portsmouth, having been wounded at Perrysville, and as we were so thoroughly broken down that we were unfit for duty, we went up the river by steamboat to that place and reported to him.

"On the boat, we met Major Ellis of our regiment, and also a Mr. Brotherlin, of Columbus, Ohio, who took a deep interest in us and our story, and requested us to call on him at his home. I did so afterward and was introduced by him to Governor Todd, who promised to remember me. He did so, subsequently, in the substantial shape of a commission as First Lieutenant, which I declined, returning it with the recommendation that our Second Lieut. be made First and myself Second, which was done.

"Col. Moore sent us home to remain until he got able to return to the regiment, when, he said, we would all go together; but I could not wait so long, and went to the front on my own motion, rejoining my regiment at Murfreesboro, only thirty miles north of the place where we left camp on the raid."

CHAPTER XXVIII.

DOWN THE TENNESSEE.—ESCAPE OF PORTER AND WOLLAM.[1]

"A T this perilous moment—the alarm in the jail-yard—we found that we had to run for our lives and every fellow for himself. We made for the woods. In a very short time the whole city was alarmed and in hot pursuit. It was a close chase. We were fired upon by the pursuing rebels, but none of us hit. Everybody was wild with excitement, women screaming, men running, bells ringing, drums beating, dogs barking, in fact a regular stampede. As I approached the woods I found my strength failing, and I associated myself with John Wollam, according to previous arrangements. We soon arrived at a clump of bushes, and as both of us were nearly exhausted we concluded to stop, though fearing recapture as it was not yet dark. We laid down, pulled some brush and leaves over us, and as everything was hurry and excitement, we were not discovered, though they sometimes passed within twenty feet of us, but were not expecting any of the party to stop so soon.

"We remained in our retreat until we heard the city clocks strike ten, and by that time the excitement had died away and we concluded to go. As the night was dark and we had no guide but the stars, we made slow progress. We arranged to travel only at night and lay up in the day-time. After three nights we were still in hearing of the city bells, but I attribute this slow progress to our reduced condition and the difficulty of traveling after night over hills and valleys in a country we knew nothing about. We had saved a morsel from the scanty rations for three or four days, to have something for this emergency. Many difficulties were to encounter, the worst being that of crossing turbulent streams, which was dangerous, as we always crossed them at night, and wherever we struck them. Generally we crossed on logs, if we could not ford the stream, in order to keep our clothes dry. We had only scanty rations to start on—less than enough for two meals—but it was all we had for the

JOHN R. PORTER. From a war-time photograph.

---

[1] Written by J. R. Porter, and slightly abridged.

first twelve days. By this time our appetites were ravenous and we were ready to run some desperate chances for supplies.

"Being in the mountainous districts of northern Georgia, where the country was only sparsely settled, we concluded to take regular turns and visit houses occasionally, one of us to be on the watch while the other presented his claim for supplies. This was hardly ever rejected, owing to a very impressive appeal that we always made, on the ground of our service to the Confederate cause. We generally made these visits about dusk in order to have the advantage of darkness for escape if suspected. Often the one who called at the cabin homes was asked to stay all night, but under the circumstances could not accept these hospitalities, although the inmates were generally old men, women and children.

"On one occasion an old man came to the door and invited me in ; I told him that I only wanted something to eat ; he asked if I was a soldier, and I told him I was ; 'Come in then, get your supper, and stay all night,' he said. I could not resist the temptation and walked in. The occupants consisted of the old gentleman, his spouse and a daughter whose husband was in the rebel army. He commenced to interview me as soon as I was seated and the young women began getting supper. I insisted on a cold lunch as I had some distance to travel that night, but it was of no avail. They must give me a warm supper, so I had to rest easy. I represented myself as a soldier belonging to a Tennessee Regiment and on my way to Bridgeport to rejoin it. The old man's son-in-law belonged to Captain Smith's company, but he did not remember what regiment, and as they had not heard from him for some time they insisted that I should look him up as soon as I got back, and tell him that his wife and baby were well, also father, mother and all the friends, and to be sure to write home as soon as he could—all of which I promised faithfully to perform. Supper being now ready I was invited to sit up and help myself. It is needless to say that I did that supper justice. I found Wollam very impatient and about to make an assault on the premises to find what had become of me, but the lunch that I had secured compensated for my long delay.

"On another occasion we came to a house while the family were absent, and as the doors were locked, we crawled under the floor, raised a board and thus got inside, where we found a supply of corn-bread and meat. At another time we tore some clapboards from the roof of a cabin in the absence of the family, and here also secured a small supply.

"As the dawn approached each morning we would hunt a place to stop for the day. After finding some dense thicket or a cave in the mountains among the rocks, we would stop and listen for the barking of dogs or the crowing of roosters. If there were none in hearing, we considered our retreat safe. Otherwise we would continue our journey for a better hiding place.

"Several times we slipped into the fields where the negroes were at work, and stole the provisions they had brought for their dinner. One night we traveled till we were chilled and weary. It was very late and we were nearly frozen, when fortunately we discovered a nest of hogs. Immediately we routed them up, and lying down in the warm retreat they had left, slept till morning.

"All this time Wollam was longing for the Tennessee River. He would speak of the rapid and easy manner in which he had passed down the stream when he and Andrews escaped at Chattanooga and he believed that if we could once reach it and get a boat, a few nights' journeying would carry us out of the rebel lines. Accordingly we kept going northwest, through the mountains, hoping to reach the river well below Chattanooga.

"Thus the time wore away, and after twenty-two days of hardship and danger, we struck the Tennessee River about twenty-five or thirty miles below Bridgeport. This was quite a relief. We intended to confiscate the first canoe or skiff that came in our way. Soon after we started down the river we saw a large canoe chained to a tree with a padlock. We twisted the chain off and sailed out, feeling that the worst was over, as we expected to make the rest of the journey by water. We ran our canoe at night and hid in the cane-brakes during the day. This was very pleasant, but it cut off our means of support to a great extent ; that is, it did not afford us the same opportunity of visiting houses as when we were traveling overland. We were compelled to adopt a new plan, and thereafter left our retreat early in the evening and returned to our boat before dark. During this time the cold caused us much suffering, as our clothes were thin and nearly worn out, while the nights were very chilly. After having been three nights on the river without anything to eat, we concluded to take some desperate chances on the following day. After our regular rest in the cane-brakes, we started out early in the evening. We soon saw a house on the island. I cautiously ventured up and found the place vacated, but there still remained several stands of bees. I communicated this intelligence to Wollam, and we concluded to have some honey.

"We hunted around, found an old crock, and raised the top from one of the hives, filling our vessel with nice honey. But this was not sufficient. We put the honey in the boat and pulled out, intent on having something more to eat. When it was nearly dark we observed a cabin near the bank of the river. We ventured as near as we thought safe with our boat, then secreted it, and both headed for the house. As we approached, a dog gave the alarm, and a lady came to the door. We told her that we were Confederate soldiers, had been sick and were on our way home to recruit, and that if she would be so kind as to give us a lunch we would travel on. She was baking corn-bread, which was nearly done, and also frying meat, which we very readily accepted. After reaching our boat we opened up our treasure, and, oh ! what a feast of corn-bread, meat, and honey ! It was either feast or famine with us on this whole journey—oftener the latter.

"By this time it was quite dark, and feeling much invigorated we again launched out and made good time that night. The next morning we stopped in a cane-brake, secreted our boat, hunted a suitable spot and fixed our bed for the day. The bed consisted principally of dry leaves and such other stuff as we could gather up, a chunk for a pillow, and our old tattered coats for a covering. Thus we rested from our fatigues. The sun was far down in the west when we awoke, got up, washed ourselves and finished up our corn-bread and meat left from the evening before, but had still a supply of honey. After lunch we took a stroll. It seemed very lonely and we had the place to ourselves, but we were free ; we breathed the fresh air. After the sun had set in the western horizon and darkness had closed around us, we again launched our gum-tree canoe and floated on down the silent river.

"At that season of the year the river was very low, and when we arrived at the head of the Muscle Shoals we had to abandon our canoe. This was quite a disappointment, but we were willing to endure almost anything for the privilege of being free, and as we had been very successful thus far, we felt able to surmount every obstacle. It was with some reluctance that we took our leave of the old canoe, as it had carried us safely for many miles and had saved us many hard nights' walk. We journeyed on foot again, but were comforted by the hope of securing another boat as soon as the river could be navigated. The country being rough, we kept near the river and concluded to travel in the day-time when we thought it safe to do so. We soon found that this added materially to our progress and comfort, and with necessary precautions was almost as safe as traveling at night, besides giving us a better chance of foraging, which we very much needed.

"I will here relate two or three incidents showing how we secured supplies. One evening we came to a field where some negroes were gathering corn. After reconnoitering we discovered a basket on a stump and two dogs lying near. We thought that the basket might contain something to eat, and we wanted it. We did not not fear the negroes, who were some distance away, but how to manage the dogs! We started in as though we were gathering corn, and gradually worked toward the basket. We got near the dogs before they discovered us, when they skulked away without a word. Upon reaching the basket we found it contained boiled beef and roasted sweet potatoes—the negroes' supper. We took possession and fell back in good order until we reached the woods, when we beat a hasty retreat to the river. We walked in the water to avoid pursuit by the dogs in case they should follow us. We then halted and feasted on the poor negroes' cold beef and potatoes, and it was a feast indeed.

"On another occasion we were traveling along the bank of the river when we came upon a small canoe that contained fishing tackle, which we supposed would be used for spearing fish that night, as there was a torch in the boat ready to light. A little distance from the boat we observed a sack placed in a tree. Wollam stepped on the fence, took hold of the sack and pulled it down. It proved to be a sack within a sack. The outer one contained some corn and the inner one provisions, which we shouldered and carried to the woods, double quick. We traveled until nearly dark, and then secreting ourselves in a cane-brake, proceeded to investigate the contents of the sack. We found a large loaf of corn-bread, four or five pounds of bacon, a tin of lard and two or three pounds of salt. We considered this a bonanza, as it lasted five days. It had been prepared for the fishing party, as there was a frying pan and other cooking utensils with the boat, and lard and salt for frying and salting the fish. No doubt we would have enjoyed joining the fishing party, but circumstances would not permit us to stop over!

"We were now near Florence, Ala., about forty miles from where we left our canoe, and at the lower end of Muscle Shoals, so that the river might again be navigated if we could obtain a boat. We arrived here a little before midnight. Everything was quiet, and we hunted until we found a board canoe made fast to a post. We loosed it and again launched out. But it leaked, and as we had no way of bailing we were compelled to abandon it and continue our journey on foot. We had not traveled far before we found a skiff which proved to be a good one. Now if we met with no misfortunes we could soon row the distance to our own lines. From that time until nearly daylight we pulled hard and made good time in order to avoid pursuit. We stopped in a dense cane-brake where we felt secure for the day, prepared a bed and were soon in the land of Nod. We slept till afternoon, when we got up and took a stroll, after which we ate lunch and felt very much revived.

"We learned from a negro the following evening that Corinth, Miss., was in possession of the Federal army, and we concluded to make our way there when we should leave the river. At Hamburg Landing, about eighteen miles from Corinth, we left our boat and started across the country. As we could hear of no rebel force in the vicinity we felt comparatively safe but we resolved to use all precautions, and not to discover ourselves until we actually saw the Union blue, and the old Stars and Stripes.

"It was now evening, and as the rain-clouds were gathering, we soon found that we would have to stop for the night. While wandering around in search of shelter we heard the bark of a dog, and discovered a house a short distance off by means of the vivid flashes of lightning. The rain was already descending in torrents. Our intention was to hunt the stable or some other out-building and remain until morning. But the dog made so much noise that we either had to leave or march boldly up to the house, and preferred the latter. We walked up to the fence; an old man came to the door and asked who was there. We said, 'Friends who desire shelter and a place of rest for the night.' He asked if we were soldiers and we answered 'Yes.' He then invited us in. They had just finished supper, and there was a nice blazing fire. We took off our coats and hung them up to dry before the fire. The old lady commenced preparing supper for us, during which time we were questioned by the old gentleman. We told him that we belonged to a Mississippi

Regiment and lived twenty miles south-west of Corinth, that we had been sick at Florence, Ala., and were on our way home to recruit, and get new clothing, which we badly needed. The old man's name was Washburn; he lived fourteen miles from Corinth, and had a son in the Confederate army; but as the rebels had been meeting with some reverses, he had lost confidence in the cause, and thought unless a change came soon, that the South would not gain independence. Supper being ready, we were invited to sit up and help ourselves. It is not necessary to say that we accepted! The demand of our appetites was great and the supply adequate. It was quite late that night before we retired, as we were anxious to learn all the particulars about the country, how we were to get around Corinth, as we lived directly beyond, and wanted to take the most direct route, and yet avoid coming in contact with the Yankees. At the same time we were careful to learn the most direct route to Corinth. As there were no Confederate troops in that vicinity we began to feel comparatively safe, and to indulge the blissful expectation of seeing our soldiers the next day! We offered to sleep in front of the fire, but this would not do. We must take a bed. We were reluctant but yielded, and slept till about four o'clock in the morning, when we were aroused by the old man building a fire, and the old lady soon got up and prepared a hasty breakfast, and we were enabled to take our departure before day-light. As we were taking our leave they asked us to stop again if we should ever travel that way, and let them know how we got around the Yankees at Corinth. We promised and started on our way.

"Owing to the rain, the roads were bad and we did not make good time, but about eleven o'clock as we emerged from a strip of woods four miles from Corinth, we discovered some men and teams. The men wore *blue coats!* As we neared the premises our joy became full. They were Federal soldiers! Our terrible captivity was past! Freedom once more!

"We were soon in the midst of a squad of the Ninth Iowa, but we still bore the resemblance of dilapidated rebels, and after relating our experience, and telling the officer where we belonged, he laughed and said that our story *might* be true, and that he would take charge of us and conduct us to the post. The officer in charge here was still more skeptical in regard to our character. He charged us with being rebel spies and said they had a place for us. A lieutenant from the Twentieth Ohio was less skeptical, and a guard was ordered to escort us to the Provost-Marshal's quarters in Corinth. Several officers were in the room when we arrived, and all eyes were upon us. The guard introduced us as escaped prisoners. There was incredulity written in the face of every officer as we were invited to be seated. I soon observed from their conversation that we were to undergo a very strict examination. They commenced in an insulting and sarcastic manner. In turn we were very independent, feeling much elated over our success in reaching the Federal lines, and were as tantalizing as they, often laughing to ourselves in a quiet way, knowing that we could easily establish our claim, and turn the tables on them. After becoming convinced that we were not to be scared they concluded to proceed in a more genteel manner. We were next conducted to General Dodge's headquarters. Upon being introduced, he bade us be seated, and we found him to be very much of a gentleman. After a short interview he recognized our true character and received a full detail of our adventures. He then ordered the Quarter-Master to furnish us with a full supply of clothing and blankets, and to report to him again. We went to the Quarter-Master's department, drew our clothing, took a ' general clean up,' robed ourselves in army blue, and felt that we were no longer fugitives and wanderers, but free men. Some of the boys soon learned where we had been, and the consequence was that we had to relate our experience wherever we went. After taking supper with the Quarter-Master, we again reported to General Dodge, who made us a present of five dollars each, and gave us an order for transportation to our regiments.

"It was now the eighteenth of November, 1862—a month and two days since leaving the prison ! We had been absent from our regiments over six months, given up by our comrades as dead, and therefore we were anxious to get back once more. We had to go by rail from Corinth, Miss., to Columbus, Ky. It was a beautiful night when we started and we fully enjoyed it. In due time we arrived at Columbus, where we took a boat for Cairo, Ill. We had a pleasant trip from Columbus to Cairo. Upon arriving we reported to the Provost-Marshal to learn the whereabouts of our regiments, but could hear nothing of them. We then got transportation to Louisville, Ky., and after a weary night's travel arrived and reported to the proper authorities, but still failed to get any tidings of our regiments. After a few hours delay in the city, we concluded to push on to Nashville, and took the first train south. Judge my surprise when we reached Nashville to find the old Twenty-first doing provost duty there, and the Thirty-third—Wollam's—regiment, in camp only eight miles away ! I will never forget the expression of Colonel Niebling, when he took me by the hand and welcomed me back. I was heartily congratulated by all the boys and there was general rejoicing. After a few days' rest I was again ready for duty."

CHAPTER XXIX.

FLOATING TO THE GULF.

AUTHORITIES: 1. Report of Wood and Wilson to Secretary of War from Key West, Nov. 12th, 1862. 2. Account published by Wood and Wilson in Key West *New Era*, Nov. 15th, 1862. 3. Conversation with Wilson in Ohio, 1887. 4. Adventures of Alf. Wilson, published in book-form in Toledo, O., 1880.

THE route followed by Wood and Wilson was at once the easiest and the least obvious. In the prison we had discussed the possibility of going south as we had all other routes, and it was agreed that it would have many advantages if we could be certain of finding some of our coast fortresses or ships. But this was so doubtful that I am not sure that any had adopted it before the break was made. Certainly none of us knew exactly the course of the Chattahoochee River or the great advantage it possessed for such a purpose. Rising in the North-eastern part of Georgia, it is already a considerable stream when it flows a short distance west of Atlanta, and then southward between Georgia and Alabama, and across Florida, till it empties into a large bay which in turn opens to the gulf. Wilson seems to have been a natural boatman, and his adventures in his first attempt to escape by the Tennessee are vividly recalled in this new effort. In no other way would it have been possible for him to have saved his friend Wood, who was too sick to put forth great exertions, but took shelter under the guardianship of his more robust companion. The bearing of this sick man safely through four hundred miles of foes was one of the most heroic deeds ever inspired by soldierly fidelity.

MARK WOOD. From a war-time photograph.

At the first, however, Wood probably saved the life of his comrade.

Wilson was engaged in trying to drive away the two guards outside of the jail-gate with bricks, so that it might be clear for exit—an unequal contest, for they were armed with muskets and were now receiving reinforcements—when Wood, who had noticed the other raiders climbing the fence at the back of the yard, called him to come quickly in that direction, or it would be too late to escape. No second call was required. They scaled the fence together, though a volley was fired, and hurried on their way. It was a terrible run, especially for Wood, but they reached the shelter of the forest, and then dodged from one thicket to another till night.

Darkness came none too soon, for a squadron of rebel cavalry in skirmish order galloped toward the place where they were. They could only dart under a pine bush and falling flat, hope that they might not be observed. Here they remained for some time hearing rapid firing and thinking that probably many of their comrades had been shot. They had already seen Captain Fry fall, and afterward reported him as dead. Infantry soon followed the cavalry, and pickets were posted along the road, one of these being within fifteen feet of their hiding place. These pickets were very free in expressing their opinion of the fugitives, and Wilson and Wood heard much about themselves the reverse of complimentary. Porter and Wollam glided by while they were still concealed, but our hidden comrades did not dare to speak to them for fear of the enemy so close at hand.

Late at night they left their hiding place and crawled between two sentinels as the only mode of passing, and getting cautiously over the fence, they crossed an open field at a full run, and at the far side found a stream of water in which they waded to break the scent should any dogs be following. Then they went a short distance up a wooded hill-slope, when being exhausted they lay down to rest.

Their emotions were of mingled character. They were glad to be out of the terrible prison, but were so weak that they feared the journey home. They now spoke for the first time, and in the tumult of their feelings prayed and swore in the same breath. The cool woods under the open sky, and the fresh night air, made such a contrast to the narrow filthy prison that it is no wonder, as they realized a hope of escaping these things forever, that they were nearly insane with joy.

But the obstacles to success,—the sleepless nights, and the days of hiding, in prospect, repressed their exultation. Wilson tried to form his plans, for on him rested all the responsibility. Wood yielding to his stronger companion with a docility that is truly touching. The idea of mountain travel about Chattanooga, where they had so sore an experience six months before, was so repulsive that Wilson formed the bold, and, as circumstances showed, the wise resolution of pushing for the Gulf of Mexico. But his knowledge of geography was slight, and to find a river that would lead in that direction was no slight task. He had heard of the Chattahoochee, but did not know where it was, or to what part of the gulf it would lead; yet he determined to reach it if possible. In no way can persons travelling at night in a strange and hostile country guide themselves so well as by the course of a river, which has the further advantage of carrying them by an enemy's picket so quietly that it must be a close watch indeed that will detect them: and even if discovered, they have the chance of crossing opposite their foes and escaping by land. There are other advantages, as will appear in the experiences of these two men; but these were sufficient to make them look diligently for a southward flowing river.

After resting and settling their plans, they arose, and finding an open place, "looked up the North star" and took their bearings. Then they set out south-west, which was a mistake, for the river running nearly in that course, it would take them a long time to reach it.

Before long they passed a railroad which they supposed to be one that led to Columbus, and judged that they were travelling in the right direction When daylight came they sought a secluded spot, and preparing clubs as weapons against any dogs that might pounce upon them, they lay down and slept. Late in the afternoon they woke, refreshed, but hungry, lame, and footsore, as might be expected from their terrible run. They ate all the rations they carried from the prison—Porter and Wollam, more provident, made theirs last for several days—and trusted the future for new supplies. They sat and looked at each other here, thinking far from pleasant thoughts. They were ragged to the point of nakedness, dirty, and haggard, so that if they met any one they were sure to be suspected; they saw the smoke curl invitingly from houses in the beautiful valley they overlooked, but did not dare to seek lodging or food. They knew that the white people were enemies, and feared to trust the negroes, with whom they had but little acquaintance. The latter would have saved them from great suffering had they been willing to risk approaching them. But as the sun was setting, Wood suggested that they ought to be travelling.

In the twilight and through the whole night they pushed on, with no adventure save the crossing a corn-field and pulling as much corn as they could carry easily. This kept them gnawing and somewhat appeased their hunger. They rested the next day, and found in the evening that it was only with the utmost suffering that they could push on. Wood wanted to die, saying that a man's life was only a curse to him when obliged to live in such agony; but Wilson encouraged him by urging that they would soon reach a river where they could travel more easily; and with great difficulty, by stopping frequently for him, he kept the poor fellow moving on. Part of the way Wood crawled on his hands and knees !

But when even Wilson was about ready to surrender, being a little ahead, he heard the gurgling of water. He had felt that they could not possibly go a mile further, but this was new inspiration: he told Wood, and so cheered him up that he put forth his last strength, and soon they stood on the bank of a small stream they believed to be the Chattahoochee. Probably, from the course they took, it was one of the tributaries of that stream which flows westward and soon falls into it. But it answered every purpose, and in describing their sensations as they looked upon it, Wilson compares their joy with that of the discoverers of the Pacific and of the Mississippi. It was the road that led home! They felt that they could follow it with confidence though it led to the South and not to the North, for was not all the sea under the old flag? They felt like shouting, but concluded to stint their joy till they reached a country less peopled with enemies; and finding the direction of the current, edged down the stream till a boat was discovered. It mattered little that the skiff was chained to a tree; neither the question of ownership or the lock troubled them, for necessity solved the one, and a stone was an effective key. In a few minutes two happy men were gliding down the smooth and narrow stream. Their wearied feet could now rest and be bathed in the cool waters, while their arms took the labor. They paddled briskly through the profound darkness caused by the closely overhanging trees, wishing to get as far as possible from the owner of the boat before daylight; and also finding this mode of progression so delightful that they hardly thought they could have too much of it. They did not realize how terribly tiresome the hundreds of miles that lay before them could yet become. As day began to break, they sought the darkest inlet of the river—it is probable that they now were on the main stream—and hid themselves and their precious boat.

But the day was less pleasant than the night. The mosquitoes preyed upon them unmercifully, and as it was now four days since they had left jail, during which time they had obtained no food but the corn, they were suffering too greatly with hunger to sleep much, and when they did, the dreams of feasting made the wakening a bitter disappointment. In the dusk they set out with the resolve to have food at any risk. Before it was quite dark they saw a house favorably situated, and, hiding the boat, went up to it. A number of small negro huts and some bloodhounds chained to the fence showed it to be a regular plantation. Their story about being sick soldiers was credited, and supper promised. While waiting for its preparation, they asked for news and heard a highly colored account of the escape of the " Engine-thieves " from the Atlanta prison. The planter said they were a dangerous set, who ought to have been hung long ago, and that some of them fought even after they were shot through the body. He was discouraged as to the prospect of the war, and gave a sad account of the composition of the Union army: but while he talked they ate, until everything cooked was devoured. They were making up for lost time! Soon they were again sweeping, with vigorous strokes, down the river.

As they passed along at a good rate of speed, Wilson found himself in a moment struggling in the river. He thought of torpedoes and every other danger, but on getting into the boat again, discovered that it was only a ferry-boat wire, which had caused the mishap: no evil more serious than the wetting followed.

Toward morning Wilson became so sleepy that he could guide the boat no longer, and entrusting the paddle to the unskillful hands of Wood, lay down, charging him to keep the boat in the middle of the stream, but was soon awakened with the terrible report that they could go no further, for they had reached the end of the river! It was only a cove into which Wood had run, under the shadow of the mountain, and when they were out, Wilson was wakened enough to keep his place at the oars till day.

They had one experience for which they were never able to account. From the first entrance into the river till they reached the rapids above Columbus, they had heard a singular noise as of something following the boat. This terrified them in the darkness; once they saw it—a large animal unlike anything they had ever encountered, and it seemed to be always at the same distance. But it finally left them. If it was their evil genius, it had respect enough for bravery and perseverance to do them no harm, and not to follow them on the lower half of the river!

They reached the rapids of the river in the night, and knowing nothing of the stream pressed on, passing places in the darkness that would have frightened them if the rocks had been clearly visible. Several times they scarcely escaped wreck. They were glad when daylight came, and on a hill-side they were not only less troubled by the mosquitoes, but the sun warmed them, and they had a fine day's sleep.

The next night they came on a mill dam, and in the darkness thought it might safely be shot; but the experience was not one they cared to repeat; then the river grew more turbulent and the country around mountainous. They kept on till they were drawn into a fearful gorge, from which it was impossible to extricate their boat. They dared not go ahead and they could not return; so they found a landing place, and abandoned to the current the boat which had carried them a hundred miles on their homeward way.

Wood was completely drenched in getting out and was nearly frozen, till, on the hillside above, the rays of the sun the next day warmed his blood somewhat. He was so weak as to be almost incapable of caring for himself and frequently had to be led, as he would stagger in walking and even be unable to see. But Wilson never thought of leaving him.

Under such circumstances the three days journey through the mountains that was necessary to clear the rapids, involved dreadful hardships to both men. They could not go much more than five miles daily. The country was rocky, and their feet were in even a worse condition than on their first setting out. They travelled in the day-time in the mountains, as they could not have made any progress at night, and the country was so thinly inhabited that this did not seem to involve any great degree of risk. But their sufferings the last day before reaching Columbus were indescribable. When they saw the town, it was a wonderful satisfaction, as they knew that the river then would be open to the gulf. But they hid in the day and made a long detour around the city. There they concealed themselves among some drift-wood covered with grape-vines, and were able to see the rebels at work on the ram Chattahoochee, of which they carried the first report to the Federals. It afterward blew up and never reached the gulf.

The next thing was to find a boat. One was seen in plain view of the new iron-clad. This was serious, as the work was being pressed night and day, and the light was also kept bright continually. But Wilson wanted the boat, and creeping up with a stick, managed to break the lock. When they got the boat after this danger, they found it to leak so as to be almost worthless. It carried them some distance down stream, when they saw a group of boats on the Alabama side of the river. They "traded" for one, but the owner arrived in sight just as they were getting off and called them unpleasant names. Not wishing to prolong a dispute which could only hurt their feelings, they pushed the other boats into the river, and as these floated down they did some rapid rowing *up* the stream. The owners got the other boats and followed. But a river with islands and dark wooded shores gives heavy odds to a fugitive boat at night, and by the time our comrades had got round an island, they were able, under the shadows of the shore, to turn down stream while they were being pursued up the river. They were exultant over their escape though nearly starved.

Wilson moralizes on the terrible deprivations which were endured in this voyage, declaring his belief that man can endure more than any other animal, but doubting whether he himself could go through the same hardships again. They could make nearly fifty miles a night down stream with their paddles and felt that the hope of reaching the gulf was good, if they could only keep from starving. But food became an urgent necessity. Their feet troubled them no more, but their stomachs gave them great solicitude.

The next evening they found a corn-field and gathered some of the ears—a very unsubstantial diet. But what was much better, they found a number of pumpkins and secured the seeds. A steamboat was safely passed, by gliding in the shadow of the shore.

The scenery of the lower Chattahoochee was found to be of a very lonely character. The river is a mere water way through unpeopled forests and frequent swamps. The trees are covered with great masses of moss, and the forests can be seen through for but a little way. This moss frequently hung from the topmost branches of great trees clear down into the water, and swayed with the sighing breeze. The moss was very useful in a way that added greatly to the picturesque appearance of the two travellers — had any one been present to observe them! They covered themselves with it from head to foot to escape the attacks of the numerous and deadly mosquitoes. The water snakes and alligators were also unpleasant company, and very abundant.

Wilson knew nothing about the habits of the latter and was very much afraid of them. They would follow the boat, or show themselves at the most unexpected places. When the poor fugitives would sleep for a short time in piles of drift-wood, they would, on awakening, often see one or more of the savage monsters lazily watching them, as if sure of their prey. Wilson had only the natural fears inspired by their formidable appearance; but poor Wood was superstitious also, and compared them to sharks hovering around a ship doomed to wreck, and felt almost sure that he and his companion would soon fall victims to these patient monsters. They found that the alligators would not fight a waking man on the shore, but scarcely ever fell asleep without the nightmare feeling that one of them might steal up and seize an arm or leg!

The fugitives had a great fright about half way down the river. They went in search of food and procured a small supply, but found to their horror on returning that their boat had been carried off! The wickedness of boat stealing was now fully realized by them! It was raining and they were in the midst of swamps. On the south, another river emptied into the Chattahoochee so that it was not possible for them to continue their journey further without a boat. It was now dark, and all they could do was to find a place slightly out of the water, where they spent a most sorrowful and hopeless night. How happy their former condition now appeared when they had a boat—that best of earthly possessions! Hunger was a small calamity in comparison with finding themselves thus cut off from all means of locomotion,—stranded in the midst of a Georgia swamp.

All the next day was spent in vain wanderings through the swamps. But toward evening they saw something across the river which looked like a boat partly sunken. To get over the stream more than half a mile wide was the next problem. A frail raft was constructed, with a grapevine as a rope for tying the pieces together, and Wilson tried the passage, sinking deeply in the water. Now was the time to think of alligators! He had a stick in his hand as an oar and weapon, but it was nearly an hour before he was able under such circumstances to get over. But when he did, there was an abundant reward—a better boat than the one they had lost!

No time was spent considering the right of property, and in a few minutes the adventurers were once more together, rejoicing over their deliverance. Rapidly they swept down the stream and made good progress that night. They also resolved that they would not both leave their boat again, starve or not; any calamity seemed less than being deprived of the boat. In pursuance of this resolution Wilson went alone in search of supplies, and in a vacant cabin was fortunate enough to find quite a prize—some fish-hooks and lines. With these he returned in triumph to his companion, feeling that they were now reasonably sure of one article of diet. The whole country seemed an immense swamp, with no corn-fields, but they had a considerable supply of corn, which disagreed with them as they were obliged to eat it raw. To their bleeding mouths and fevered stomachs the raw catfish, of which they caught a good quantity, was far more nu-tritious. Had they been able to provide fire they could have feasted.

Wood had suffered so much from hunger and excessive fatigue that his comrade feared at times that he was losing his mind. He began to talk wildly and would frequently call out in pain and terror while asleep. This added greatly to the cares of Wilson, who hardly dared trust him alone for a minute. But the raw catfish, of which he ate large quantities, seemed to help him. For two weeks after leaving Columbus they had but four meals in addition to their raw corn. Wilson believes that the fish saved their lives.

From day to day the stream widened, showing that the river journey was almost ended, but that their greatest danger lay just ahead. The rebels might have picket-boats, or the swell from the gulf might prove too much for their frail craft. They soon saw signs of a large town ahead that proved to be Appalachicola

But before trying to pass this point they resolved to take some risk in order to gain a little necessary information, and possibly some cooked food as well. So they landed not far from a cabin, and Wilson, taking a pipe he had fortunately found in the last boat, went up to the house to beg a light. The cabin proved to be inhabited by an old Scotchman, who gave some matches, and told him that the blockading fleet was off the mouth of the gulf. Wilson returned to the boat elated with his success, and found that Wood had been equally fortunate. He had discovered a negro canoe not far away with some sweet potatoes on board, and at once appropriated a good number. They were now rich indeed, and resolved to recruit a little before entering upon the dangers still before them. Accordingly they rowed to the opposite side of the river, and from thence up a small tributary to a place which looked as if it had been solitary since the days of Adam. There they built a fire in security, roasted their potatoes, broiled their catfish and ate! The only drawback was the fear of Wilson that such a banquet, after their long starvation, might prove too much for their endurance; so, when they had eaten *enough for eight men*, he persuaded his comrade to suspend for a time, and began to cook supplies for the remainder of the voyage. This done, they ate again, slept the day out, and in the night once more set forth in much better spirits and condition.

They pulled down stream till the city was passed on the opposite side of the river, then crossed to the same side, and worked along the shore. It was still fifteen miles to the mouth of the bay, and they were naturally anxious to trade their boat for a large one before going out to sea. But in this they did not succeed. They fell in with some large fish they could not name—possibly porpoises—and were a little afraid of the damage they might do. At length they resolved, as it was nearly day, to land and rest till morning. This they were enabled to do, and in a dense thicket, took what they hoped would be their last sleep on shore till they were under the Stars and Stripes.

In a wild orange grove, with tropical trees around them, they awoke the next morning, cheered and refreshed with a wild hope stirring in their breasts.

They expected great things that day ! The beauty of their surroundings had no charm for them, or rather they were too busy to attend to it. As soon as they sought their boat they noticed a singular phenomenon. They had hidden it at the water's edge; but now it was far up on the shore ! Wilson was surprised; but Wood who had crossed the ocean and knew something of the ways of salt water, told him that it was only the tide, and that they would either have to haul the boat down or wait some hours for the water to rise. They preferred the former, and were soon once more afloat.

But in the morning light they looked in vain for the white sails of the blockading fleet. They could see the spires of the city far to the north, but seaward there was nothing to cheer them. Yet they bravely—in partial ignorance of the dangers they encountered—pushed right down the bay. The cold but cooked food left from the day before, had given them strength, and they rowed steadily. They staked everything on being able to find the blockading squadron. If they failed in that, they were ready to die on the ocean, rather than go back to rebel prisons. A fishing boat coming down from the city passed not very far from them and caused no small degree of uneasiness, but it did not care to meddle with them—or perhaps they were so low in the water that they were not seen at all. On and on they rowed, right south, and toward what seemed to be the open sea. There was no sign of ships, and the waves began to be longer and to toss them about in no small degree. The land to the right and left sunk out of sight and nothing but water was all round them.

In a very plain and matter-of-fact manner, as if there was nothing especially remarkable about it, Wilson narrates this voyage out to sea in a skiff in search of the blockading squadron. The venture was hazardous. It was quite possible for the ships to be missed till night came on. Or a November breeze might very soon raise a sea that would destroy their frail craft. But unconsciously—shall we not rather say under the guidance of a better pilot than mortal man—they pushed forward.

About the middle of the afternoon, long after they had been out of sight of land, and when they began to feel somewhat lonely and uneasy, they saw to the southward an island. They gladly rowed toward it, for it would be pleasanter to rest upon it, if they failed to discover the blockading squadron in the few hours of daylight that remained to them, than to toss about on the waves all the long night. But they found that it was much further away than they had supposed and they had hard work to reach it. The island seemed to be small, and as they had no idea of stopping on it till dark, they were discussing as to which side of it they had better pass, when they discovered to the left of it, and a long distance away out at sea, a few dead trees. This was strange, but supposing them to be on some large island which would afford a good place from which to seek the American ships, they rowed diligently toward them. Before long they found a line of sand in the way, a regular bar, which seemed to grow longer as they got nearer, and they could not see how to get across it. They paddled along in the direction which seemed to lead nearest the dead trees, and at length found a narrow passage leading through it. They struck into this, and, as they passed, Wilson had another surprise. For now Wood was the leader in knowledge; at least he put out his hand and picked up what looked to Wilson like a muddy stone, and with their piece of a knife opened and ate something out of it. Wilson got the inside of the next " stone," and found it the sweetest oyster he had ever tasted ! They had struck an oyster bed; and were not in too great a hurry to stop and make another feast. These were wholesome as well as delicious.

But as they were eating, another glance at the dead trees electrified them with a new sensation, the most overwhelming of this glorious day. There were smoke-stacks among the trees! The oysters were dropped in a moment and with the vigor of assured hope they again set forth, and it was not long till the trees took the shape of masts, and best of all, there was the old flag streaming over them !

They stood up and screamed and shouted with delight. Wood was almost disposed to jump overboard to swim to them; but they considered it better, as soon as their rapture had moderated, to sit down and row steadily toward the largest. There were three of the ships, and rowing up to them with the knowledge that they were safe from their foes, was one of the happiest experiences in their whole lives.

They were crossing the bows of a little gunboat which seemed almost too insignificant to be noticed, when they were brought too so roughly that for a moment they half thought that perhaps it might be a rebel cruiser under false colors. The officers and sailors were almost equally struck with their odd appearance. They were covered with their old moss, and so starved, that it was not hard to make the commander believe that they had passed through a rough experience. When within

At Sight of the Old Flag.

speaking distance, Wilson had to tell who they were, and his story of being from the camp of Gen. Mitchel—who was no longer in Tennessee—puzzled them still more. In helping Wood up the ladder, the commander pulled off the moss with which he was covered, and the nakedness and emaciation of the poor fellow excited both wonder and compassion.

They told their story, which made the commander—Lieut. J. F. Crossman, of the gunboat *Somerset*—terribly indignant at the cruelties the men had suffered. With a sailor's generosity he gave them every kindness. They received brandy, a good wash, new clothes, and plenty of

ood food. So hungry were they that it seemed almost impossible " to
et filled up." As they wished to report the whole matter to the War
)epartment they asked the commander to signal the cruiser, which was
ne large ship they had first seen, and send them to Key West. The
ruiser very appropriately bore the name of the *Stars and Stripes.*

On this voyage two or three days passed like days of paradise, and
hen Wilson was prostrated with fever, the result of his terrible exposure
nd hardship, and for several days knew nothing more. Then recovering
nd arriving at Key West he sent a report, which is still filed in the War
.ecord office at Washington under date of November 12th, of the whole
xpedition. A still fuller account was published in the *New Era* of the
5th, and these are the first reports given of the raid by any person en-
aged in it. These two were first of all the refugees to arrive at their
estination, notwithstanding their journey of more than four hundred
hiles.

CHAPTER XXX.

THE RECAPTURED PRISONERS.

THOUGH the story of those who were retaken presents more of variety and vicissitude of fortune than that of the escaping prisoners, yet it is darkened with the shadows of prison life, and the continued apprehension of rebel scaffolds. But it is not without a bright side, as indeed no human lot is. At the very first our fears were not realized. We expected worse treatment than before—perhaps some of the horrible punishments which were too possible in the South, or a speedy trial and the death of spies. But on the contrary, the mildest and most humane treatment we received during our whole sojourn in the Confederacy followed. Col. Lee no longer thought the jail a safe place, and ordered us to be taken to the barracks of the rebel troops, which were in the middle of the city, but two squares from the depot. Here we were in the midst of soldiers who had the exclusive custody of us, and their commander was held directly responsible for our safe keeping. Any attempt to escape would meet with far greater difficulties than at the city jail. We did not cease to discuss the subject of breaking out, but the opinion was general that it was hopeless.

The building was large and airy, with full-sized windows, and with only one room in addition to our own, devoted to prisoners. These were Union people of Atlanta, and while closely guarded were not especially ill-treated. Our room was perhaps twenty feet square and was situated on a corner, with two windows overlooking the main street of the town, and two others opening on a side street. To make it almost a paradise in comparison with our former quarters, it had a gas burner which was kept lighted all the night and a large open fire-place with a generous hearth, on which blazed a heap of logs. There was but one door to the room and this was never shut; a soldier stood continually in it with orders to notice all that was going on.

I love to linger on the details of this prison room, for when we were expecting a dungeon like that of Chattanooga or death, it came as a blessed relief,—a place where we could rest for a little time from the storms that had swept over us. Our hearts were very grateful to God and we felt that He had not deserted us. To sit in the window and watch the tide of human life flowing past was an unending delight. Scarcely less pleasant was it to gather round the fire, especially in the evening, and watch the play of the flames. We carried in the wood, always with an armed guard for an attendant, while the wood-pile itself was guarded carefully, as well as every entry of the building. In fact the obstacles to escape here were not bolts and prison walls, but the eyes of watchful sentinels, for we were always in the midst of a little army. We did not think the latter less of a hindrance to escape, but we well knew which was pleasanter. We were soldiers, and to be in the midst of this army-life was far better than any prison, though we would greatly have preferred another flag!

We were also sometimes taken out of the rooms for our meals, which was a grateful change in the routine of the day. It is true, we always marched under a strong guard, but to stand around a rude board, and have the not very abundant food, consisting of bread and meat with the addition of a little soup, placed before us in a way that suggested a table, made this change very acceptable indeed.

Much of the improvement in our treatment we attributed to the Commander of the barracks, Major Wells, or Jack Wells as he was familiarly called, who had been an old United States Army officer. One of the traditions of that service was to treat prisoners well, and he had none of the pro-slavery feeling, which in the South too frequently over-rode all sentiments of humanity. He would come around to our room and talk with us by the hour—telling us great stories of his adventures and receiving as great in return. Very often he was half drunk, for temperance was not among his virtues, and sometimes he did not stop at the half-way point. But the habits of a soldier were so completely grounded in him, that no one could take advantage of such occasions to violate any military rules. When in particularly communicative moods he would tell us that he did not care a cent which side whipped—that he only held his present position to avoid being conscripted. But his superiors knew him to be so faithful and vigilant, and he could so readily control the rude mass under him at the barracks, and enforce so stern a discipline, that they readily forgave these little slips of the tongue. When aroused he had the regular officer's sternness in punishing, as one or two incidents will show.

There were six left of our original party. These with the four prisoners of war and ten Tennesseeans made twenty in all who were confined in this room. One morning our provisions, which were very scanty, were brought up in a tray, each man's portion being set by itself. Mr. Pierce,—called "Gun-barrel" from the wound in the head,—thought his portion too small, and without a word, but with a look of contempt, threw it back into the tray. In a few minutes a guard came up and seized the old man—he was over sixty—and taking him out into the cold hall, tied his hands before his knees with a stick over his arms and across under his knees in a way soldiers call "bucking." He was left there all night! We all felt very indignant, but could do nothing.

One of the guards was a malicious fellow who delighted in asking the prisoners how they liked being shut up, playing checkers with their noses on the windows, etc. Another Tennesseean, named Barker, told him that he need not be so proud now, for when the North conquered them he would have to work like a slave in the cotton fields to help pay the expenses of the war. The guard reported the "treasonable" remark to Wells. Poor Barker was taken to a room where punishments were usually inflicted, and there hung up by the heels till he fainted; then taken down till he revived, and hung up again; this being continued till they were satisfied. Then he was put in a little room, with no light, the smallest in the house, and kept there with nothing to eat for twenty-four hours!

Here I had many opportunities of seeing instances of the iron tyranny to which Union men in the South were continually subjected—a tyranny which explains the apparent unanimity shown in the cause of secession. The strictest espionage was maintained in every order of society. The spies of government would pretend to be Union men, and thus worm themselves into the loyal societies which existed in Atlanta as well as in many other cities. Clear proof of disloyalty to the Confederacy was not needed, for a strong suspicion would do as well when only a military order was needed to keep a man imprisoned in a horrible dungeon for months. In one day, seventy men and twelve women were arrested and sent in irons to Richmond! Afterward we had the opportunity of seeing some of these very persons in that town. In spite of this persecution there existed at this time a society of loyal people who were ready to extend to Union prisoners all the help in their power. From some of them I received—as did many of the other prisoners—supplies of money and other needed articles which were of the greatest value. This help was given at great personal risk to the donors, for to give a Union soldier money was considered a serious offense. One man who was in the opposite room, and with whom we managed to hold some communication, had been in prison for four months on the suspicion of having given some help to a few of the Shiloh prisoners who were taken to Atlanta. A physician named Scott visited us on the plea of mere curiosity, and on learning that if we wanted a doctor they would allow one to be sent for, we managed to get him into our room and talk with him a little. He gave me some money to be used for the benefit of the party, which was of great service. A year afterward when I found him a fugitive in the North, I was able to partially repay the favor by vouching for his loyalty, and putting him in the way of gaining a livelihood.

A little further lightening of my own confinement came about in a very singular manner. I was making some short-hand notes on the margin of a book one day, when Wells walked quietly into the room and saw me at work before I noticed him. I feared that he might be displeased, and take the book from me, but it was too late to retreat, and I wrote on. He watched for some time with great seeming interest. "What kind of crowtracks are you making there?" he asked. I explained as well as I could, and showed him particular words that I wrote. He then wanted to know why I was writing. I did not think it safe to tell him that I was making notes in the hope that they might be useful some day, and therefore told him a part only of the truth; that I wrote to pass the time, as I had more of it on my hands than I knew what to do with. I then read him a few isolated sentences which did not strike him as having any treason in them, and he left me indulging the hope that nothing bad would come out of his discovery. I did not wish to give up the little book in which I was writing, but as a precaution copied some of the notes into another place.

The next day he came again, seemingly in a very pleasant humor, talked cordially with different members of our party, and then turning to me asked if I would be willing to do some writing for him? I said I would, very gladly, if it was for him only, but I did not wish to do any for the Confederacy. He laughed at the distinction, and asked me to go into his office. There he showed me the report he had to make each day of the number of prisoners, and the rations drawn for them as well as the requisitions. The blanks were not very well filled, for he was not a good clerk. I asked him why he did not have some of the soldiers do the work. He answered in terms very uncomplimentary to them, and it was a fact that literary ability was not as common in the Southern army as in the Northern. I could not see that my making out this return daily would do any harm to the United States, while it might be the means of giving me valuable knowledge or even of effecting the escape of myself and comrades. The work was nothing—I would rather do it than not; but I told him that I could not make out such a sheet properly without the use of desk, paper, rulers, etc. He said that I should have them, and could come right into his office for the purpose. This was exactly what I wanted, but the boon was not as great as I hoped, for he went to the sentry at the door of the prison room and instructed him to let me pass to the office when I wished, but nowhere else; another sentinel who stood at the head of the entry within a musket-length of the office door was told to watch that I never passed further than the office. In short all the guards were informed of the new arrangement and charged to watch me especially! This did not leave me much chance for getting away, but it was possible that their vigilance would relax with time, while it was certainly pleasant occasionally to change from the prison room to the office and to get many an item of intelligence which I could carry back to my comrades. But this work gave me the heartfelt pleasure of helping a man to escape death,

While writing in the office one day, a person dressed in the uniform of a rebel officer was brought in for confinement in the barracks. He appeared to be very drunk, but remonstrated so hard against being put in the room where the remainder of the rebel prisoners were kept, that Wells consented to let him stay for a time in the office, especially as regular charges were not yet brought in against him. He was carefully examined, but when his money was taken he was so furious that Wells, first showing me how heavy it was, for there were some five hundred dollars in gold, allowed him to retain it. He was very drunk, and Wells, in the absence of other charges, supposed that he was arrested only for that—an offense with which he had a great deal of sympathy. Soon after Wells went out to attend to some business, leaving Sergeant White with us. He too soon went out and I was alone with the drunken officer. Of course there was a sentinel in the entry, not a dozen feet away. I was busy writing, but looking up I saw the stranger approaching me. There was no trace of drunkenness about him. I watched his movements attentively. Soon he was standing by me.

"You are a prisoner.'

I nodded assent.

"One they call ' engine-thieves,' " he continued.

I answered in the affirmative.

"I know you," said he. "I know all about you. I was here when your comrades were hung. Brave men they were, and the cruel deed will yet be avenged. I am not afraid to trust you. They do not yet know who I am, but they will learn to-morrow, and then if I am still in their hands I must die, for I am *a spy from the Federal army.* Can't you help me to escape?"

I was astonished at this revelation, and for a moment doubted his story, and his object—thinking he might wish to betray me for a selfish advantage. I put a few hasty questions to him to test his knowledge of the Federal army. The answers were satisfactory. I hesitated no longer, but determined to help him if in my power.

"What can I do for you?"

He answered, "Can't you write me a pass and sign the Commander's name to it?"

"That would hardly do you any good," I said, "for it would probably be detected. I think I can put you on a better plan. Take that overcoat," pointing to one belonging to Wells that was lying on the foot of the bed, "put it around you, and just walk past the guard as independently as if you owned the entire establishment. It is now nearly dark, and the chances are that you will not be halted by the guard at all."

"A good idea," said he; "I'll try it."

At once folding himself in the coat he bade me a hurried but grateful adieu.

Eagerly I sat in the deepening twilight, listening for any sound that might betray the success or failure of the scheme. But all was silence. I have since learned that the guard, seeing the familiar coat, supposed of course that the owner was in it and allowed it to pass unchallenged.

I have often been asked why, if I could help another man off in this manner, I did not avail myself of the same opportunity. The circumstances were very different. I had no thought of it at all till it came as a sudden inspiration, when the man asked for help. Besides, I was especially watched, and was several inches taller than the officer and the commander, much more slender, and always wore my glasses, without which I would have been helpless. If all these obstacles had been surmounted, I would not have been, like this officer, dressed in rebel uniform, with abundance of money, but destitute and nearly naked. The chances were in every way a hundred to one in his favor.

A moment after Sergeant White came in, and I engaged him in conversation, inducing him to tell some good stories to keep him from missing my companion, and to allow him as much time for a start as possible, before the inevitable alarm was given. I succeeded perfectly for five minutes, for he did not make the least inquiry, probably thinking, if he gave the matter any thought, that Wells had taken him over to his natural place in the barrack-room opposite. Then Wells entered, and throwing an uneasy glance around, exclaimed at once,

"Sergeant, where is that officer?"

The sergeant protested that he knew nothing about him; that he was not in the room when White entered.

Wells then turned to me and demanded,

"Pittenger, where is that officer?"

"What officer?"

"That officer I put in here."

"Oh! that drunken fellow?"

"Yes; where is he?"

"The last I saw of him he picked up his coat, and said he was going to supper."

"Going to supper, was he! Ho! I see it! Sergeant, run to the guard, and tell them that if they let him out I will have every one of them hung up by the heels."

This was rather a useless punishment considering that the prisoner was already far away.

But the sergeant departed to muster the guard. Shortly after Wells, who had resumed his seat, said in a meditative tone,

"Had he a coat?"

"I suppose so, sir," I returned, " or he would not have taken it."

"Where did he get it?"

"Off the foot of that bed."

Wells sprang to his feet as quickly as if a serpent had bit him, kicking over the chair on which he had been sitting, and exclaimed:

"My coat, sure as ——! worth eighty dollars! The villain!" Wells pressed his hands to his head and sat down again; then, as if thinking better of it, he ejaculated, "Well! if that is not a cool joke," and burst into a loud laugh which ended the scene.

While here we secured the most delicious article of diet we had yet received in the South. Sweet potatoes were cheap and very abundant now, and the money kindly given by Union men sufficed to buy a considerable quantity. These we roasted in the ashes of the open fire that made our room so comfortable, and if there is any thing more palatable than these potatoes were, I have not found it. But for the want of clothes and the opportunity of being clean, and the vermin which it was impossible to banish in any of these prisons, we might have been reasonably comfortable.

We still continued our devotions morning and evening. The fact that the door was open, so that we had no privacy at all, and that a guard was always looking on, did seem like a hindrance at first, but in time we grew used to it. A few of our own number seemed to think that now we were not in a dark cell, and were treated more as regular prisoners, there was no need of so much prayer. But the majority clung to the good resolutions made in darker hours. When one of the guards once interrupted us with coarse remarks we appealed to Wells, and he stood our friend; saying that he did not like praying himself and would stay away from it, but if we could get any good out of it we were welcome, and the guards should not interfere. Thus we were enabled to pass many hours most pleasantly and profitably.

But our friend MacDonell came no more. When taken back to our prison room after the flight, I tore up the note I had written to him, finished reading the books we had, and returning them, asked for more. But they did not come, nor did he ever enter our prison again. Two reasons were sufficient to account for this. The Provost Marshal was disposed to attribute the success of our effort in breaking from the jail to outside aid; and those who were very friendly would more likely be suspected than any others. It might not be safe for MacDonell to come to us again for awhile; besides his term for serving that church in Atlanta was nearly ended, and removal is an event in a Methodist preacher's life which for some weeks overshadows everything else.

But as I knew nothing of this and he did not come, I resolved to go to him. Accordingly I asked Wells to send a guard with me to go to see a preacher. I might as well have some indulgence for the amount of help I was giving him as not. He asked me a good many questions and made me tell the whole story of our acquaintance, but then consented, and ordered a soldier to take me to the minister's house and not to let me get out of his sight for a moment; to treat me well, but to promptly shoot me if I tried to run away. Then he made me, in addition, promise that I would make no effort to escape.

Thus I set forth through the streets of Atlanta guarded by *only one man!* It did seem almost tempting Providence to go forth in this unsafe manner, and as I gazed at the soldier, it looked an easy thing to take the musket from him and march off. He imagined that he was keeping me most safely by being close to me; the very position in which his musket would have given him no advantage. But I resisted the impulse as I had no intention of trying to escape on this occasion.

I felt a good deal ashamed on entering the fine parlor of Mr. Mac-Donell in such shabby dress and with the armed guard behind me, but the kindness of his reception put me at ease and we talked for quite a long time. I tried to make him understand how grateful I was for all his kindness, and gave him my copy of *Pilgrim's Progress*, with an inscription, as the only remembrance in my power, informing him of my resolution to devote myself to the ministry in case my life should be spared. He prayed for me most fervently and we parted. He promised to come to see us in the prison again, but we were shortly moved away, and I have seen him no more, though since peace came we have often exchanged letters.

An effort was made at this time to get some of us to enlist in the rebel army. Whether our party would have been included we could not tell, and had no desire to provoke an investigation, though at an earlier period we would have been delighted with the opportunity. But others were strongly urged, especially the two regular soldiers. These, with the two from the 10th Wisconsin, were still with us, and it did seem as if they were being held simply for our sake. Yet this was scarcely worth while any longer, for it was certain that if any of our number escaped, they would carry the news of our adventures through the lines. These four were, however, kept with us for some time longer. One of the regulars, Geo. W. Walton, became a great friend of mine, and contributed not a little to lighten the dreary days we still had to endure.

At length we heard what seemed to us fearful news—a new court-martial was ordered! We waited the result with the greatest anxiety, and my comrades charged me to try all I possibly could to get from Wells any intimation of an intention, if such should exist, to try us. In case anything of that kind occurred we were resolved to make an effort for freedom even if the result would only be to die with muskets in our hands. That would have been almost inevitable. We could capture the three guards—one at our door, one in the entry, and one at the head of the stairs; but not less than twenty were in reserve in the guard room, and the whole building was surrounded with soldiers constantly on their beat. We could only have had the six of us to count on, for none of the other prisoners would make such an attempt; but we would have preferred dying in this manner to being led powerless to the scaffold. A week of sickening suspense passed and no summons came for us. Then the court adjourned and we breathed easier. It seemed as if they did not intend to persecute the feeble remnant of our party any further, and we passed from the extreme of fear to that of hope. We began now to talk of exchange, as if that was the most probable ending for us.

It may be interesting to inquire whether the alarm under which we were led to break out of the jail was well grounded, or a mere scare. At the time we had no doubt. To escape or die was the dread alternative we saw. To the present day, nearly if not quite all of our number are perfectly convinced that it was only this bold attempt that saved the life of a single member of the party. The chain of evidence was very strong, as previously stated, and until recently I did think that we would at least have been summoned before a court-martial, and to us this had but one meaning. But since examining so many of the records I am disposed to doubt whether anything had been discovered that would have led to a new trial. It was weeks after the escape before any court met, and then no one of our party was summoned before it, which would have been the most natural course if information was their aim. In Col. Lee's report of the escape, he still complains of a want of knowledge in regard to the facts of our case. But he says nothing about the summoning any court. The regular soldiers in the front room, from whom came the most alarming reports, were very anxious that we should attempt to escape and succeed, though they had determined beforehand, as we learned afterward, not to go with us; but they were anxious to be separated from us, believing that they would then be exchanged. But the strongest evidence is the simple fact we were not tried. There had not been time for any representations from our government to be made. Possibly our enemies were weary of following us further. But it is on the other hand sure that charges against us were not dropped, and we were not treated as ordinary prisoners of war.

The weeks rolled by. Few things occurred worthy of note. The same monotony which makes prison life so dreadful robs it of interest when recorded. We would rise in the morning from our hard beds, which consisted of a few fragments of carpet, with our shoes and coats for pillows, and wash ourselves—for we had that great privilege here—pouring the water over each other's hands; then eat our scanty breakfast; then put in the dreary forenoon as best we could. We no longer had the regular employment which had marked our life in the jail. When dinner-time came we would eat either in our own room or taken out by the guard to the common mess-room, which we much preferred; then read or talk, sit round the fire, roast potatoes or do anything else that could be found till dark. Then the gas was lit—not for our benefit, but that the guard might see that we were working no plot for escape. This hour, as in the jail, was the most pleasant of the day. Then came prayers, and with the dull fire light and the brighter gas, we again lay down to sleep.

Thus days glided into weeks and weeks into months. The golden hues of autumn deepened into the sombreness and death of early winter, and still we were in Atlanta barracks. Our weak faith could scarcely conceive that we should ever be anywhere else! A mental and spiritual numbness, like that which follows the long-continued infliction of physical pain, took possession of us. We almost ceased to hope!

But at the close of November there was a startling change—a day of rejoicing! A number of officers came up to the barracks and inquired for the room occupied by the Yankees. On being shown to our room they called us all into line and said, with great manifestations of friendliness, that they had glad news for us:

"You have all been exchanged, and all that now remains is for us to send you out of our territory."

Then they came along the line and shook hands with us, offering us congratulations, and wishing us a safe journey home.

These were some of the leading officers of Atlanta, and we could not doubt. I am now disposed to think that they believed what they told us. The correspondence about exchange in the War archives shows that at this time arrangements were in progress to bring to Richmond the prisoners scattered through various parts of the Confederacy for exchange. The only fault with which our visitors can therefore fairly be charged, is that they spoke as if that was already accomplished which they simply believed would be done.

But we had no doubts, and our feelings may be better imagined than described. There was an overwhelming rush of emotions too deep for utterance. We were very happy, but there was a deep undertone of sorrow. Oh! if our seven comrades had only been with us to share the joy of this unexpected deliverance! And we were uneasy about the eight who had fled. Of only two had we heard anything, though we had tried every source of information. From a rebel paper containing an extract from the *Cincinnati Commercial*, which I had seen in the office of Wells, we learned that Wollam and Porter had arrived at Corinth in almost starving condition. Of the others we had no reliable information. Lee had told us that three of them had been killed in the woods, but we could not

24

meet any soldiers who had done the killing, or had seen them dead, and we doubted. This suspense had caused us many unhappy hours.

But when all allowance was made for these things, the prospect of gaining our liberty was enough to make our hearts overflow with gratitude to Almighty God for his wondrous mercy. I was so agitated that when Wells asked me to write the requisition for rations for our journey, I was not able, and had to transfer the work—to more steady hands. It was in the forenoon that we received the news, and we were to start for home,— *via* Richmond,—at seven the same evening.

We spent the intervening time in arranging what clothing we had— little enough—and preparing for the journey. As the time drew near we rejoiced over the prospect, but were not without a little of that strange reluctance to leave a familiar place which has so often been considered one of the marvels of human experience. We lit the gas and built a roaring fire, the ruddy blaze of which was itself an emblem of cheerfulness, and took a farewell view of the place where we had spent so many not unhappy hours. Often afterward did we think of that bright fire and still brighter hour of expectation during the cold and dreary months that we were still doomed to suffer.

We had here quite a number of pieces of carpet which served us very well for blankets, but we were forbidden to take them along; being told that we were going where good blankets were made, and would soon have plenty. We managed, however, to secrete two very small pieces, which were afterward of the greatest use. I had one article of dress in which I took a good deal of pride—a very fine hat. Nearly every other part of my apparel had been worn out long before. But one day, when he was somewhat intoxicated, Wells was pleased with the neat look of a report I had made out for him, and showed his approval in a characteristic fashion. He made a grab at my cap and threw it on the floor, saying that I was too good a man to wear a cap like that, and putting his own hat, a soft, new felt of finest quality on my head, said, "Wear that." I could afford to let him have his way in the matter, though I did not suppose the trade was to be permanent, and therefore, when he was gone, carefully gathered up and secreted the old cap, expecting that he would take the hat. I was almost sure he would before we left for Richmond, but had no disposition to call his attention to it.

All was now in readiness, and we took our last look at Atlanta—as a rebel town. We were not tied for the first time in our removals. This was truly remarkable and afforded strong confirmation of our hopes. The guards fell in on each side of us and we wended our silent way along the dark streets. Wells, drunker than usual, accompanied us to the cars, where he hiccoughed an affectionate farewell—and did not ask me for his hat ! Sergeant White, who was with me when the spy escaped, commanded our escort, and we could not have had a more agreeable companion. Like his superior he cared little which side came out best in the war, so long as he was not hurt! There were but ten guards, while with the Tennesseeans we were twenty,—a great falling off from former precautions. We were crowded into box-cars, and began to suffer severely with cold, for the night air was most piercing. It was now the 3rd of December, and we had nothing but our ragged spring clothing—not a single article of any kind—always excepting my hat!—having been supplied by the Confederates. I had also the well-worn coat left by Capt. Fry, without which I could scarcely have endured the journey. At three o'clock in the morning we arrived at Dalton, from which place we were to change to the road direct to Cleveland. We were not to go through Chattanooga, and I saw that famous town no more for a score of years. The stars were sparkling in light and frosty brilliancy when we stopped. The other train had not yet arrived, and as we stood on the platform the keen icy wind seemed to cut through us. We could find no relief from its piercing breath and shivered for an hour, when the other train arrived, and finding it a little more comfortable we managed to doze away the time till daybreak.

In the morning we found that there was hardly enough of our three days' rations, which were to last us to Richmond, for breakfast. We ate what there was, hoping to be able to buy more with the remnant of money given us by our Union friends of Atlanta. When that was all spent we had another resource that never failed—the endurance of hunger!

During the day we quietly discussed the question whether it would not be best at night-fall to try to make our escape, as we were now within forty miles of our own lines. It would not be difficult, for the guards were careless and we could at any instant have taken more than half the guns. They sat on the same seats with us and slept. Frequently those guarding the doors would fall asleep, and we would wake them as the corporal came on his rounds, thus saving them from punishment. The most complete security prevailed with them, showing that they were sure that they were taking us for no other purpose than that of exchange. Once Sergeant White laughingly told us that we could escape if we tried, for we had the matter in our own hands; but that he thought it would be more pleasant for us to ride around, than to walk across the country on our own responsibility. This very security lulled our fears and made us the readier to believe what had been told us in Atlanta. There were formidable difficulties in the way of escape, outside, in the terrible task of trying to travel almost naked and barefoot over the rough mountains, and in the snow which now began to appear.

In the afternoon we passed the beautiful village of Knoxville, with its sad memories of the court-martial, where we had been so terribly deceived, and then the town of Greenville, which we noticed as being the home of our heroic companion, Captain Fry; then into the lower part of Virginia. It was nightfall when we entered this State, and a beautiful night it was. The moon shone over the pale cold hills with a silver radiance which made the whole landscape enchanting. On, on, we glided, over hill and plain, and leaning back in the car seat, with the guards and most of my comrades asleep around, I seemed to see in the shifting scenery of the unreal-looking panorama without, a type of the fleeting visions of human life—like us, now lost in some dark and gloomy wood, or walled in by the encroaching mountain side, and now revealing a magnificent view of undulating landscapes far away in the shadowy distance. Thus through the silent night we journeyed on, and morning dawned on us still steaming through the romantic valleys of Virginia.

But there soon came a change in the weather that added very much to our discomfort. The day became wet and dreary. Our car leaked, the fire went out, and all day we had nothing to eat. At length the miserable day wore to the evening, and found us at Lynchburg, which is literally a city set on a hill. Here we discovered that we had missed connection, and would have to lay over for twenty-four hours. We were very sorry to hear this, for though the journey had come to be intensely disagreeable, we were in a great hurry to get to our own lines, and had been talking all the way about what we should do in Washington. But there was no help for it, and we marched up to the great empty barracks with as good a grace as possible.

We found an immense room in which was gathered some of the refuse of the Confederate army,—prisoners for various infractions of Confederate law. There was a great stove in the centre of the room, but as it had no fire, it was not so welcome to us in our wet and chilly state as it might have been made. We resigned ourselves to another night of freezing, with the consoling thought that we would not have many more of such to spend. I was always sensitive to cold, and therefore could sleep but little; so I paced the floor till nearly morning and witnessed a good many amusing incidents. Some of the Confederates were drunk enough to be disposed to mischief. One of them diverted himself by walking around on the prostrate forms of those who wanted to sleep. Soon he came to Bensinger, of our party. He endured the infliction patiently the first time; but as the drunkard came the second time, Bensinger was on the lookout, and springing up, planted a blow on his ear that not only laid him out, but disposed him to be quiet for the rest of the night! Some of his comrades rushed forward to resent the well-deserved punishment; but when they saw the company that gathered around Bensinger, they concluded to press the matter no further.

In this Virginia prison we met the most determined and bitter rebels we had yet seen. One prisoner said that he had long advocated raising the black flag, declaring that if it had been done at first, the war would long since have been over. I assented, adding that the whole Southern race would have been exterminated in a short time. This way of ending the war had not entered his mind, and now he did not enjoy the suggestion.

All the next day continued cold and gloomy. About noon we succeeded in obtaining wood for the big stove with permission to light a fire, and a genial glow was gradually diffused over the room in time to warm us thoroughly before we started for Richmond.

From this point to a junction we had good cars, the most comfortable we had yet enjoyed. But in a little while we were obliged to leave them, and as the other train had not yet arrived, we built a large fire to defend us against the bitter cold. We were allowed to assist in gathering wood, and even went some little distance into the dark woods after fuel. Not to escape under such circumstances seemed almost a sin; if so, we repented it afterward as heartily as a sin was ever repented of. But we were on our way to be exchanged, and when we stopped to think of it—though instinct urged us to go—it was surely better to go on by rail and steamboat to Washington, than to dare the rigors of the Virginia mountains in winter, without clothing enough to defend us from perishing, through a single freezing night. But in this case instinct was right and reason wrong. Soon the other train arrived, and a few hours more of rapid and eventless travel placed us in the rebel capital.

CHAPTER XXXI.

RICHMOND AND ITS PRISONS.

RICHMOND is a beautiful city, but we saw little of its attractions on this first visit. There was still the same sparkling moonlight and the same intense and piercing cold which marked our journey on setting out from Atlanta; we were instantly chilled when we had left the shelter of the cars, and started through the streets of the sleeping town. Everything looked grim and silent through the frosty air, but we cared little for such a cool reception if we could speedily go through on our way to the place of exchange.

We had hardly passed a dozen squares of this historic city, when suddenly it occurred to Sergeant White that he did not know what to do with us. He had been ordered to bring us to Richmond, where he had never been before, and was given a letter to the Provost Marshal, but had no idea of where to find him. This was an unfavorable hour for inquiries. He preferred entering upon the search alone, rather than with a crowd of prisoners at his heels. For a time we were left in charge of a subordinate, and endeavored to shelter ourselves as well as we could from the unbearable cold, which threatened to prove fatal. Necessity is rich in expedients. We huddled into the angle of a brick wall and spread the two pieces of carpet over our heads. It was astonishing to notice how much more comfortable this made us—especially those on the inside of the pack where I happened to be. In perhaps an hour the sergeant came back and conducted us to the Marshal's office.

Our company, thirty strong, threaded several of the principal streets that would have been entirely dark but for the moonlight, and entered the office, which, to our further discomfort, was fireless. We stood for an hour or more in the empty room, looking at the grim portraits of rebel generals that stared at us from the walls, until the Marshal himself entered. He did not deign to speak to us, but opened the letter handed him by the sergeant and read that ten disloyal Tennesseeans, four Federal soldiers, *and six engine thieves*, were hereby forwarded to Richmond by order of Gen. Beauregard. We had hoped that this title of which we had become heartily sick would now be left behind; but it still clung to us and seemed ominous of further suffering. The Marshal then gave his orders and called a guide, under whose direction we again marched off.

By this time it was daylight, December 7th, 1862. Richmond looked still more cheerless by the cold beams of daybreak than it did before.

We marched along several tedious streets until we came near the bank of James River, where we halted in front of a most desolate-looking, but very large brick building, surrounded by a formidable circle of guards. A sentinel summoned an officer, and we learned, as we had already surmised, that this was none other than the far-famed *Libby Prison.* We had expected to enter it, having heard that it was the place to which exchanged prisoners were sent on their way to City Point, where exchanges were completed. We hoped that we would not be obliged to stay here a whole

Libby Prison. From a photograph.

day as we had done at Lynchburg. We went up a flight of stairs, and on turning to the right entered a vast open room, with a multitude of people in it, and saw almost for the first time since leaving Mitchel's camp, the old familiar United States uniform. We were at once in the midst of comrades.

At first our greeting was not very warm, as we still wore the citizen clothes, or rather rags, which had done duty day and night for the past eight months. Our appearance was not prepossessing, being now more emaciated and tattered than the average of prisoners; but our story was an unfailing passport, and we were soon heartily welcomed. There was only one small stove in the midst of the great room, and as many of the inmates as possible were closely huddled around it. But they generously made way for us and our blood was thawed, after which we had leisure to look around us.

The room was bare, and with the exception of great rafters, was open clear to the roof. The window sashes had been removed and the wind whistled in from the river far more sharply than was consistent with comfort. The stove was kept red-hot, but this only warmed a limited circle, and had no effect on the general temperature of the room. But with all these discomforts and many others of which we heard, we rejoiced to be here. It was to our simple minds the sure pledge that our enemies had not been deceiving us in their promises of an exchange; for the soldiers in whose company we were placed expected to go north in the next truce-boat, which was now due, and they had actually been paroled for that purpose. Our hearts beat high as we talked with comrades from different parts of the country, and thought that, after drinking the bitter draught of bondage and persecution for eight long months, we were at last to taste the sweets of liberty. What wonder if our joy was too deep for words, and we could only turn it over in our minds, and tremble lest it should prove too good to be true! But there came a swift and terrible shadow, and our dream of delight vanished!

From our companions we learned many interesting items of news. They told us of the battles in which they had been captured, and of the varying fortunes of war. But strangest of all to us, was what they said about the existence of a large party at the north who were opposed to the continuance of the war, "because," as my informant explained, "they feared if it went on that they might be drafted and have to take part in the fight."

After we had been here an hour or two, an officer entered and called for the men who had been last put in. Expecting that we would receive the regular exchange parole we promptly responded. They took us down to the entrance hall and called over our names, identifying each one. The four soldiers and one of the Tennesseeans were put on one side and we on the other. The first party were then conducted back up stairs again, while we were put into an immense, dark, and low room on the opposite side of the first floor.

These changes were too significant to be easily misunderstood. It was an awful moment. We knew that we had been deceived, and our hopes at once fell from the highest heaven to which they had soared, down to perfect nothingness, and misery and despair overwhelmed us. To be thus separated from our friends seemed like parting the sheep from the goats, and could be only because they still held charges against us which would prevent our being treated like other prisoners. No wonder we looked at each other with pale faces in the dim light, and asked questions that no one wanted to answer. But only for a few moments were we crushed under this unexpected blow. Soon we again sought avenues of hope.

Perhaps they did not recognize us as soldiers and only wanted to exchange us as citizens—a matter of indifference to us, so long as they exchanged us at all. We looked around to see what foundation we could find for this pleasing conjecture.

There were even more prisoners here than upstairs. Civilians were gathered from all parts of the South. Some of them had been in prison ever since the war broke out, and a few even had been arrested for supposed anti-slavery principles before that event, and had lived in loathsome dungeons ever since! This did not look much like a speedy exchange! A few only of our soldiers were here, and they as punishment for attempting to escape.

In the meantime breakfast was brought in. It consisted of a small quantity of very thin soup, and a very scanty allowance of bread. To our delight the latter was made from wheat instead of corn, which had been the only food further south; and all the time we remained in Richmond we received good bread, though but little of it.

Just as we began to feel somewhat at home in this dark, dirty, and gloomy place—which, however, was not so chill as the room above—an officer entered and inquired for the last fifteen men who had been put in. We answered with great alacrity, for hope instantly suggested that there had been some mistake which would now be corrected and we taken up stairs again. But no such good fortune was in store for us. There yet remained a lower deep to sound. We were taken out of doors and found a guard waiting to receive us. Any change to another prison was almost sure to be for the worse, and we were probably the only band of prisoners who were sorry to leave Libby!

We crossed the street and halted at another desolate-looking building which we afterward learned was "CASTLE THUNDER," the famous Bastile of the South. Like "*Libby*," it had been a tobacco manufactory before the war. We were conducted through a guarded door into the general office, where we had to wait for some time. A fierce-looking, black-whiskered man, whom I knew afterward as Chillis, the prison Commissary, walked through the room, and looking at us for a moment, as if his eyes would pierce us, said:

"Bridge-burners, are they? they ought to be hung, every man of them; and so ought every man that does anything against the Confederacy."

I could have better agreed with him in the state of feeling I now had if he had said "for."

The guard returned and we were ordered up stairs. At the first landing we passed by a howling and yelling multitude, who made such an outrageous noise that I was compelled to put my hands to my ears. As we came in view, more than a score of voices screamed with all the energy their lungs could command, "Fresh fish! Fresh fish!" I found that the same exclamation greeted every new arrival.

On the floor above this there was another howling multitude. We were taken into a little office at one side and searched, to see if we had anything that might be useful to a prisoner. They took some pocket-knives from the Tennesseeans, which they had managed to keep secreted till now. But I had been learning prison arts, and when it came my turn, I slipped a large knife which I had obtained at Atlanta up my sleeve, and managed by turning my arm when they were feeling for concealed weapons to keep it out of the way.

When this examination was over I thought they were going to put us into the imitation mad-house we had passed; and they did not do much better, for they took us down the centre of the room about half way, and then placed us in a stall beside it. I call it a stall, for no other word so well describes it. It was one of a range about the size and shape of horse sheds, partitioned off from the large room in which were the noisy miscreants, and from each other only by long boards nailed to uprights with cracks between wide enough to let the wind circulate freely. Most of the window sashes of the large room were out, which greatly increased the cold. Our stall was only eight or nine feet in width, and possibly sixteen in length. Its entire furniture was a low shelf, about six feet long, which served to support a bucket of water placed on it.

In this cheerless place our party, six in number, and the nine remaining Tennesseeans—fifteen in all—were confined during the months of December and January. There was a window in the shed which afforded abundant light. We did not suffer from crowding or lack of air as in the Swims dungeon; but other evils endured, especially cold and hunger, were scarcely less tormenting than the inflictions of that vilest of all dens.

The first day of our imprisonment here our spirits sank lower than they ever did before. All our bright hopes were dashed to the ground, and there seemed good reason to fear that we were doomed to this dreary abode for the remainder of the war, even if we escaped the dreadful fate of our comrades. It was too disheartening for all our philosophy, and that day was one of the blackest gloom. We seldom spoke, and when we did, it was to denounce our folly in suffering ourselves to be deluded by fair words to Richmond, when we might so readily have escaped. Our lot would have been much easier to bear but for the conviction that it was our own fault. We ought not to have believed them. It was only the shrinking from the bitter cold on the mountains that disposed us to credit their promises and kept us from leaving; and now the promises were all broken—and from the way we had begun it seemed likely that we would suffer as much from cold as we could have done in making our escape—an anticipation abundantly verified.

But it was no use lamenting, and all we could do was to register a solemn vow that we would never be deceived again in that manner, and that we would never let another chance of escape pass. It is easy " to lock the door after the horse is stolen."

But when night came we found the help we might have had sooner if we had not been too angry with ourselves to seek it. At the time of evening worship, we read our chapter, sung our hymn, and then knelt in prayer. I tried to roll all our cares upon the Lord, and felt a sense of his nearness that was inexpressibly precious. Rising from my knees I was comforted in the assurance that whatever happened we had one friend, " mighty to save." The more I thought of Him the more it seemed that this prison hardship was not an unendurable evil as long as He gave us the light of His countenance.

The next morning all awoke still cheerful, and nerved for any fate that might follow.

The prison life here was somewhat broader, and presented a little more variety than we had formerly known. The door of our stall was unlocked in the morning, and we had to go down the two flights of stairs in the charge of a guard to the open court-yard, for the purpose of washing at a hydrant; then we were immediately taken back to our stall and locked up. But the principal part of our difficulty proved to be the want of fire. The early part of this winter was exceedingly severe, and as we had virtually neither clothing nor bedding, this was a source of continuous suffering, and effectually prevented all those pleasant fireside chats which had done so much to make our condition endurable in the Atlanta barracks. There were but few pleasant employments in which we could indulge through the day, and the night was even worse. The gas jet in the corner of the room was lighted by the guard as soon as it began to grow dark, that we might still be under observation by any one looking through a crack in the partition. We had abundance of light, except when the awkwardness of the gas managers left the whole city in darkness, which happened more than once. In the colder weather we did not attempt to sleep till sleep became a necessity, pacing back and forth till exhausted; we would then pile down on the floor as close as we could possibly lie and spread over the whole mass the two thin bits of carpet which constituted our only bedding. Here we would sleep, till awakened with cold, when we would arise and pace the floor again. A little device to which we resorted will gave an idea of how much we wished warmth. We found part of the blade of an old spade in our room, and by putting it diagonally over the gas-jet it diminished the light but made the heat available; two or three were always toasting their hands around this feeble furnace.

We never omitted our devotions. For a time the deserters outside, who were composed of the very scum of Southern society, many of them guilty of the worst crimes, tried to interrupt us by every means in their power; but finding their efforts of no avail, as they could only shout derisively at us through the planking, they grew weary of such tame sport and left us to pursue our own way in peace. We afterward found, when we were put among them for a short time, that they respected us all the more for our perseverance.

A few days after our arrival we noticed a great stir among the prisoners at *Libby*, which was in plain view from our window, and not more than two squares distant. Inquiry among the prisoners outside, who were as much interested in anything of the kind as we, revealed the fact that the expected truce-boat had "arrived from the United States," and that a large load of prisoners was being started North. The mode of effecting an exchange was first for the two Commissioners to agree on the terms. This had been already done so far as the great mass of prisoners against whom no charge was made on either side was concerned. Then a large boat-load would be sent up the James River to City Point, a few miles from Richmond; here they were met by Mr. Ould, the rebel agent, and a similar number was ordered by rail from Richmond. These were received at City Point, and the exchange completed there. The boat, which was under Union command, (for the Federal government took the greater part of the trouble in the matter of exchange,) would carry these troops northward, and bring back another load of prisoners from the North, when the process would be repeated. All prisoners before starting signed the oath of parole, not to serve against the opposing government till declared regularly exchanged. This was to prevent all attempts at escape while on the way. The arrival of a truce-boat was therefore a matter of the deepest interest to all prisoners who were each hoping that his turn would be the next.

Soon we saw a band of United States soldiers under rebel guard come up the street, on their way toward the cars to take passage for the truce-boat. Our five friends with whom we had spent so many days in Atlanta and on the way to Richmond were in the same company. As they passed our window they looked up, saw us, and waved their hands. We could not shout messages in the presence of the guards, but wished them a happy journey. My especial friend Walton did write to my father as I had before requested him, but the letter was not received. He afterward wrote again telling the story of our acquaintance, and adding a rumor of our subsequent execution. Fortunately another and later letter from a different source had reached home first, and friends were spared the suffering this would have occasioned.

This parting was bitter. We were glad that some who had been near us were to be delivered from the power of our enemies. But it seemed so much like fulfilling the Scripture, "The one shall be taken and the other left," that we turned away from the window feeling again the gloom which darkened the first day of our arrival. A sense of utter desertion and loneliness swept over us, which only disappeared when we once more sought help from God.

In the dead and wearying sameness which again settled upon our daily life we had one delightful half-hour nearly every day in reading the Richmond papers. In this prison we were not debarred this privilege, for the first time in the whole of our imprisonment. Among the hundreds of prisoners outside there was always some one who had money enough to buy the dailies, and charity enough when read, to pass them to us through a crack. Frequently we would get several in one day. As compared with the Northern papers they were very meagre, contradictory, and unreliable; but they were a wonderful treat to us. As soon as one came to hand all the party would gather round while I read all the news and editorials aloud, and these would afford much food for fresh conversation. It was only by carefully putting together different accounts that we could gather a fair idea of the real situation. The advance of Burnside preceding the Fredericksburg battle was wonderfully exciting. When we heard of his crossing the river, we prayed most sincerely for his success and his continued advance toward Richmond. In this case we would either have fallen into his hands or have been removed; and we had firmly resolved never to be moved again without a desperate effort to escape. But the sad news of his repulse destroyed all the hopes we had been indulging.

We were not able to borrow books as in Atlanta, and felt the need severely. Among other expedients I managed to sell the fine hat which Wells had placed on my head for three dollars and a half, and bought another in better keeping with the rest of my apparel—that is, long since worn out—for half a dollar. A Union man who visited the prison not long after, secretly gave me five dollars more. As I was now wealthy, I tried to procure a book which would stand a good deal of reading and lighten the long hours for us all. I entrusted the money to the corporal who attended the prison with very strict instructions. He kept the money for several days; and then, as I began to think that he was likely to imitate Swims, he gave it back, not having been able to get the book desired. I next tried a higher officer, but met with no better success. Determined not to be baffled, I dropped the money through a crack in the floor to a lady prisoner in a room below who was sometimes allowed to go out in town; but in a few days she sent it back, saying the book was not in Richmond. Encouraged by the honesty of those who so faithfully returned me the money, I made sure of succeeding in another manner. I wrote the names of several books on a slip of paper and gave it to Commissary Chillis;—the man who wanted us hung when we first arrived, but who was by far the kindest officer in the prison; he likewise returned it after due effort, telling me that none of these books were to be had in Richmond. My hope for obtaining any reading matter beyond the daily papers was completely extinguished; and I therefore devoted the money to the next best purpose to which it could be applied—the purchase of food, which was sorely needed.

Our bread was very good and the meat better than any we had so far received. One complaint only was made against the food, but that was most serious. There was too little of it. At no time would it have been difficult for one of us to have eaten two days' rations at a single meal! We had a standing argument as to whether it was better to eat all that was given —for a whole day's allowance was given at once—or divide it into two or three portions. In the first case, there was a fair meal, but a terribly long interval followed; and the sight of others eating in the afternoon was not calculated to appease the hunger that would be raging by that time. If it was divided, each portion was so small that it seemed hardly worth while to begin. We were also permitted, if we thought good, to exchange our allowance of meat for a certain extra quantity of bread, by notifying the Commissary. We tried this also, and could not see that there was much if any profit in so doing, for these changes left the *amount* of food about as before and were therefore of little service. Here each man received an allowance that had been apportioned for him, and the prisoners were not entrusted with the work of distribution, which was well, for some in the prison outside would have fared badly had they been in the hands of their companions. But those who had funds could purchase without restriction from the Commissary. The price, however, was such that a small amount of money was soon exhausted. A small cake or biscuit, such as sold at our homes for a penny, was ten cents at first, but soon rose to fifteen. One of these added to our daily allowance made a great improvement. But our money was soon gone, and thus this resource was cut off. Besides, we had to take a good many rebel postage stamps in change, and if these became torn or cracked, which frequently happened, they were not received, and we lost a considerable portion of our precious currency in that manner.

That we were very restless and discontented here, and revolved all kinds of desperate plans for escape, will be understood without statement. The prospect seemed hopeless, for we were in the third story, and could only succeed by passing through successive relays of guards, all of whom had reserves ready to co-operate with them in case of need; or we might *think* about getting down from our window into the street with a guard especially watching that window, and no possible means of descent, so far as we had yet discovered. Our room had some offices on one side, and a stall on the other. The latter was occupied by a number of Federal soldiers, some charged with being spies, and others with murder.

One of the latter in whom I became greatly interested was a young and handsome man called Captain Webster. On one occasion he had been sent to capture a notorious guerrilla, Captain Simpson, who was then in hiding within our lines. Webster found him and summoned him to surrender. Instead of doing so, he fired his pistol and started to run; but Webster also fired and mortally wounded him. When Webster was afterward captured by the Confederates, he was charged with the murder of Simpson and confined in the room next to us. Some time afterward he was hung, but the charge as published against him was changed to that of violating his parole.

Webster was now tired of confinement, and knowing that revenge on the part of Simpson's friends endangered his life, he was ready to make the boldest strike for freedom. In concert with a large number of men who were in the room outside, and with some citizens in the room below, we decided to make the attempt. The midnight before Christmas was fixed upon as a favorable season. As in Atlanta we had opened secret communication with all the rooms of the great building, and we decided after full consultation to let the citizens initiate the movement. This they were to do by giving the startling cry of "fire." This is liable to arouse panic at any time, and coming late at night in a building crowded with prisoners it would be doubly alarming; probably those not in the plot would rush also to the doors. About a hundred and fifty were in the secret, and as soon as the signal was given, we were to cry "fire" at the top of our voices and rush upon the guard. There were only thirty soldiers actually on duty in the building, and I had no fear, if the plan was fairly carried out, that the first stage of the game—that of getting out of the building, which was not the hardest part—would be immediately successful.

On Christmas eve everything was in readiness and most anxiously did we await the signal. The hours rolled slowly on; midnight came and went while we listened. But no cry was heard. We afterward learned that the citizens failed in courage at the decisive moment. Thus they defeated a plan which would in all probability have succeeded—at least so far as the escape of some of those engaged,—and which would have startled rebeldom not a little by breaking open its strongest prison in the city of Richmond itself.

We resolved to try again the next night; and that no faint-heartedness might interfere we appointed Captain Webster our leader, knowing that he would not falter. Again we prepared, and this time more carefully than before. The locks on all the side rooms which contained the prisoners most dreaded by the enemy were carefully drawn, except our own, which was so close to the guard that it could not be removed without imminent danger of discovery.

There were some who did not wish to make the desperate attempt, but these were very kind to the men who intended to go, supplying us with serviceable shoes, and taking our worn-out ones in return.

At length every thing was in readiness and we again waited only for the signal. In our room we were perfectly ready. The board which supported the water bucket was taken, and four of us holding it as a battering ram, did not doubt our ability to dash the door into the middle of the big room, and seize the guard before he could make up his mind as to the nature of the assault.

As the critical time drew near, the small rooms were all gradually vacated. The inmates of the large room stood before the guard and the door in such a manner as to hide the movement from his view.

For an instant all was silent. We listened breathlessly for the word "fire"—a word of double signification, for we had often waited for it with beating hearts on the battle-field, but never with greater anxiety than now. We lifted up our hearts in a mental appeal to God that he would be with us, and preserve us through the coming strife, and if consistent with his own righteous will, permit us to win our liberty.

But what can cause the delay? Every added moment is a source of danger and the guards may see our preparations and get the great advantage of the first stroke. Minute after minute passes and the dead silence is only broken by the throbbing of our own hearts. We stand with the board ready, and our spirits eager for the coming strife which shall lead us to grapple with naked hands the shining bayonets of the guard. We do not doubt the issue, for the hope of liberty inspires us.

But what terrible surprise is this? We see our partners in the desperate enterprise creeping back to their rooms! We are enraged and freely use the word "Cowards;" but soon hear the explanation which makes us feel that the Lord is indeed keeping watch over us.

Just as Webster was about to give the signal, a comrade pressed hurriedly to his side, and whispered that we were betrayed, and that an extra guard of over eighty men were drawn up before the street door. He said that he had been told that their orders were to shoot all who came out while another detachment was to close in behind and make the slaughter complete. Webster was at first incredulous, but slipped to the window and *saw the soldiers for himself.* He was convinced, and the word was passed around which resulted in a rapid but quiet return to the cells, and as complete a repair of damages as it was possible to make.

I confess that when I first heard the story of the soldiers in ambush, I thought 't the invention of some faint-hearted individual who feared the danger of the assault. But it was all too true. The next day the Richmond papers gave a full account of the affair, and Captain Alexander, the tyrant who commanded the prison, threatened to have all who were engaged in it tied up and whipped. But nothing more serious than putting a few who had been prominent into close confinement for a week was actually done. One prisoner, however, did secure his release by this affair—the man who betrayed us! He was placed in the prison for some slight offense, and secured his own pardon by informing the authorities of our design. Where so many were engaged in a plot this was a very natural outcome.

25

CHAPTER XXXII.

LAST EXPERIENCES IN REBEL PRISONS.

W E had made repeated efforts to send letters home to inform our friends that we were still alive, but all had hitherto been unsuccessful. The Confederates professed to allow letters to pass the lines if so written as to be inoffensive and to give no military information; but though we had written and been very guarded in our expressions, no letter was ever received on the other side. We could not write without betraying our existence, and it is possible that this was the very fact to be concealed ! Robt. Ould, rebel commissioner of exchange, under date of February 9th [1] says to Col. Ludlow, the U. S. Commissioner, '' Not one of the Tennessee and Ohio men referred to in your letter is now in Richmond. If they are elsewhere, they will be delivered to you." As we had been in Richmond two months when this was written, and were all the time prominently reported in the prison returns, and special pains were taken to keep us safe, this letter has a strange sound; and when coupled with the neglect to send any of our letters through the lines, it certainly looks as if there was a motive of some kind in keeping our own people from knowing anything of us. Secretary Stanton told me that when he remonstrated with the rebel government about the hanging in Atlanta, they denied having any knowledge of the matter—and this, too, not long after the death sentences of the seven had been sent to Richmond in response to inquiries, and were then in the war archives !

But we now had a providential opportunity to get word through the lines without thanks to the Confederates. Prisoners captured at the battle of Murfreesboro' were brought to Richmond in such numbers that there was no room for them in Libby, and the surplus were for a day or two confined in the basement of Castle Thunder. I wrote a note on the flyleaf of *Paradise Lost*, tore it out, and when we were taken around to wash in the morning, I slipped away from the guard, and got to the door of the room where the western prisoners were confined. Here I saw a generous-looking Irishman, and having little time to spare, I handed it to him with the request that he would try and get that through the lines for me, and send it to its address. This was asking a good deal of a stranger,

---

[1] War Records, Series III.

but I had a strong conviction that he would do his best to carry out my wish. The note was written very closely, and carefully directed. As these men were exchanged in three or four days, it was but a short time till it was mailed and reached its destination. To the dear ones at home it was like a letter from the dead. My father received it at the post-office, and was so overcome when he perceived its nature that he could not read further, but had to leave that office to a friend. It was published at once in the *Steubenville Herald* and widely copied. The light tone assumed was intended to comfort those hearts that I knew must be sorely wounded. The letter is here reproduced just as it was written, with the exception of one paragraph addressed particularly to my mother, in reference to religious experience, and also, I think, to the fact that if I got home again, I would feel it my duty to be a preacher, and thus fulfill a long-cherished hope of hers—a paragraph which from motives of delicacy was left out of the newspaper publication and which I cannot now recover.

"RICHMOND, VA., January 6, 1863.

"DEAR FATHER:—I take this opportunity of writing by a paroled prisoner, to let you know that I am well and doing as well as could be expected. I have seen some rather hard times, but the worst is past. Our lives are now safe, but we will be kept during the war, unless something lucky turns up for us. There are six of our original railroad party here yet. Seven were executed in June, and eight escaped in October.

"I stand the imprisonment pretty well. The worst of it is to hear of our men (this refers to the Union army) getting whipped so often. I hear all the news here; read three or four papers every day. I even know that Bingham was beat in the last election, for which I am very sorry.[a]

"The price of everything here is awful. It costs thirty cents to send a letter. This will account for my not writing to all my friends! Give my sincere love to them, and tell them to write to me.

"You may write by leaving the letter unsealed, putting in nothing that will offend the Secesh, and directing to Castle Thunder, Va. I want to know the private news,—how many of my friends have fallen. Also tell me who has been drafted in our neighborhood, who married, and who like to be. Also, if you have a gold dollar at hand, slip it into the letter,—not more, as it might tempt the Secesh to *hook* it. I have tried to send word through to you several times, but there is now a better chance of communicating since we came from Atlanta to Richmond.

"No doubt you would all like to see me again, but let us have patience. Many a better man than I am has suffered more, and many parents are mourning for their children without the hope of seeing them again. So keep your courage up, and do not be uneasy about me. Write as soon as you can, and tell all my friends to do the same.

"Ever yours,
"WILLIAM PITTENGER."

"To THOMAS PITTENGER,
"New Somerset, Jefferson Co., Ohio."

We remained in the little stall until about the first of February. Then all the other rooms in that range were wanted for hospital purposes. The

---

[a] The Congressman from our home district.

garret over the big room had been used for this purpose, but now it over-flowed, for the ravages of disease among the prisoners had become fearfully great. Our rooms were not well fitted for the purpose, but they were no worse than the garret above, where it was believed that all who went were sure of death.

Small-pox had broken out among the prisoners in a virulent form, and the whole town was alarmed. Stringent orders were issued for vaccinating all who were in danger of infection. This was a fearful hardship in connection with the kind of treatment received, and under such circumstances, was not infrequently fatal. None of our party of six took the disease, but several of the Tennesseans did, and men were dying around us every day. All of us were vaccinated, and several were very sick from that cause. But the sickness had at least the one good result of securing our removal from the pen in which we had been confined now for nearly two months.

From this time the loneliness and isolation of our prison life was at an end. We were always in the midst of a great company, with the unending sources of interest opened by numbers. At first we were taken into the noisy bedlam before described, which was by no means a desirable location in itself, but it made a change that was agreeable for a little time.

The space here allowed was so great that it seemed like freedom by contrast. Almost the whole of the upper floor was in the one room, and there was an extension to one side only a little smaller than the great room which must itself have been much more than a hundred feet long. It was more like the upper story of a great barn than an apartment in a house. There were pillars as supports, and great beams and rafters visible overhead; there were a great many windows, from all of which could be seen the gleaming muskets and bayonets of the sentries. Best of all, there was a large stove. It did not warm the immense loft, but there was a chance of occasionally getting near it and being warmed once in a while —a luxury that only one who has been *freezing for two months* can properly appreciate.

The amusements of our new friends were *striking* if not elegant. There was a great deal of very rude practical joking. One prank never seemed to lose its charm for some of the roughest of the lot, among whom a number of robust Irishmen, who seemed to cling together, were conspicuous. When a dense crowd was seen around the stove, a company would mass themselves at a distance, and one giving the word, "Cha-r-ge, me boys," they would rush upon the unsuspecting crowd about the stove, striking like a solid battering ram. Men would be knocked down and scattered in all directions, and sometimes serious hurts inflicted. As tempers were naturally irritable in such a place fights were of no infrequent occurrence. It only needed the addition of intoxicating liquor to make the prison a perfect Pandemonium. But fortunately this was strictly forbidden. At night restraints were less than in daylight, and in spite of the efforts of the guards, there were robberies—not of anything of much intrinsic value—and terrible beatings. On two or three occasions men were found dead in the morning under such circumstances as to lead to the belief that they had been murdered. But we had nothing to fear, for the advantage of perfect union and perfect trust was such as to defend us against much more formidable danger. As it was understood that we would take care of ourselves and of each other we were not often molested at all, except sometimes in the general tumble around the stove.

But not infrequently under the inspiration of the gaslight the worst ruffians would put on a new character. When the day's turmoil was over and all who had blankets had gone to rest, a group would gather around the stove and begin to tell stories. I would sometimes join them, and listen for a great part of the night to some of the finest fairy tales and most romantic legends it has ever been my fortune to hear. But the approach of day took all the romance out of them, and restored them to their original unlovely character.

All of this was endurable for a time as a contrast to former life, but we soon wearied of the perpetual noise and ferment. We had learned that on the ground floor there was a large room occupied mainly by Southern Union men. We would have much preferred being put with our own soldiers, but as that could not be, it seemed more natural to be with those who held the same allegiance as ourselves, and we petitioned to be removed—without, however, much expectation that our wishes would be heeded. But after some delay we were taken down as we had desired, and then began a more pleasant part of our Richmond experience. The new room was not half so large as the last, and was well filled, indeed crowded, having about eighty men in it. It was dark, and very low, with many supporting columns, causing it to look more like a cellar than a room. The windows being on the level of the street or a little below, were not only covered with bars but with woven wire as well. The refuse stems of the tobacco manufactured in this building had been thrown into the room till they covered it to the depth of many inches. This was considered a great prize by the chewers and smokers of our party. The dirt mingled in nearly equal quantities with it did not daunt them !

But as an abundant compensation for all these disagreeable accompaniments of our new apartment, it had a stove and was warm throughout, so that now the terrible suffering with cold, which those only can appreciate who have endured it, was near an end. There was also comparatively good society here, men who were intensely patriotic, and not a few of a higher order of intelligence. In talking with these men and hearing their adventures and opinions, I passed many a pleasant hour and gained a great insight into the character of Southern Unionists.

Among others I became much attached to a Scotchman named Miller. When the war opened he was living in Texas and witnessed the manner in which that State was driven into secession. The first part of the plan upon which the conspirators wrought was to excite rumors of a negro insurrection in a certain neighborhood; next they would place poison and weapons in certain localities and find them as if by accident. This would be done several times till the public mind was in a perfect panic. Then some slaves were selected and whipped till the torture made them confess their own guilt, and also implicate the leading opponents of secession. This was enough. The slaves and Unionists were hung together on the nearest trees, and all opposition to the rebel cause thus crushed out. The alliance between slavery and treason, as in this instance, was the most natural in the world. Miller himself was captured on suspicion of being a Union man—he had settled in the country but a little while before this time—and narrowly escaped hanging with his associates. He was sent east to be tried for treason. Twice he made his escape, once travelling over a hundred miles, and each time, when captured, telling a different story. Finally he represented himself as a citizen from New York. When brought before the magistrate in Castle Thunder he merely said,

" I told you all about my case before."

The Judge, who had been drinking, thought that he must have examined him before, and sent him back without further question. Several other times he was brought out, but gave the same answer, and was at length exchanged.

I was here also much interested in Charles Marsh, a young and adventurous scout from the Potomac Army, whose history was most remarkable. Since the close of the war I once met him at Steubenville on his way to Chicago to unearth some revenue frauds. He was then serving as a government detective. He recognized me at once, and recalled with a great deal of pleasure our adventures in Castle Thunder. But it was less pleasant to endure than to remember them !

Marsh had been sent into the Confederate lines to burn a certain bridge, if he found it unguarded, and to collect important information. He succeeded in the latter only, and while on his return was captured, with papers in his possession, which clearly established his character as a spy. He was immediately started toward Richmond with a strong guard. On the way the sergeant in charge got a chance to indulge in liquor, a practice to which he was addicted, and became so careless that while Charlie was not able to escape on account of the men who were not drunk, by watching his opportunity he managed to slip from the breast pocket of the sergeant the packet of papers containing the charges against him, with directions for his disposal, and dropped them into a pond over which the road ran.

Great was the perplexity of the sergeant on arriving at Richmond to find his papers gone. He could only report that he was to bring the prisoner, who was a bad customer, to Richmond, and that he had lost his papers on the way. The authorities arrested him for neglect of duty, and put Charlie in the room with us, sending back to the sergeant's regiment for information. It was only a day or two till the evidence arrived, and the commanding officer promptly entered our room with a guard and called

Taking a Dead Man's Name.

in a decided manner for Charles Marsh. The latter well understood what this meant and made good use of his last chance for life. It so happened that a man had expired in the night—an occurrence not at all uncommon —and Marsh at once answered in a careless tone

" O, that fellow died last night," and pointed to the corpse.

The purpose was instantly recognized by all in the prison who knew either of the two men, and no one wished to interfere. The officer was thrown completely off his guard.

" Died has he ! the rascal ! We'd 'a hung him this week, and saved him the trouble," growled the baffled magnate, and took his departure

It is needless to say that Charlie took care that when the morning report was made out " Charles Marsh, died " was duly entered, or that he continued to answer to the dead man's name. He was finally exchanged under the name which he had thus fortunately borrowed !

A few occurrences about this time throw a lurid light on the design of the Confederates in making the hardships of prison life as great as possible. There was always the endeavor to persuade the prisoners to enlist in the rebel army. As all able-bodied men outside were conscripted, it was but natural that every means should be employed to make the Union prisoners and others enlist also. On their part, however, the threat of imprisonment had lost its power, for it was already realized; but their sufferings might be aggravated till the sufferer was willing to yield and go out to fight the battles of his tormentors. Of course such allegiance was not very hearty, but abundance of military executions were relied upon to maintain discipline and prevent desertion. One day a Tennessee Congressman visited our prison and made an address. He employed the usual arguments, appealing especially to the Tennesseans to return to their rightful allegiance and enter the army; promising that all would be forgiven, and that they would share in the glory of Southern independence; adding that the North was weary of the war, and that a powerful party there was favoring peace, so that it could only be a short time till the war was over, and then if they were still prisoners, they might expect to be hardly dealt with. To men who had spent from six to eighteen or even more months in prison, these were powerful arguments. In one room over twenty yielded; how many of them were afterward shot for desertion, or escaped to their native mountains, I do not know.

But the great majority remained faithful to the Union, saying that they could die for the old flag if they could do nothing more. This " obstinacy " excited the ire of the authorities and caused us even greater suffering; for while it is probable that our band would not have been permitted to enlist if we had applied, we were sure to have part in all general prison inflictions. One penalty was that we should all be put at menial work like the negroes,—forced under bayonets to do all the disagreeable service of the prison. The work itself was nothing to most of us, but the manner of it made the infliction most galling, especially as it was confined to Union men, while rebel deserters and criminals were exempt. Some obeyed; but others would not. The former complained of the latter, and to remedy the inequality, a list was made out and every man given his turn. One of the first called under the new arrangement was a fine young Tennesseean named McCoy. He answered boldly: .

" I'm not going."

"What's the matter?" demanded the sergeant.

"I didn't come here to work; and if you can't afford to board me without, you may send me home," replied the fearless man.

"Well, well, you'll be attended to," growled the sergeant, and proceeded with his roll. Four others likewise refused and were at once reported to Captain Alexander, who commanded them to be put into "the cell." This was a dark place by the open court, only about four feet wide and six or seven high. It had no floor except the damp earth, and was utterly destitute of light and air, except a little that came under the door. Here they were informed that they should remain till they agreed to go to work, or died.

But we found another alternative for them. There was a piece of file and a scrap of stove pipe in our room, and watching our opportunity—after buying a piece of candle from the Commissary,—we slipped to the door when taken out to wash in the morning, and passed the articles under the door. When they received these, their objections to working vanished at once! They began to dig most faithfully at a tunnel, and the second day toward morning they broke upward through the crust of ground outside the prison walls. The one who had been selected to lead, worked his way through the tunnel, and coming to the open air, glided noiselessly away. As he was never heard of again the presumption is that he reached the Union lines, for if any prisoner was recaptured, it was sure to be published by the authorities very prominently in the prisons. The next man was just coming out also, when the barking of a dog that happened to be prowling around called the attention of the guard that way. The man seeing he was discovered, managed to dodge back into his hole, and thus escaped a shot from the guard. No more could go; but this incident prevented the confinement of any others in that cell.

They were not yet ready to give up the design of showing us off in the character of servants. I happened to be on the next list that was prepared, and this time it was rumored that the task was to dig in Captain Alexander's garden, which we would have been obliged to perform with an armed guard standing over us. I would have greatly liked to be for a little time in the open day, and would not have at all objected to digging a little. In fact, if volunteers for work had been called for, they would have been promptly furnished from those who wanted to get out of prison for a brief time. But to be forced to labor seemed to the Southerners among us to be putting them on the level of the negroes, and against such degradation their pride of race revolted. This was not so sore a point with us, but as regularly enlisted United States soldiers, our professional pride was wounded just as badly; so it amounted to the same thing and we were equally determined. When we discussed the matter and counted the cost, the refusal became general, and the issue was thus made up. Nothing was done that day, but the next we were ordered out in the yard as a punishment. The place was perfectly bare, about fifty by twenty yards in size, having the prison for boundaries on two sides and high brick walls on the other two. Sentinels were on the tops of these walls as well as in the streets outside.

The full cruelty of this infliction is only apparent on consideration of all the circumstances. We were enfeebled by long confinement, and many actually sick. Our clothing was very slight and thin. There was not an overcoat in the party, with the exception of the old one given me by Captain Fry, which had been used for seat, quilt, and bed by day and night, till little of it was left. I was perhaps about the average in dress, and I had a flannel shirt, a pair of pants somewhat torn, and a thin spring coat. The latter had a dog bite taken out of the skirt as a reminder of the beginning of our adventure at Shelbyville. All had been worn out some months before. I had also what passed for shoes and stockings, but the soles were completely worn out of the shoes and the feet out of the stockings. It was a cold day in February, and was raining. The wind also was high, laden with torrents of rain, and swept around the corners of the jail with such drenching force as almost to rob us of sensation. There was no shelter, and in a few moments we were wet through. Water stood everywhere an inch or two deep in the yard, and was soon tramped up into mud, through which we splashed as we paced to and fro in a vain effort to keep warm. We could scarcely prevent ourselves from being actually chilled to death !

Here we remained from early in the morning till late in the evening. We were told that we would have to stay there till we froze to death or agreed to work. The latter we resolved never to do. The former was prevented by relief from a most unexpected source.

It was our intention when night came to try to force the guard on the top of the wall, though there was little hope of success from want of means to get up promptly; but we had nothing to lose, for as night came on the rain turned to sleet, and it was unlikely that any of us could have survived till morning.

But the old Commissary, who had been so harsh on our arrival, and who continued to grumble at us on all occasions, but who was really almost the only one of the officers who was ever known to do a kindness, came out to see us several times; each time he went back shaking his head and grumbling; we were still so much influenced by his rough ways that we thought he was gloating over our sufferings. But in the evening he went to the tyrant Alexander and remonstrated with him in most profane and energetic fashion, exclaiming:

"If you have anything against those fellows bring them to trial and hang or shoot them—no doubt they deserve it; but don't keep them out there to die by inches, for it will disgrace us all over the world."

This logic, enforced in various forms and with much profanity, produced a good effect, and the order was given to send us back to our room, which, with its warm fire, never seemed more pleasant. But the result of that day of terrible freezing did not soon pass away. The grateful warmth of the room produced a stupor from which most of us awoke sick. Several died very soon after. I have always since been on the list of invalids, and vividly remember that day of exposure as about the severest merely physical suffering of the whole year.

When we first entered this room we had a striking instance of the manner in which the way of duty is often opened to those who are willing to follow. While up stairs in our own little apartment we had continued the custom of having prayers, morning and evening. But when we were put with the great crowd here it really seemed as if there was no opportunity, and that the better way would be for us to do our praying and Bible reading privately,—at least we did not find it hard to persuade ourselves that as we probably would not be permitted to pray or sing publicly, we might as well save ourselves from ridicule by not attempting it. But the most profane man in our party of Tennesseeans prevented this cowardly surrender. Mr. Pierce, whose split head made him such a noticeable figure, climbed on a box toward evening, and, by dint of vigorous calling and stamping, succeeded in getting the attention of the assembly. He then told them that he wished to propose a matter of general interest. All listened. He continued:

"We have a number of preachers among us who are accustomed to sing and pray and read the Bible out loud every morning and evening. Now I do very little"—with a great oath added—"of that kind of thing myself; but I like to see it going on; and I think it will do none of you any harm if we have them do the same thing down here. I propose that we invite them to go ahead."

The motion was instantly seconded. In the prison almost anything which breaks the terrible monotony of the long days is likely to be popular, and when Pierce put the question, there was a very hearty "Ay," and no negative voices at all. Then he stepped down and, turning to me, said :

"You hear the decision. Now go ahead."

If I had wished to evade duty there was no opportunity. But I did not, for I had felt far from easy at the thought of abandoning a practice which had been taken up in the darkest hour of our history, and which had done so much since to support and comfort us. There were no "preachers" in the technical sense with us, but wasting no time in discussing that question, we formed a group near one of the windows and read, sung, and prayed. During the first and second we had no difficulty, and I even thought there were manifestations of interest and pleasure on the part of many; but when it came to prayer, there were a good many very fervent responses at wrong places, and the disposition to turn the whole matter into sport on the part of a dozen or more of the rudest characters. However, as the way had been opened for a beginning we did not despair of being able to continue. In the morning there was the same trouble, but I carefully marked the one who seemed to be leading in the irreverence, and determined to make an effort during the day to bring him to a better disposition.

This young man had quite a history. I had noticed him before we had been brought down stairs as the victim of one of the merciless punishments that were fearfully common. Being guilty of some infraction of prison discipline, the lash was ordered. He was tied up to a post in the larger room, that all the prisoners might have the benefit of the lesson, and the work began. He never uttered a word. His silence—so different from the shrieks and groans and pleadings that were common, especially among the negroes,— seemed to touch the barbarous Alexander, and he suspended the lash before the full number prescribed had been given. When the young man was untied, he turned to the officer and said, "Are you through?" The other said he was. "Then, sir," he continued, "I want to tell you, it is a shame for you to whip a man in that manner." So far from being offended at this plain and truthful speech, the latter seemed to be pleased by such marvellous powers of endurance, and exclaiming, "Well, you *are* a man, anyway," sent for some better clothing, which he gave to the sufferer, who was afterward put in the room below, a few days earlier than ourselves.

During our second day in the lower room I managed to get into conversation with him, learning something of his history—he was but a little more than seventeen, though large and strong—and in return gave the history of our raid, which never failed to command the fullest attention. Then he told me that his relatives were all dead, except one sister who still lived in Canada, his native place. He was very much affected when he spoke of his longing to see her once more. He took occasion after that to talk of us to his comrades, and we had no more disturbance from them. I wish I was able to tell of his conversion and release, but I lost sight of him soon after.

Having formed a beginning for religious worship in the prison, quite a number of persons came to us, each one of whom had supposed that he was alone, and was cherishing his religious convictions in secret. There were two or three Methodists, nearly the same number of Baptists and one or two of nearly all the principal denominations. A Roman Catholic was as devoted and brotherly as any other. In the presence of the fearful wickedness by which we were surrounded, both within and without the prisons, distinctions between the followers of Christ seemed very slight indeed, and many a dreary hour was cheered by Christian fellowship

CHAPTER XXXIII.

HOMEWARD BOUND!

ONE day there was a capital sensation in our prison. We were ordered into line, and the names of our railroad party, with a few others, were called over and checked. Instantly our thoughts were busy in conjecture. Were we to be moved again, and thus given an opportunity of repairing our mistake in not escaping before? or was it worse than that? One of the men whose name had not been called made so bold as to ask what this roll was for, and was still more mystified by the apparently frank reply:

" We can't tell, for this list comes from Yankee-land."

From " Yankee-land !" Then we were not forgotten, but our own government in some way was asking after us ! It was good to be remembered even if nothing came from it. Why should a list with our names be sent from the North? The whole prison was in a ferment !

It was just about this time that Commissioner Ould assured our Commissioner, Ludlow, that we were not in Richmond. Did he answer in this manner after he had made the inquiry, or give a positive answer first and inquire afterward? This is the dilemma in which he seems placed. We gave no equivocal response. Two subjects are mentioned in the following extract from his letter: " Not one of the Tennessee or Ohio men referred to in one of your letters is now in Richmond. If they are elsewhere they will be delivered to you. The clothing, etc., have been received, and your directions have been complied with."

The item in relation to exchange seems to indicate that in response to a special demand our release had been promised, *if we could only be found !* But from events it seems that we were too well hidden ! The matter of clothing, however, soon assumed a tangible form.

In the prison a rumor came that a general exchange of political and civilian prisoners was in contemplation. We talked of it day by day, but as one truce-boat after another came and went, and nothing more was heard, hope died out so completely that at length every one who happened to use the word " exchange " was greeted with a burst of derision.

Yet we now know that the general arrangement for this purpose was concluded as early as February.[1]

One day an officer came into the room and asked every man who claimed United States protection to fall into line. We did not know how much protection the United States could give in Richmond, but were determined to lose nothing for want of claiming, and gathered promptly into line. He took the name of each one, and said we were to get a suit of clothes from the North. This would have been most welcome news if we had believed it. We had a strong conviction that if anything of the kind was sent from home it would be turned over to the use of the Rebel army. But it was pleasant to find that the thought of the government was turned toward its unfortunate citizens imprisoned in a hostile capital. I presume there were seventy-five names reported as in allegiance to the Federal Government —we were in the minority in this prison—for Captain Turner writes[2] to Captain Alexander:

"I send you according to Gen. Winder's orders boxes marked as follows: Seventy-five trousers, seventy-five pairs boots, seventy-five flannel shirts, and seventy-five jackets."

These did not all reach us. What became of them is more than I can venture to say. All of the Federal prisoners got something, but none of them a full suit. One man would have a new jacket, another would revel in a new shirt. My own share was a pair of boots. I was rather dissatisfied at first, but still had the remnants of Fry's overcoat, and on reflecting how extremely useful the boots would be if we should happen to escape, I became reconciled. The prison officers also helped to comfort those most in need by declaring that more would soon be received and make out a full suit for us all. No more ever came; but the new clothing, however inadequate, was a source of wonderful pride. It was the token that our nation had not forgotten us.

But better things were to come. There is one day of imprisonment which stands out amid all the miserable prison record like a lily in a cluster of weeds or a diamond in a heap of garbage. Through all the mists and winters of twenty-five years it shines in clear and unfading radiance. To have made sure of such a day any of our number would gladly have sacrificed a right hand or led the van in the deadliest charge of the civil war.

A little before dark on the evening of the 17th of March, we were sitting around the stove lazily but not indifferently discussing the siege of Vicksburg, and my friend Miller had just shown with all a Scotchman's intelligence and positiveness how the town might infallibly be captured, when an officer entered in haste and gave the strange order, "All who want to

---

[1] War Records, Ould to Ludlow, Feb. 2d, 1863.
[2] War Records, Feb. 19th, 1863.

go to the United States, fall into line, and come to the office." There was a rush and a scramble to the middle of the floor, and no line was ever formed more promptly. We all believed that we were already in the United States, but had no objection to be still more so; and it was with a good many questionings that we marched out of the guarded room-door, across the court yard, and as many of us as could enter at once, into the office. The proper officers were prepared with blank forms, and they filled out rapidly the oaths of parole binding the signers not to serve against the Confederacy until regularly declared exchanged. Even when I saw this good work progressing I feared that the opportunity was only for citizens of our company. To test the matter, I pressed forward and gave my name, fully expecting to hear "The engine thieves can't go;" but to my surprise no objection was made. For a moment a delicious hope thrilled through my veins—a vision of happiness and home, dazzling as a summer flash of lightning, shone forth before my eyes—but it faded in the remembrance of our Atlanta deception.

I called others of our party, who also went in, gave their names fully, and still no objection was made. The work of paroling went forward, those who had signed being sent back to the prison room between the files of guards, who looked far more pleasant than usual, as if they sympathized— and I have no doubt many of them did—in our deliverance. We were told to be ready to start North at four o'clock the next morning. We could have been ready in four seconds ! But we were unutterably glad to have it brought even that near. At times a deadly misgiving arose for a moment that all this might prove a delusion; but the actual signing of the parole was a strong anchor to our hope. The obligation in such cases is mutual: the soldier is not to serve until a man has been obtained for him, and the government is to send him to his own lines as soon as practicable. In previous deceptions they had never gone so far as this; and we felt assured that they would not now, if they had intended to cheat us.

As might be expected, that evening was one of wild excitement. Nearly everybody in our room acted like men bereft of reason. Some danced and bounded over the floor, embracing each other and pledging kind remembrances. Others shouted till they were too hoarse to shout any more. Others sat down and wept The deliverance was so great that they were completely overpowered. But a few were, for some reason, not permitted to go, and we deeply pitied them. I remembered when we had been left by our comrades on our arrival in Richmond more than three months before, and my heart bled for these forsaken ones, as they sat cheerless and gloomy, seeming amid the general joy to be more wretched than ever. It is one of the contradictions of prison life that while comrades will nearly always help each other to escape, and rejoice in the prospect of deliverance for any one, yet when either actually takes place, those who are left behind suffer the keenest pain—feeling an emotion not unlike the pangs of jealousy.

It was near midnight before we became calm enough to offer up our usual evening devotions. But when the roomful of excited men were still at last, wearied out by the very excess of joy, and overcome by the quietness which ever follows powerful emotion, we knelt in prayer. Many more than the usual number assembled with us in the corner of the room where we always gathered. The prayer was one of overmastering thankfulness. When we remembered all the sufferings through which we had passed it seemed as if we could never cease to be grateful. We asked God for strength to bear every trial; but we also implored with a fervor and sincerity which few can realize, that he would not allow our bright and vivid hopes to be disappointed and us to be dashed back from the paradise of liberty. And we asked with no little solicitude for strength to continue in His service when no longer confined within prison walls, so that the precious possession we had found in the darkness might not pass away in the light; thus composed we lay down to sleep and await the event.

How was our most unexpected participation in this exchange brought about? It seems from the correspondence that there was an inquiry from the Federal side and a strong desire that we should be embraced in the provisions of any cartel of citizens. A denial of our presence in Richmond followed. But the matter was still urged from the North. Finally the exchange was completed with the promise that we should be included, *if we could be found.* The tone as well as the wording of the correspondence reveals a strong wish on the rebel side to complete the exchange but to get all possible advantage of the Federals in it. That the policy of the Confederate government was in all things relating to exchange of prisoners highminded and honorable, would hardly be the opinion of one who carefully reads the following remarkable letter. There are many others in the War Records that breathe the same spirit. It was sent to Richmond on the afternoon of March 17, and immediately there began a great search by the Confederate authorities for the necessary number of prisoners in order to prevent some of their own men that they were very desirous of securing from being carried North again. It is the opinion of the writer that our party would hardly have passed through if the whole affair had been more deliberate; but as the orders were for starting at four in the morning and it was then evening, there was so little time for consultation that we slipped through their fingers.

"CITY POINT, March 17th, 1863.

"BRIG. GEN. WINDER,

"SIR :—A flag of truce boat has arrived with 350 political prisoners, Gen. Barrow and several other prominent men being among them.

"I wish you to send me at four o'clock Wednesday morning, all the military prison-

ers (except officers), and all the political prisoners you have. If any of the political prisoners have on hand proof enough to convict them of being spies or of having committed
other offenses which should subject them to punishment, so state after their names. Also
state, whether you think, under the circumstances, they should be released. *The arrangement I have made works largely in our favor. We get rid of a set of miserable wretches
and receive some of the best material I ever saw.*

"Tell Capt. Turner to put down on the list of political prisoners the names of Edward
B. Eggling and Eugenia Hammermister. The President is anxious they should get off.
They are here now. *This, of course, is between ourselves.* If you have any political
female prisoner whom you can send off safely to keep her company, I would like you to
send her.

"Two hundred and odd more political prisoners are now on their way.

"Yours truly,

"ROBERT OULD."

"ENDORSEMENT :—Send all called for in this unless they are charged with criminal
offenses.

"JOHN H. WINDER."

It was not easy to sleep that last night in Castle Thunder. Fancy was
too busy, peopling her fairy landscapes, picturing the groups that awaited
us beyond that boundary which for nearly a year had frowned before us,
gloomy and impassable as the river of death. But even as we are revel-
ling in these anticipations, what unbidden fears spring up to darken the
prospect and dim the brightness of our hopes! How many of the friends
at home whose love was our life may be no more! For a year not a
whisper had been heard of private news, and we trembled as we thought of
the ravages of time and of battle. These and other thoughts whirled
through our brains, whether sleeping or waking, the whole of that ever-
memorable night. They were only banished at last by the commanding
officer, who stepped into the door long before the morning light and gave
the thrilling order, "Get up and prepare for your journey!"

At his call we hurriedly thronged to our feet. All doubts and
fears vanished. It was true! Freedom once more! Our terrible captivity
was passed. O joy! Joy! Joy inexpressible!! almost too wild and
delirious for earth!

There was a hurrying around in the partial darkness left by the glim-
mering of lights; discordant calling of names, a careful inspection of each
man to see that none went but those intended; and then we formed in the
courtyard for the march toward freedom. We fell into two lines and when
all was ready we passed again into the office, were carefully counted, and
then with bounding hearts passed outward through the dreaded portals of
Castle Thunder—the same portal through which we had passed *inward* more
than three months before, and which we had not passed in the interval—
passed outward into the cool but free night air. There was all the hope
of the morning in it, and by contrast it vividly recalled the chill morning
of our arrival in Richmond.

26

We halted on the pavement until perfectly formed, and then, with the guard, marched through the muddy, unlighted streets for many squares. There were a number of sick in our company, but not one of them was willing to be left behind; and as the rebels had neglected to provide any conveyances, we helped them—each being supported between two stronger men, and thus with encircling arms they were able, at the expense of much suffering in some cases—to accompany us the weary distance. Two or three had to be almost wholly carried, but the burden was cheerfully endured, upborne as we all were on the wings of hope and exultation. After we were seated in the cars, as the train did not start for a little time, we managed to get the Richmond *Dispatch*, and were especially interested in a reference to our departure containing the following words :

"Included in the list of citizen prisoners" (who were to be sent North) "are also a number of renegades from Tennessee and Kentucky, some of whom were arrested for bridge-burning, *engine-stealing*, and similar crimes in the states named. The departure of these prisoners will relieve the Confederate Government of a considerable item of expense."

The name applied to us no longer had the old sting, and we could not help thinking that the expense might have been saved much sooner.

Seated in comparatively comfortable cars, with no ropes or irons upon us, and but a weak guard, and with deep peace and content in our hearts, we glided out of the rebel capital—to see it no more till all that belonged to rebel power had passed away "Like the baseless fabric of a vision, and left not a wrack behind." The unwonted light dazzled our eyes. The motion of the cars was an intoxication. This was happiness indeed. The thoughts and visions of the night were lived over again in scarcely less bright waking dreams. I tried to look at the country though which we were passing, but my pre-occupation was so great that I could scarcely see it at all.

Petersburg was reached with no notable incident; then after but a short detention we moved on toward City Point. As the train passed around a curve about eleven o'clock we saw a very large steamboat in the river; we had seen none such since we left our own Ohio. But there was something still better than the boat. Waving over it in the morning breeze was the "Flag of the free." I seized the comrade next me and shouted, "Hurrah, boys; hurrah ! there's our flag !" They needed no prompting; indeed some in the forward cars had seen it sooner than I had. Cheer after cheer went up from the whole train; some of the guards were discontented at our vehemence, and said, "Stop that noise; there's no use in making such a fuss;" but we did not stop. This was the first time we had seen that banner for eleven long months, and it meant to us life, home, freedom, country—everything which men love; and tears and shouts intermingled till we drew up at the station, but a short distance from the truce-boat.

The grossest frauds were often practiced in these exchanges. The letter of Col. Ould quoted above shows how they were willing to secretly put down names for the mere purpose of making count, and how they gloated over the miserable condition of those who were exchanged in comparison with the good material they got instead. There were many striking instances of the manner in which this principle was carried out. One which occurred here will well illustrate this: A rebel soldier was wounded in the head at the first battle of Manassas. It affected his brain so that he was incapable of performing his duty as a soldier even after he seemed physically well. He was confined for a short time in Castle Thunder, which was the reason of our knowing the case. He was then taken to Camp Lee near Richmond, but did no better. Then they exchanged him to the Federals and got a sound man in his place !

From the side of the truce-boat, *State of Maine*, there issued a long procession of prisoners, three hundred and fifty in number, who did not seem to be nearly so much rejoiced as we were. They were formed in line, and there was a calling of names, a checking of lists, with inquiries about special officers, for what seemed to me a long time. Some disputes arose, but I could not catch the words, and could only pray silently that all might be arranged in such manner as to prevent the necessity of our going back to Richmond. I remember somewhat dimly the picture presented immediately at the wharf under the sunshine of that March day. Of more distant objects I have no recollection whatever. In fact, I was dazed almost equally by the brightness of open daylight and by the happy coming of this long-expected day. I could not, however, help being struck by the contrast between the sets of prisoners; those on the Confederate side looked strong and hearty; their clothing was whole, neat, and clean: But on our side all the number were in the last extremity of raggedness, dirt, and emaciation. I have totally forgotten whether any breakfast was given us that morning or not—a matter of small moment when we had been obliged to starve through so many longer journeys. I do not wonder that Col. Ould said: "We get rid of a set of miserable wretches, and receive some of the best material"—that is for conscription into the rebel army—"I ever saw." He was perfectly right as far as appearance went; but some of that motley crowd did manage to do effective fighting, and with very good will, against their former oppressors !

While the wrangling over details was still going on, without, as far as we could see, any prospect of termination, the order was suddenly given us to "Go on board." Most promptly it was obeyed. It meant a great deal for us to be actually under the *Stars and Stripes* once more. But not till the boat cast loose and swung into the stream did we count ourselves

truly out of rebel hands. Then we felt as one who has awakened from a
hideous nightmare dream to find that all its shapes of horror and grinning
fiends have passed away, and that he is in the wholesome sunlight again.
The blue sky above us was heaven indeed, and the sunshine pure gold.
Our hearts beat glad music to the threshing of the wheels on the water,

Eating in the Engine-Room.

knowing that each stroke was placing a greater distance between us and
our enemies.

Then, too, the hearty welcome with which we were greeted; the good
cheer, so different from our miserable prison fare; and the kind faces smil-
ing all around, showed in living colors that we were free men again.

In regard to food, the great difficulty now was one we had not encountered for a year past—that of over eating ! A year's famine had made

us terribly hungry, and we received cautions on this point which were not unneedful. Nothing in the rations was more enjoyed than the tin cup of good coffee-which was given to us all—genuine coffee was something which we had not tasted inside the rebel lines—and I am not certain that I have found any as good since! No place on the boat seemed more homelike than the engine room, where I sat and ate slowly for a long time. The working of the strong machinery which seemed impelling us lovingly homeward was about as much company as I wanted for the time.

Down the river we went in a delirious dream of rapture! We were scarcely conscious of passing events. Probably no emotion on earth has greater sweep and intensity than the wild throbbing sensations that rush thick and fast through the bosom of the liberated captive.

Then I roved all over the boat in the mere luxury of liberty. There was no guard following me with gun and bayonet, and I caught myself several times looking round to see what had become of him! To walk boldly up to a door, open it, and go out, was so enjoyable that I did it much oftener than was necessary, for the triumph of doing it! Our party talked little with each other, for we had done enough of that in prison, and wished rather to do other and new things; but in passing one would say to another, "Not much like Castle Thunder!" "Going home are you?" "Did you get enough to eat!" etc.

CHAPTER XXXIV.

REPORTING TO SECRETARY STANTON AND PRESIDENT LINCOLN.

I HAVE forgotten almost everything that took place on this homeward voyage, if indeed, my mind was not too nearly in the condition of a waking dream to receive definite impressions. I saw some of the great gunboats as we drew near the mouth of James River, which looked like grim sentinels guarding the avenue to rebellion. We were furnished comfortable blankets, and when tired out, as much by unwonted emotions as by physical effort, we could lie down and slumber at our will. When rested and disposed to talk, we found abundance of admiring auditors ready to listen to as much of our story as we chose to tell. Thus the time passed on very pleasantly.

But at last a dispute arose between us and our new friends which threatened for a short time to mar our harmony. One of our party—Buffum, I think—came to me and said, "Pittenger, they insist on our going to Annapolis instead of to Washington. Is that right?" I said "No; we must go to Washington, and it will be better to do it at once." Soon others came saying, "The word is that all the soldiers must go to Annapolis, while the citizens can go there or to Washington as they please." I inquired as to who gave the orders and learned that it was one of the officers who was making arrangements for the transfer of released prisoners to a boat for Annapolis. I went to him and asked why it was that he wanted us to go to that point. He told me that the government had a camp for paroled prisoners there where they were made very comfortable, and that as my comrades had said we were soldiers, that was the place for us to go; adding that if we had any business in Washington we could easily get a pass and run over. I saw no use in this delay, and asked for the commander. When I found him, he told me that his orders were positive to send all soldiers to the parole camp at Annapolis. I told him that our case was altogether exceptional; that we had been on a peculiar expedition and felt it our duty to report directly to Secretary Stanton. General Mitchel, who had sent us disguised into the enemy's country, was dead. Our leader also had perished. We had long been thought to be dead ourselves, and had escaped only as by miracle. There was no place so appropriate for us to go with the whole story as to the War Department at Washington. I added as a perfectly conclusive argument that for a whole year the rebels had refused to treat us as prisoners of war, and had contended that our case was exceptional; that they had wished to put us to death on that theory; and that it was only right if we should be a little different from other soldiers now. He smiled, saying that our case seemed a good one, and gave orders that we should continue with the citizens from Castle Thunder. I had then to give him the outline of our adventures which impressed him greatly. He promised to do all in his power for us when we should arrive in Washington.

After this episode I remember nothing more with distinctness till we approached Washington. The snowy Capitol with its towering dome never looked so attractive before. I had last seen it in the summer of '61, when war seemed little more than a holiday parade. What mighty changes for the nation had occurred since then! and for us who had seen the sternest of possible experiences, who had felt what famine, prison, and the scaffold meant, the change had been still more momentous.

We arrived in Washington late in the afternoon—I know not of what day, but presume it was the day after leaving Richmond, which would be Thursday, March 19th. I made no memoranda at this time, such as I had not failed to keep during the darkest parts of our history. In my own experience joy was more overpowering than sorrow.

The trials of our East Tennessee friends, with whom we had been associated since leaving Knoxville, were not a little amusing at this point. They knew nothing of Washington circumlocution; and now that they were delivered from the power of the tyrannous rebels they believed their troubles ended, supposing that as a matter of course they would be free, the moment they placed their foot on Northern soil. When Pierce especially began to grumble, we comforted him, Job-fashion, by declaring that he expected the President to come down to meet him with a band of music! Instead of this we were all marched off to the barracks, put into a bare room, and then to crown all, and to make the Tennesseans boil with wrath, a guard was placed over us! Pierce declared that he was going South the first chance he had, and would tell Jeff. Davis that he gave up his opposition and would now be as good a rebel as any of them. Bensinger said laughing, "Then he'll put you in Castle Thunder again." Buffum teased him still further by asking, "How are you going to get past the guard to start?" We who had been subject to the discipline of a soldier's life and had always slept encircled by guards in our own camps saw nothing strange in this. One of our number—probably Reddick—suggested that as we looked like poor people from the country they had only employed these guards to protect us from city sharpers! But soon the matter was made a thousand times worse for Pierce and his companions by the arrival of an orderly who handed the officer in charge orders to let those six Ohio men go out and in at pleasure and have every privilege! This was adding insult to injury; and the other Tennesseeans were as much irritated as Pierce. But the men were loyal and true as steel; they had suffered every thing for the cause of the Union, and I could not endure to see them suffer. So I explained to them that time was necessary to examine into the different cases and determine what should be done with them; that we were only put here for shelter for the night, and that soon some better disposition would be made of us; they would soon see the difference between the old government and the Confederacy. Then we offered to do any thing we could for them, as we were going out for a short walk. This comforted them a good deal, Pierce saying that he supposed it must be all right, but it went hard for a man who had been locked up by the rebels to come North and find that the first thing they did there was to lock him up again! It did seem hard; but soon there was a good deal of comfort. A huge roll of blankets were brought in—two or three apiece, and real blankets, not pieces of carpet. Bensinger could not help saying, "Which of the Southern prisons was it, Pierce, where you got a couple of big blankets like these?" He was pleased as a child and answered that he could not just exactly remember! The blankets were hardly shared out when another grateful surprise occurred. The door leading to an adjoining room was opened and a guard called, "Come out, and get your suppers." On going, we found a long table on which were great plates of excellent soft bread, good boiled beef, and other articles in unstinted supply! while the odor of strong coffee filled the room. We were always hungry, and could now eat without fear of injury. Even Pierce's countenance relaxed as he grumbled, "Well, if Uncle Sam does shut a fellow up, he feeds well, which is more than Jeff. does." The guard who conducted us in, said hospitably, "Help yourselves. There's plenty. I guess you did'nt get much to eat among the Johnnies."

After supper we took our proposed walk. But once outside in the dark street, with no place to go, no money to buy anything, and nobody to care for us, it began to seem rather lonely. We had been objects of so much solicitude for so long that we could scarcely help feeling a little neglected now that nobody prevented us from wandering in any direction we chose! Beside we were not yet provided with new clothing—though I stood on better footing than the rest, for I had good boots; a regular requisition would have to be filled before our wants in that direction could be supplied; and the air was chilly. After wandering aimlessly for a few squares, looking in at the shop windows and staring at the great buildings, we turned and were soon back in the barracks. The room, which was well warmed, seemed very cosy; it was easy to consider the sentinels at the door as placed there to guard us from an enemy, a view of the case which greatly pleased the Tennesseeans. I really think we slept better than we would have done if at a first-class hotel !

The next day, as I had expected, all restrictions were removed from the movements of the other members of the party as well as our own, but we were invited to make our home at the barracks, receiving regular soldiers' rations, as long as convenient.

But we did not remain long. A wealthy and patriotic lady, Mrs. Fales, had fitted up a pavilion in her yard in excellent style and kept it filled with convalescents from army hospitals, to whom she gave the attendance and most of the comforts of home life. Hearing our story, she applied to the Ohio State agent, Mr. J. C. Wetmore, to secure us as lodgers. He visited us and carried the invitation, which we gratefully accepted. Here we lived in fine style, with all a soldier's comforts, and excellent company. We had now received a suit of army blue, and made a bonfire of our rebel rags; so that we were a little less afraid of lady visitors. Our hostess made an evening reception for us, where there was an abundance of ice cream and cake,—which did not remind us of anything in Dixie;—and had the privilege of telling our story to many sympathizing ladies. Having a very comfortable ambulance always at command—the property of our kind friend, and kept with a driver for the use of her soldiers—we were able to visit the places of interest in Washington with great comfort. Of this privilege we freely availed ourselves when business permitted.

On one of these occasions when we were at the Smithsonian Institute, I saw a tall man of striking appearance, and at once said to my comrades, "That is President Lincoln." But they were incredulous. They had read in Southern papers that he never stirred out of the White House without a heavy guard, and at once they said, "You are trying to deceive us; don't you see there is no guard here?" I was a little puzzled by the absence of the guard myself, for I had believed the story, and there were only one or two civilians with him; but I had seen Lincoln, who could not easily be forgotten. I said, "Come up and let us speak to him !" A little reluctantly they advanced, and when we were near enough, I said, "Can you tell us the name of this animal?" pointing to the skeleton of the geologic monster he was looking at. He smiled in his kind, sad way, and answered, "That's its name written on that card; but I won't undertake to pronounce it. I don't know much about such things." One or two remarks were exchanged on indifferent topics, and then we parted. When at a safe distance, Buffum said, "Pittenger, you can't play such a joke on us as to make us believe that a man who will speak to common folks in that off-hand way is President Lincoln !" But I only replied, "We'll go and see him at the White House soon and then you can judge."

The first Sunday of our stay in Washington we gladly accepted the invitation of our kind hostess to accompany her to the Baptist church of which she was a member. This led to a great trial on my part. The preacher seeing five men in soldier uniform in her pew—Mason, who was sick, had remained at home—came down and spoke to her. A few words were exchanged, and then he turned to us and said, "Can't you tell us a little about your strange deliverance? I will make the other services very short?" The faces of all our party turned toward me; and the minister, taking the hint, repeated his question to me individually. Something like this was just what I had feared when I had the great mental struggle in the Atlanta prison; I wished to refuse, but in the freshness of deliverance, such a refusal would have seemed little less than a crime. The minister was better (or worse) than his word, for he did not preach at all, but on finishing the opening services, spoke a few words about the great sufferings and lessons of the war, and the hardships and temptations of the army; then added that there were some men present that morning whose experiences were marvelous beyond the common lot of men, and one of them had consented to give some account of them.

I rose at the call and scarcely could have felt worse for a moment if in Swim's prison! But the people were so kind and attentive that embarrassment vanished. I passed very lightly over the military part of the enterprise, and spoke especially of God's goodness in delivering us. When I told of the sudden death of our comrades in Atlanta, and of their regrets for not being better prepared for death, there was weeping over all the church. At the close of the meeting kindly greetings and proffers of service were showered upon us.

We lodged in the tent provided by this lady during the whole of the ten days of our stay in Washington. Many other persons wished to take us as guests, but she refused unless we would say that we wished to leave her care. Mason, being sick, was taken into the house, and nursed most carefully.

On our first morning in Washington I had written a note to Secretary Stanton, giving notice of our arrival and of my belief that the government, if not already fully informed, ought to know our story. I supposed this would produce more speedy results than to let the matter pass through the ordinary channels, and was not disappointed. He gave written orders to Hon. Joseph Holt, Judge-Advocate-General, to investigate the whole matter, provide for all our wants in the meantime, and then bring us to him. Hon. J. C. Wetmore, of the Ohio Military Agency, attended to procuring us the legal allowance for commutation of rations during the time of imprisonment, so that we were soon in funds. Regular pay with arrearages could not be given till we returned to our regiments.

Our first visit to Judge Holt was a merely friendly one, and the time was spent in familiar conversation. Major-General Hitchcock, Commander of the Post at Washington, was also present.

The next day we went again, being taken each time in a government carriage, and found Justice Callan of Washington ready to administer an oath, with a phonographer to take down the testimony as given. I was examined first, telling the whole story, and then each one followed, confirming what was said, and adding omitted particulars. The evidence that we were telling a true story was far stronger than we knew. The story of Wood and Wilson had been on file in the department since the preceding November. It was far less full than ours, but confirmed it in every essential particular. Indeed the certainty that we were giving a simple, plain, and uncolored narrative was so complete that no one save General Buell ever called any part of it in question; and he only because of ignorantly confounding the first and second expeditions.

While talking I watched the phonographer with much interest. To see a pencil move as if it was animated by my tongue rather than by the hand of the man who held it was a new experience. I tried to speak at different rates; the pencil quickened or went slowly with no more appearance of effort in the one case than in the other. I was more interested because of writing shorthand myself, but at no such speed.

The testimony thus taken was published in the next issue of the Army and Navy Gazette, as also in the Washington papers of April 4th, and copied over the country. The examination before Judge Holt was on the 24th of March, and his report to the Secretary of War based upon them was dated the 27th. The latter is given in the Supplement.[1]

The following day was fixed for an interview with Secretary Stanton. Agent Wetmore and Major-General Hitchcock accompanied us. Generals Sigel and Stahl, with many other distinguished persons, were in the anteroom waiting, but as we were there by appointment, they continued to wait, while we were at once admitted. Stanton had long resided in my county town of Steubenville, and I had seen him, and knew him well by reputation, though I could then claim no personal acquaintance. We were seated, after he had shaken each one of us warmly by the hand and uttered words of greeting and compliment. We talked for a considerable time, not so much on the subject of our expedition—for I took it for granted that, lawyer-like, he had looked over the evidence in the case and made up his mind about it—as upon general topics, such as our impressions of the South and the Union men in it, and of our hope and feeling about the war. I was especially struck by his asking us how we had liked Gen. Mitchel as a commander; and when we spoke of him with unstinted enthusiasm he seemed greatly pleased and said, " That's the way all his men talk

---

[1] Chapter XL.

about him." He told us that he had been aware of our expedition at the time, but had no accurate information of the fate of the party. His impression was that all had perished at first. On the escape of the eight in October he had made official inquiries of the Confederate government about us, but had been answered that they had no information of the hanging of any of the party. (The papers which were forwarded to Richmond in response to inquiries much earlier, show this statement to be ab-

Face of Medal.   Back of Medal.

[Fac simile of medals of honor given to surviving members of the Andrews raid and to the families of those who perished. Natural size].

solutely incorrect). He had then threatened retaliation in case any more were put to death, and had endeavored to effect our exchange; he was very glad indeed that these efforts had succeeded, and surprised us by saying, "You will find yourselves great heroes when you get home;" then added many kind words about the high appreciation of our services by the government, which, coming from the Secretary of War of a great nation, to private soldiers, was most flattering. Stanton seemed especially pleased with Parrot. He was the youngest of our number and of very quiet and simple manners. Stanton gave him the offer of a complete education if he would accept it—I understood him to mean at West Point. Parrot answered that while the war lasted he did not wish to go to school; but would rather go back and fight the rebels who had used him so badly. At this Stanton smiled as if he greatly approved his spirit, and then said to him, " If you want a friend at any time be sure to apply to me." Then going into another room he brought out a medal, and handed it to Parrot, saying, " Congress has by a recent law ordered medals to be prepared on this model, and your party shall have the first; they will be the first that have been given to private soldiers in this war ! " Later all the survivors of the party received similar medals. Then he gave us a present of $100 each from the secret service fund, and ordered all the money and the value of arms and property taken from us by the rebels to be refunded. Finally, he asked us about our wishes and intentions for the future. Finding that we were all resolved to return (as soon as health permitted) to active service, he offered us commissions as first lieutenants in the regular army. We expressed a preference for the volunteer service, saying that we were soldiers only for the war, and would wish to resume our usual pursuits when peace returned. He promised to request Gov. Todd of Ohio to give us equivalent commissions in our own regiments. Then with a hearty good bye we left him.

We had been invited to call upon the President the same day; and General Hitchcock accompanied us on this pleasant mission. My companions had done a little jesting about being able now to show me how much I was mistaken in trying to impose the tall, plain stranger of the Smithsonian upon them for the great and good Abraham Lincoln, which I enjoyed as much as they did. A still greater crowd than at the War Office was awaiting admission; but as we came by appointment, we had the preference and were conducted immediately to the private office of the President. We did feel some little embarrassment, but this scarcely accompanied us over the threshold.

The office was very plainly furnished. There was a long table and some chairs, but scarcely anything else. Lincoln met us at the door, greeted us warmly, and told us how much he had been interested in hearing of our adventures, and how glad he was that we had at last escaped from the hands of the enemy. We answered as well as we could. I remember telling him that we were very glad to see him, though we had been hearing a great many things not complimentary about him for the past year. He smiled, saying, " Indeed, there are a good many people up here that say about as bad things of me." I also mentioned the reason that the other members of the party would not believe that he was the man we had met in the Smithsonian, which caused him to laugh heartily and ask if we really imagined he went everywhere with a great guard parading after

him, and if the people of the South believed all the stories printed in their
papers? While talking he did not keep one position, but shifted from place

Interview with President Lincoln. " A little luck with the battles now ! "

to place, going from one to another of us, as he addressed each one with great
courtesy. I specially remember part of one remark and his position while

making it. Something had been said about political matters, and our joy that the Union party was now gaining in the country after the great defeats of the fall before. " Yes," Lincoln said, as he stood in a stooping position by the fire-place with his elbow resting on the end of the mantle-piece, " if we could only have a little luck with the battles now, all would soon be right and the war be over." The quaint phrase, a " little luck with the battles " made an indelible impression on my memory, for we had been having very ill-luck in that direction for some time. We did not wish to be tiresome, and ourselves made the first motion to leave; the President took the hand of each in both his own, saying again how thankful he was that we had been spared, and that he hoped we would find all our relatives living, and well when we reached home. We left him, exceedingly proud of the honor the greatest man in the nation (or the world) had conferred upon us. We had now nothing further to detain us in Washington, and were most anxious to be in Ohio again. A furlough for sixty days was given to us and an order for government transportation to our homes.

The father of William Reddick had heard of his being still alive and in Richmond, and possessing abundant means, had started for Washington at once with the intention of trying to get a pass to go South in search of his son. He arrived the day befor. we were ready to start for Ohio. His surprise and joy in finding his boy free and safe were indescribable. We almost envied our comrade the happy meeting, but hoped for others like it soon. Reddick greeted each of us only less cordially than his own son, and was in our company in the journey over the Baltimore and Ohio Railroad as far as Wheeling.

Of the thoughts that filled our hearts as we bade adieu to Washington, where we had spent such a pleasant week, and of the kind friends who had seemed to try to make amends for all we had suffered by rebel hands, I have no time to speak. The freshness of liberty had by no means worn off, and each change was a delight in itself, yet the hours that intervened between us and *home* seemed all too long. Mr. Reddick insisted upon buying for the whole party everything we could possibly eat on the way, and in the delight he took in the presence of his son we could see reflected what was in store for the rest of us.

When Bellaire on the Ohio river had been reached, our homeward roads parted and I was left alone—the first time for many months. We had promised to write to each other and never forget our prison association —a promise which has been kept. The others took the road which carried them into the interior of Ohio, while I had to go up the river to Steubenville. But our train was behind time, and the one with which it should have connected had long since gone. There was no other till afternoon. The waiting among total strangers for some hours seemed very wearisome; but getting a newspaper, reading the news and thinking of the meeting before me caused time to pass pleasantly. A party of travellers ordered a special dinner and I joined them, and then enjoyed a new sensation—that of *paying my own bill*—a thing I had not done since the supper at Dalton the preceding April! The price was a dollar, and I was correspondingly impressed. When the cars came I met no acquaintance. The run to Steubenville was short, and at that point I found that I would have to wait some two hours before I could go on up the river to the station nearest my father's. I preferred walking up to the office of the *Herald* the paper with which I had corresponded ever since the opening of the war, to remaining in the depot, and started. Before I had gone half a square, some one recognized me, and another, and another. I could hear them calling out of shop doors, "There's Pittenger!" The forming of the crowd was fully as prompt as in some of the southern towns through which we had passed, but the expressions were very different. I could scarcely get to the newspaper office; and I could not get away from the town till I had given my promise to Dr. John McCook, father of Anson G. McCook, my first Captain, to come back and give an account of my adventures to the people the second evening from that date, which I did to nearly the whole town, very much to my financial advantage.

The journey over the old familiar hills about which I had dreamed in Southern dungeons, the tearful welcome of father and mother, the surprise and joy of the little brothers and sisters,—all these are beyond description. For the first time in its history a public supper was given in honor of an individual in the little village of Knoxville. It was held in the village church, and the whole surrounding country was represented. The next Sunday I attended the Methodist church in New Somerset and had my name enrolled as a probationer. The vow I had made to God in the hour of trouble was not forgotten.

After the expiration of my furlough, though still in very feeble health, I returned to the 2nd Ohio, which was then encamped at Murfreesboro —the very town from which I had planned to go South the year before. Here my reception was not less hearty than at home, and I realized how much more pleasant it was to wear the blue and face the foe, even in all the hardships and dangers of war, than to be cooped up in dark prisons.

CHAPTER XXXV.

SUBSEQUENT HISTORY OF THE RAILROAD ADVENTURERS.

Brought down to 1893.

CAPTAIN DAVID FRY lived at his old home in Greenville, Tennessee, until the 21st of August, 1872, when he accidentally fell in stepping from a moving train of cars and received injuries from which he died a few days afterward.

The eight of our party who escaped from prison in October, 1862, after enjoying short furloughs, returned to their places in the ranks. All of them were commended for good conduct, but received no special promotion until after the exchange of their six comrades in March, 1863. Several of this first number participated in the battle of Stone River, and the fortunes of war carried all of them a second time over nearly the same ground they had previously traversed. In the spring of 1863 all who made proper application to Gov. Todd of Ohio, as directed by the Secretary of War, received lieutenants' commissions, though several experienced technical difficulties and delays in being duly mustered as officers.

It may be of interest to give a brief account of the individual fortunes of the raiders, dwelling a little more fully upon those which present features of special interest.

Wm. Knight served as a private soldier with great credit until his final discharge on the 28th of April, 1864. He spent a year in California, then came east again, resuming his old occupation as railroad engineer, and residing in Logansport, Indiana. When he married, he exchanged the locomotive for stationary engines. He has lived in Ohio, Wisconsin and Minnesota, returning at length to Stryker, Ohio.

Wm. Bensinger, at the expiration of his furlough, applied to Gov. Todd and was promised a commission as soon as a vacancy should occur in his regiment. But as there was considerable delay, he decided to accept instead a captaincy of colored troops. Here he rendered excellent service, finding the negroes, after sufficient drilling, to make good soldiers. At the battle of Nashville his company fought with desperate gallantry, losing more than half of their number.

Bensinger married a Southern lady during the continuance of the war, under very romantic circumstances. He was mustered out of service on June 11th, 1866. After the war he farmed for some years, but the burning of his house entailed so heavy a loss that he could not replace it, and he left the farm, first for mercantile business, and afterward to serve as railroad engineer. After spending eight years in this latter occupation, he purchased a farm in Hancock Co., Ohio, on which he still resides.

J. A. Wilson, from whose published "Adventures" frequent citations have been made, served in the army till the expiration of his period of enlistment. He has since resided in Wood Co., Ohio.

John Wollam returned to the army and was recaptured at the Battle of Chickamauga. He had the misfortune to be recognized as having belonged to the Andrews raid, and was taken to Atlanta, and compelled to wear a ball and chain for three or four months. From this severe treatment he managed by his own shrewdness to escape, and being now quite familiar with the best modes of travelling through an enemy's country, succeeded in piloting two companions who shared his fortunes through to the Union lines. He returned to the ranks and finished his period of service, after which he resided in Ohio and Illinois until 1877, and in Kansas until September 25th, 1890, at which time he died.

W. W. Brown. From a recent photograph.

Robert Buffum had a sorrowful history. He was as brave a man as ever lived, and had been prominent in the Kansas struggles which preceded the great civil conflict. After exchange he received his commission and served with honor till the close of the war. But unfortunately he had formed the habit of hard drinking, and when in liquor was uncontrollable. His mind, also, was a good deal unsettled by reason of the terrible hardships he had endured, and this grew worse instead of better. Before he reached home, after being mustered out of service, he was arrested in Kentucky for some offense the nature of which I never clearly knew, and very summarily condemned to the State Penitentiary. Receiving a letter from him in which he professed innocence of any grave fault, while admitting that he had been drinking heavily, I secured the signatures of Judge Holt and Secretary

Stanton to a petition for his pardon. This Gov. Bramlette—probably thinking the man had been hardly dealt with—very promptly granted. When released, Buffum returned to Ohio, signed the temperance pledge, and lived a perfectly sober life for some months. During this time he engaged in the sale of the first edition of *Daring* and *Suffering*, and met with such success that he was not only able to maintain his wife and three children, but also to pay for a home. But there were so many persons who wished to "treat" that he yielded to temptation and was soon in trouble. In a heated discussion in Minerva, Ohio, a man used violent language against President Lincoln, declaring that he ought to be hung. Buffum's latent insanity blazed out in a moment, and procuring a revolver he shot the man in the face. For this he was imprisoned in Canton, Ohio, but the wounded man recovering, and leaving the State without appearing against him, Buffum was soon released.

The worst still remains. A few years after, he shot a man dead in Orange Co., New York, in a fit of insanity, and was condemned to the Penitentiary for life. He was received at Sing Sing, May 23rd, 1871, where his derangement was so evident that only six days after he was transferred to the Asylum for Insane Criminals at Auburn. In this

JACOB PARROTT and D. A. DORSAY overlooking Chattanooga. From a war-time photograph.

refuge he remained until July 20th, when he cut his own throat and died —a pitiful ending after his marvellous escapes !

Mark Wood received his commission, and after the war settled in Toledo, Ohio, where he died, in the year 1867, on account of disabilities contracted by prison hardships.

W. W. Brown received his lieutenant's commission, and was afterward promoted to a captaincy. He was wounded, and for the disability thus incurred received a pension.   He has been a farmer in Wood County, Ohio, for a number of years past.

Jacob Parrott and D. A. Dorsey both received the commissions to which they were entitled, and accompanied the army to Chickamauga, in which battle they bore their full share, as also in the battles of Lookout Mountain and Missionary Ridge.   The accompanying photograph, in which they are represented as looking over the scene of so many adventures and sufferings, was taken just after the capture of Lookout Mountain by Gen. Hooker.   Dorsey resigned on account of broken health before Sherman reached Atlanta.   He farmed for some years in Illinois and Nebraska, and afterward settled in Kearney, Nebraska, where he has been admitted to the bar, and devotes most of his attention to legal and real estate business.   Parrott received command of a company, became a universal favorite, and was especially commended for gallant conduct.   He was finally discharged at Savannah, Jan. 3rd, 1865.   He soon married and has since resided at Kenton, Ohio, meeting with a very good degree of prosperity.

E. H. Mason received a Captain's commission when he returned to the army, and at the battle of Chickamauga found himself once more a prisoner.   He also, like Wollam, was recognized and received much ill treatment on account of the former adventure.

WILLIAM H. REDDICK. From a recent photograph.

Martin J. Hawkins returned to the army and served till the close of the war as a private.   He was wounded in the back of the head by a shell at the battle of Chickamauga, but managed to keep with his comrades when they fell back to Chattanooga, as he had no wish to try the rebel prisons again.   He was given a commission in the regular army in 1866, but was physically unable to be mustered.   When able to work at all he served as engineer.   For a time he resided in Kansas, and died at the residence of his son-in-law, B. D. Higgins, in Quincy, Ill., Feb. 7th, 1886.

William H. Reddick and J. R. Porter have each furnished me a sketch of their adventures, which is appended in their own words slightly abridged. Reddick writes:

"DEAR COMRADE:—Yours at hand. Would just say that I served through the war as 2d Lieutenant of Co. B, 33d O. V. I. I was discharged July 12th, 1865, at Louisville, Ky. After discharge I returned home, but for two years did not do anything, on account of disability contracted while a prisoner. Finally, all my financial resources failed, and I was obliged to go to work. I have since the war fished a little in all kinds of labor, from farming to chopping cord-wood, making railroad ties, peddling notions, book agencies, and township clerk, back to honest clod-hopping, which latter I hope will furnish me my daily bread for the remainder of my life. I am a living encyclopedia of all the aches and pains that flesh is heir to—a used-up man, from the treatment received in that abominable hole of Swims', in Chattanooga. This is all the story that I can give of myself. If you can get any interesting facts from it, you are welcome to them.

"Yours as ever,
"WILLIAM H. REDDICK."

Porter says:

"Shortly after I rejoined my regiment we started on the Stone River Campaign, which terminated in the capture of Murfreesboro. After a few months we again marched for Chattanooga. At Cave Spring, Alabama, I was commissioned and mustered as 2d Lieutenant of Company G, 21st Regiment, O. V. I., August 29th, 1863.

"I was captured at the battle of Chickamauga, September 20th, 1863, in company with many more of the 21st. It was after dark, and we had run out of ammunition and were surrounded. Fearing that I might be recognized by some that knew us during our first imprisonment, I unbuckled and dropped to the ground my sword and belt, and having no other marks of a commissioned officer, gave a fictitious name, and passed as a private soldier. We were marched some distance from the battle-field that night. The next morning, after finishing our supply of rations, we were marched to Dalton, and after nine days' ride by rail, brought up at Richmond, Virginia. Here we were put in what was called Pemberton building, on the corner of Nineteenth and Carey streets, nearly opposite Libby. The Pemberton was a large four-story building, and contained about fifteen hundred prisoners. After we had been there a short time we concluded to investigate the contents of the basement; we cut a hole through the floor, and found that it was stored with salt. This was a scarce commodity in the Confederacy, and very expensive. After we had supplied our own side of the building, we cut holes through the doors in order to supply our fellow prisoners in the opposite side. They also concluded to see what their side of the basement contained. They struck sugar, and undertook to supply us in the same way that we had been supplying them with salt. A few of us concluded to try a flank movement by cutting a hole through the brick wall in the basement. We succeeded admirably. After we had cut our way through into the sugar, we filled our haversacks and returned to our respective places. There were very few found this out for three or four nights, for we knew that as soon as it became generally known it would cause a stampede. And there was a grand rush made for sugar one evening, which continued until late at night, causing so much excitement that it attracted the attention of the guards, who came into the building, called us into line, placed guards and kept us standing until the next morning. We had supplied ourselves with plenty of sugar, which we expected would be taken from us, but they allowed us to keep what we had, but moved the balance from the building. After the sugar riot, we remained in Richmond until November 15th, 1863. From Richmond we were taken to Danville, Virginia, where we remained during the following winter.

"Here we made several efforts to escape, but all to no avail. We tried tunneling several times, but never succeeded. The prisoners in one of the other buildings succeeded in tunneling out, and quite a number made their escape, consequently they were always on the alert for such things, and searched the buildings every few days. During our stay here the small-pox broke out among us, and owing to bad treatment, many died.

"We took our departure for Andersonville, Georgia, May 15th, 1864. When at Black Stock Station, South Carolina, I succeeded in making my escape from a train of cars in company with T. W. Harrison, a member of the 10th Wisconsin Regiment. After being out three days and nights we were re-captured, taken back to the station where we made our escape, put aboard the train and sent to Columbia, South Carolina, where we arrived May 20th, 1864. During our stay here we tried tunneling again, but without success. On the morning of June 29th we were again started for Andersonville. We arrived at Augusta, Georgia, in the evening, and changed cars for Millen. Soon after leaving Augusta it became dark, and as there was no light in the car, we proceeded to cut our way out by sawing a board off in the bottom of the car with a table knife with teeth filed in the back for a saw. We succeeded in getting a hole through the bottom of the car large enough to crawl out at by nine o'clock, and as soon as the train stopped, we rolled out.

"This was the night of June 29th, 1864, and we had a long journey before us. We concluded to travel in a north-western direction and strike the Federal lines somewhere between Dalton and Marietta, as we knew that General Sherman had reached Marietta, on his way south. When we escaped we had two days' rations, this being the only bread we had during the journey, which was twenty-six days and nights. We travelled altogether at night, subsisting upon blackberries and Irish potatoes. We were provided with a blanket each, one gallon coffee-pot, and plenty of salt and matches, so we could boil our potatoes. We once got a goose, and had a regular feast on goose and potatoes. Occasionally we would get some apples for dessert. Having a pretty good knowledge of the country, we made good time. During our journey we encountered but one person. We had many streams to cross, but they were low at this time of the year.

"On the morning of the twenty-sixth day out, we hunted a resting-place as usual, expecting to hide for the day. We had been in our retreat but a short time, when we were aroused by the whistle of the iron horse, and soon learned that we were near the railroad. Our joy was unspeakable! We gathered up our traps and were soon upon the railroad, and, as the train had passed a short time before, we started for a water station that was in sight, and found one man whose business it was to pump the water. We remained here until the first train came along, which we boarded for Marietta, where we reported to the Provost Marshal, who ordered us to General Thomas at the front. It was now the 25th of July, and Sherman's army was in front of Atlanta. The next train that came south we boarded, and were soon at General Thomas' head-quarters. After giving a full detail of everything concerning our capture, imprisonment and escape, he promised us each a furlough to go home. After this interview we started for our regiments, and were soon with our old comrades again after an absence of over ten months. Many of the noblest, bravest and best who were taken prisoners with us, were never permitted to return.

"After a leave of absence for thirty days I again joined my regiment (after the capture of Atlanta), participated in Sherman's Campaign to the sea, having charge of a party of foragers on that march. I was mustered out of service at Goldsboro, N. C., March 31st, 1865. Returning home, I remained two weeks, then spent five years in Kansas, Idaho, Oregon, and California; after which I returned to Ohio and spent seven years. Moving to Arkansas, I remained eight years and again returned to Ohio. During the last fifteen years I have been engaged in mercantile pursuits."

My own history may be briefly told.

After returning to the army I participated in the march from Murfreesboro' toward Chattanooga, but finding my health so seriously im-

paired that I was unable to endure the hardships of the campaign, I
accepted a discharge for disability on the 14th of August, 1863. Before
this time I had received a commission but was unable to be mustered as
an officer. Returning home I finished writing the story of the railroad raid
in very brief form, which was published under the same title as the present
work, by J. W. Daughaday, Philadelphia, in October of the same year.
Before and after this time, I gave numerous war lectures, which may not
have been altogether without effect on political sentiment. In March,
1864, I was admitted to the ministry in the Pittsburg Conference of the
M. E. Church, at its session at Barnesville, Ohio. I was stationed success-
ively at Minerva, Massillon, Cadiz, and Mount Union, all in Ohio. In
1869, Fowler and Wells of New York published for me " Oratory, Sacred
and Secular, or the Extemporaneous Speaker," which had quite a large
sale. In 1870 I was transferred to the New Jersey Conference, where I
have ever since been in the active work of the Ministry, being stationed
at Woodbury, Vineland, Burlington, Princeton and Bordentown and
Haddonfield; and am at present (1893) in my first year at Colton, California. In 1871 I published through J. B. Lippincott & Co., Philadelphia, "Capturing a Locomotive," which has been widely circulated.
Since 1876, I have also been connected with the National School of
Elocution and Oratory in Philadelphia, lecturing, and giving weekly
lessons on Shakespeare, and Extempore Speech. Three other volumes
of mine have been issued by the publishing house connected with that
school: " Extempore Speech," " How to Become a Public Speaker," and
" The Debater's Treasury."

In the spring of 1889 I procured a transfer to the Southern California Conference, and continue in the active work of the ministry in
this winterless climate, being pastor of the Methodist Episcopal Church
in Colton, San Bernardino Co., at the present writing (1893). I have
gained much in health from this California residence.

PENSION BILL.

Passed July 7th, 1884.

The opinion was often expressed in influential circles that the extraordinary hardships endured by this band of soldiers deserved some
substantial Governmental recognition in the shape of a pension. Several
of the adventurers had obtained pensions for disability under the general
law, but for a number of years no legislative action was taken. On
one occasion, Buffum saw General Grant, then President, and called
his attention to his own condition, being at that time sick and in very
straitened circumstances. General Grant was well acquainted with the

facts, and disposed to give help. He referred the matter for advice to Judge Holt, who returned an opinion that while Buffum's story was true, and the case meritorious, the executive department had no authority to do more than to call the attention of Congress to the subject, which was done. A bill granting a pension to the survivors, so far as they were then known, was afterward introduced and readily passed the House, but in the Senate it was adversely reported and dropped. In the next Congress the bill was revived, the rate made $24 per month, and after a full investigation it was passed. The only arguments made against its passage were based upon the criticisms of Gen. J. B. Fry, Chief of Staff to Gen. Buell. These will be considered in Chapter XLII. It is sufficient here to say that they had their origin mainly if not entirely in a difference of opinion between Generals Mitchel and Buell as to the manner in which war should be carried on. The bill became a law by receiving the President's signature, July 7th, 1884.

CHAPTER XXXVI.

THE SOUTH REVISITED A QUARTER OF A CENTURY AFTER.

TO go freely amid scenes of peace over the ground associated only in my mind with war, prisons, and chains, was a pleasant experience which only limited means and the duties of an exacting profession long prevented me from enjoying. At length, in 1886, the wished-for opportunity came. Six years earlier, just before writing "Capturing a Locomotive," I had made a hasty and partial visit to the South, and the information then received made me anxious to fully explore this field. My resolution was to search all records at Washington; to seek at Atlanta for the place of execution and the grave of Andrews; to look over the War-files of Southern newspapers; to visit Flemingsburg and seek out the early life of J. J. Andrews, about which many readers were curious; to meet the survivors of the party; and generally, to get information from every source which the passage of time had rendered available for completing and illustrating the history of this railroad raid, which, in the estimation of many critics, is the most curious and romantic event of the Civil War.

With time, letters of introduction, and financial resources provided by friends, there was every reason to anticipate success. I was not disappointed.

On Nov. 2nd, 1886, I arrived at Washington accompanied by Mrs. Pittenger. Mr. F. G. Carpenter introduced me to Mr. Spofford, Librarian of Congress, Col. Scott, Editor of the Official War Records, and Mr. Hodgeson, of the Records Department of the War Office, and I was soon hard at work. My only grievance was the shortness of working hours—from 9 to 4. Mr. Spofford, whose memory of books seems infinite, produced for me everything having any bearing on my subject, and also the files of rebel newspapers for 1862-3. From the books I secured no new facts, but several valuable illustrative items. The newspapers also were somewhat disappointing, for they were meagre in historical material as compared with Northern papers. Their facilities for gathering news were not great, the censorship as regarded military matters was rigid, and their use of the telegraph so limited, either by their poverty or by its pre-occupation for

Government purposes, that the most important events of which one paper would be full at a certain date would not reach another paper a few hundred miles away for three or four days! But the time spent in looking over the papers was far from lost; nothing could have so fully recalled the spirit of those days—their intolerable prejudices, bitterness of feeling, and the strange illusions under which the war was fought.

But a far deeper interest was evoked in Col. Scott's office. The Government is there selecting from the official reports and correspondence of the war, those papers which possess permanent interest, for the purpose of publication. Already some twenty-five great volumes are in print, and the work is scarcely one-third done! No history of any part of the Civil War will have any standard value in the future which does not rest on these reports. Of that not yet printed, much has been classified by months and subjects, becoming thus easily accessible. I was kindly supplied with a writing table, and all the bundles of MSS. I called for were furnished, with only the reasonable condition that I should rearrange each as I found it. Never have I had a more fascinating employment, though it sometimes became positively painful. To be thus groping among the very papers and orders which once had the power of life and death to thousands, to see the secret springs of events with which I was familiar on the outside and often wholly misunderstood, to feel that at any moment I might come across a great historical prize, to see the very hearts of men laid bare, often with startling revelations of greatness and meanness—all written in dim and almost faded characters which were never intended for the public eye—this caused the hours to pass by almost like minutes. Nothing that I found had a deeper personal interest than the letter expressly marked as private of one Confederate to another which led to my own release from a rebel prison. There was also—far earlier—the order to proceed with the trial of spies,—an order which, but for the intervention of Providence in the form of an advance of Union armies, would have led to my own death on a scaffold.

But I was not satisfied with examining those documents which had been selected for publication. There might be others among the great mass in the war archives of even higher interest. Of one document published in the official reports, Vol. X, I wished to see the original, and secure a photographic copy for lithographing. This was the petition of our party which reached Jefferson Davis and received his endorsement upon it—a hard, merciless command which neither he nor his friends could take any pleasure in reading!

The war records are guarded very carefully, as is right. The Government does not allow its own archives to be used as a basis for sustaining claims against itself. And there is always the possibility that some valuable but compromising paper might be abstracted. A letter to the Secretary of War, however, removed all obstacles in my own case. I not only found the document wanted, but another letter of even greater interest, written by myself to Jefferson Davis on a terrible day. I think I was able to get hold of everything in the department which could throw any light on the raid. Then a swift journey over the Piedmont Air Line carried us to Atlanta.

In this city, Mr. F. J. Cooke not only rendered me most valuable services, but had by careful inquiries almost if not quite accomplished one of the chief objects of my visit—the location of the grave of Andrews. The evening was cold and gloomy when Mr. Cooke, Mrs. Pittenger, and myself drove to the spot. The beautiful houses of Peachtree street—the finest avenue of Atlanta, or perhaps in the South—were passed with but little comment, until the thin and scattering woods beyond were reached. The land here is said to be all laid out in town lots, but held at high valuation and not yet built upon. The story told to all who had inquired about this grave was that the execution and burial had taken place in what was now a compactly built portion of the city, where the grade had been changed, and that the site was hopelessly lost. I presume that we were not intentionally misled, but in the confusion occasioned by the burning of the city, some one had made this statement, and it was caught up, as such things are, without investigation, and repeated until generally believed.

In a little ravine two or three hundred yards from the roadside we were shown where the scaffold had stood. Here the sad, pathetic tragedy of twenty-five years ago had culminated. The poor fugitive, who had been twice hunted down by dogs and men, was alone in the midst of his enemies. His feet, on which cruel fetters had been riveted by an unwilling blacksmith, probably still clanged with the chains from which they were not released even by death. How the calm, pensive soul of that brave man must have thrilled in view of the near lifting of the mists on the other side of that " Jordan " whose secrets he said he had often wished to behold ! He had tried to serve his country, had nobly failed, and was now about to pay the penalty. We can well believe that no word of insult—such as had often been heaped upon him a few days before—was spoken then; that even those who were most determined on shedding his blood were awed into silence; and that outside of the rope barrier there may have even gone up prayers for him from Christian hearts. All seems to come back now; and when the cold face was laid in the earth, with no barrier of boards to protect it, the cup of earthly misfortune for one human heart had been filled to the brim. It seems little that at last his bones should be coffined and honored. But perhaps in the life beyond, toward which the sufferings of prison and the swiftly nearing shadow of the scaffold had turned his heart, the little we now can do will not be without its value.

The spot pointed out as the grave was some three hundred yards from the scaffold, on a little dividing line between two small ravines, and some dozen yards away from a large stone. The latter was spoken of at the time as a convenient landmark and still remains.[1]

Very quiet and still were we as we drove back again to the lighted city.

The next day the same party, with the addition of Captain Fuller, the conductor whose train had been captured, visited the Atlanta city cemetery and stood upon the site of the still greater tragedy consummated on the 18th of June, 1862. Capt. Fuller was among the spectators then, and took the dying message of Ross, heard the wonderful speech of Wilson, and when all was over carefully marked the seven-fold grave. Long since the remains were removed to the National Cemetery at Chattanooga, and appropriately marked, but this spot will ever remain hallowed ground.

There was a meeting of the prison Congress in Atlanta presided over by Ex-President Hayes, which opened the evening of our arrival. A number of addresses were made by prominent citizens of Georgia, full of sentiments of hearty devotion to the Union. These with the enthusiastic cheers they evoked from the inhabitants of this beautiful and prosperous Southern city sounded strangely after the memories of the afternoon. But surely no one can rejoice more over manifestations which prove the reality of Union than those who suffered to preserve it; and the prophecies of Wilson have been abundantly fulfilled in the city with which his name will ever be associated.

The war files of Atlanta newspapers had an especial interest. The Young Men's Library possesses these almost complete, and I found again a copy of the *Southern Confederacy* of April 15th, which I had purloined from the jailor Swims in Chattanooga and carried home. This contained the first full account of the capture of the train, though several partial ones had preceded it, and the paper contained many deeply interesting references to the subject during six months afterward.

It would have been no small degree of melancholy pleasure to have once more visited the prison in Atlanta—the scene of that darkest day when loved comrades were hurried to death, and where the consolations of religion first arose to overcome the fear of death; but the prison had been twice destroyed and rebuilt. There was first a temporary log structure in place of the brick building we knew, and then, when Atlanta became a state capital, and a large city, a more pretentious jail took the place of the old, having, at the time I visited it, some 120 inmates; but its larger size and greater accommodations did not in the least supply to me the place of the rude house kept by Turner! The surrounding country looked familiar, and I could realize very vividly the scenes of that day of suspense, when

[1] On April 11th, 1887, the government authorities acting on information secured during the above visit, and by the direct orders of the U. S. Secretary of War, disinterred the remains of J. J. Andrews for burial with his comrades at Chattanooga.

we believed that our only hope of escape from speedy death lay in wresting their muskets from the armed guard by whom we were surrounded. The old jailor, Turner—a kindly and merciful man—had long been dead. The present incumbent in his office, a Major Poole, greeted me kindly, saying that it was always a great treat to meet a Union soldier and talk over the war times; adding the quaint remark, " We whipped you and whipped you until we naturally wore ourselves out at it and had to give up !" This was really an epitome of the whole contest from a Southern point of view; or, as engineer Knight says in describing the battle of Brown and himself with bloodhounds, " We won the victory but evacuated the ground !"

I would have also liked to visit the building used as barracks, where a season of comparative peace was enjoyed by the recaptured remnant of our party. But this portion of Atlanta was burned by Sherman, and afterward completely changed.

A still greater interest, perhaps, centered in the road we had striven to destroy. I visited Jos. M. Brown, son of Ex-Governor Brown, General Passenger Agent of the Western and Atlantic Railroad, who received me with every courtesy, and gave me fine maps of the road, together with a month's free pass. After going twice over the route alone and examining every point, I secured the company of Capt. Fuller, who took great pleasure in making me understand every detail of the wonderful pursuit on the part of himself and Murphy; and some matters which had always seemed mysterious became perfectly plain when thus explained on the ground. It was hard to realize as we chatted amiably together that we had been engaged on opposite sides in deadly strife, and had sought by every resource in our power to destroy each other on this very line !

I found Chattanooga marvelously changed—even more than Atlanta. The latter is a beautiful and stately city, but still distinctly Southern in type; while Chattanooga is largely built up by Northern enterprise and has become a great manufacturing centre. Little care is in it for the stories or traditions of the past ! In every sense it is progressive, filled with the rush of improvements, as speculative and pushing as a Western metropolis, and looking for a great future. Scarcely any landmarks of the old city are left, as is but natural when we remember that it had but three thousand inhabitants in 1862, and now, 1887, has 30,000 ! I found the site of the old Swims jail—long since destroyed,—on Lookout Streets between Fourth and Fifth, near what is called Brabson's Hill. Mr. Allen, Chief of Police, an old Confederate soldier, located it exactly for me. There was only one thing that I remembered belonged to the surroundings—the house of William Lewis, a colored man—which had been pointed out to us because it stood close to the jail and was a very good house for that day.

I had an amusing experience regarding the location of this Chattanooga prison, which illustrates the old truth of how easy it is to be mistaken. Fuller asked if I would like to visit my former prison. I assured him that few things would give me greater pleasure, but that I had little hope, as I had been told that it was long since demolished. He declared this to be a mistake and offered to conduct me to the spot. For a moment, remembering the error about the grave of Andrews, I did indulge the hope that I might be able to see the old place once more, especially as Fuller was perfectly confident of having visited us there and being acquainted with all the surroundings. He took me without hesitation down Market Street to the corner of Fourth, and stopping before the old city building, said, "Here it is." I remained silent for a moment, for the thought was struggling in my mind, "How can such a mistake be made by a man so intelligent and well informed?" He continued, "And now, Pittenger, you see how mistaken your party were in thinking they were put underground in a dungeon. None of the rooms in this building are below the level of the ground; but I do not wonder that you thought they were, for you first went upstairs and then down a ladder."

The building was one of fair proportions, and none of the rooms could have been very dark or close. Was it possible that all the story of the terrible hole of Chattanooga was only an evil dream? I had no doubt, but some curiosity; so I said, "Are you sure that this is the place?"

He said he was, but to make it still more certain, asked several of the bystanders if that was not the building where the Union prisoners were kept? They all answered in the affirmative; but noticing that I did not appear satisfied, he called to a very old man who just then came across the street, asking him how long he had lived in Chattanooga; and was told that he had been there constantly for forty years. "Then," said Fuller, "you can tell us what building the Union prisoners were kept in." "Why in this one right by you, of course; both Federal and Confederate prisoners were in the same place," was the confident answer.

Fuller clapped me on the shoulder and said, "You will have to give it up, Pittenger." "Not yet," I answered; turning to the old man, I said, "Was there any other prison in town?" For a little time he reflected, and then said, "I think there was; a very small one; but not many prisoners were kept in it." "Where was that," I asked. "About three squares from here," he replied "It was so small that it would only hold three or four at a time, and was kept by a man called Swims."

Fuller looked less triumphant, and I asked further, "Do you remember about some bridge-burners captured in the early part of the war?" He said that he remembered them very well. "Where were they put?" I asked. He answered at once and very confidently, "Oh! they were put in Swims' jail."

I clapped Fuller on the shoulder and asked him if he was ready to give it up. He very frankly acknowledged his mistake, saying that he supposed that he had visited other prisoners here so often and visited us at other places, and thus got the locations confused.

I have been the more particular to relate this, because almost any person in Chattanooga would make the same mistake if not put on guard.

I had a strong desire to go out to the village of Lafayette, where I had my first experience of prison life. No railroad led to the place, and the roads were muddy, so that all conditions of country and weather were very like those of former times. Two days were required for the trip. On the way out we found as dinner time drew near that we were far from any hotel. With the hospitality of the South in view this seemed to give us only a better opportunity of seeing the present state of the country. Probably we made a mistake in asking at the best-looking houses; but it was not until after seven applications that a dinner was found! The refusals were kindly enough, good excuses being offered, but the people were fearfully unanimous! Our hostess was a lady of nearly sixty, who had lived in that one place all her life with the exception of about a year and a half, when "there was too much fighting along the road." She said that she did not feel like leaving home on account of the soldiers, until one day a cannon ball came through her sitting-room door and broke her fire shovel; she thought it then time to move off.

At Lafayette I was rejoiced to find everything in almost exactly the same state I had known it. The old hotel where I had been cross-examined for four hours was unchanged, and I had the privilege of lodging in it for the night. But I inquired with still deeper solicitude for the jail with its iron cage. I was gratified to learn that it still stood, and was unchanged, except that port-holes had been pierced in its brick walls by Union soldiers during their fight with cavalry in 1863. Myself and friend found the jail standing open, and were enabled to explore it without hindrance. For years no prisoner had been placed in it. The explanation given of this strange state of affairs in a county jail was that "local option ruled." For a few moments I was again in the iron cage that I had known so long before, but there was no crowd of men and women outside to comment on my appearance! My companion, Frederick Gates of Chattanooga, who made an accurate sketch of the building, was soon an object of more interest than myself, for the people learned that he was a land agent and were anxious to sell their farms at prices that seemed to me exceedingly low. But there had been scarcely any sign of improvement for the last twenty-five years and they were almost cut off from the outside world. No doubt a railroad would work a great transformation.

The editor of the Chattanooga *Times* suggested to me that it might be well to visit Gen. E. Kirby Smith, who was located at the University of Sewanee, on the summit of the Cumberland Mountains, some sixty or seventy miles from Chattanooga, for the purpose of seeing any war papers he might have. General Smith received me cordially, but assured me that he had no document whatever bearing in any way upon the raid, or the trials of the men engaged in it. I had hoped to learn from him whether it had been the intention of the Confederate authorities, of whom he was chief, to try the whole party or only the twelve who were sent to Knoxville; also the reason for not proceeding against the survivors, after their removal from Knoxville. On the first point he could give me no information. All papers had been turned over to his successor and probably lost, while he did not remember anything definite. In regard to the second, he said that he had no doubt that "the case against you had gone by default and many of us were glad to have it go that way." But while I received little direct information, the conversation with a well informed and acute-minded man, who had held my life in his hands for months together, was wonderfully interesting and stimulating.

The next point visited was the beautiful metropolis of East Tennessee, Knoxville. Here I met Judge O. P. Temple, one of the attorneys who had defended our party on the memorable Court-Martial which had terminated so tragically. His kindness was unbounded, but I regretted to find that Judge Baxter, our other benefactor, had been dead for a year past. I had the pleasure of tendering to the son the thanks which could no longer be given to the father. Judge Temple showed me the old courthouse (now used as a school) in which the trials took place, but the jail was gone.

From Knoxville I started to Flemingsburg, to ascertain all that I could about the life of Andrews previous to his embarking on the Railroad Adventure. This I regarded as the most difficult and important part of my journey. The character of Andrews and all that was known of his early life tended powerfully to stimulate curiosity. I found that Flemingsburg was on the branch of an obscure railroad, and was only reached after much zig-zag travelling. Arriving a little before noon on Saturday, I found the town, though a county seat, to be a mere village, only a little less rural and out of the world than Lafayette, Ga. How such a man as Andrews, with his powerful genius and refined manners, could have settled down in such a place was a problem to be solved if possible; or had we, through the glamor of suffering, misfortune, and time, completely idealized him? Within an hour I was relieved of at least one great fear—that of not finding any knowledge of Andrews in the place. It is a strange feeling one has when making inquiries for persons who have been dead for a quarter of a century—very much like stumbling about over forgotten graves! But Andrews was not forgotten, and I soon learned that there were very especial reasons why he should be held in remembrance, and a portion of his history at least spread upon the county records. The landlord, in response to my inquiries, said that he had only been eleven years in the place but that he would introduce me to an Andrews in the bank who could probably tell me what I wished to know. I knew he could not be a relative, but thought it as well to begin at that point as at any other. Judge of my surprise when Mr. Andrews of the bank told me that he had just written a letter to a gentleman of Missouri on the same subject! In a few minutes H. C. Ashton, the postmaster, and James B. Jackson, a retired hotel keeper, both old friends of J. J. Andrews, came into the bank, and we were asking and answering questions as fast as we could find words! They had never seen one of the prison comrades of Andrews; I had never seen any person who had known him before the war; and there was so much to say that the important business of dinner was almost forgotten.

In the afternoon I was shown the letter of Andrews written just before death. The old tattered original is preserved with scrupulous care; and a careful copy has been placed among the records of wills. The reason for this is furnished by a remarkable financial transaction of which he was the centre. Andrews deposited twelve hundred dollars in gold in the Flemingsburg bank, and gave a check for that amount with interest, to Mr. D. S. McGavic, a trusted friend, with instructions, in case he should perish in the critical business in which he was engaged, to draw it out and lend it on good security, and give the interest as a perpetual legacy to the poor. In his last letter written just before execution, he says to his friend, "in regard to other matters, do exactly as instructed before I left." The receipt of this letter containing certain intelligence of the tragic fate of their townsman, of which they had only heard rumors before, filled the little village with sorrow; and this was greatly intensified when the touching bequest of all his savings to the poor in their midst was made known. Many references to this universal mourning were made; one gentleman saying to me, "Nothing that ever happened, not the darkest calamities of the war, ever cast such a gloom over our town as this news from Mr. Andrews." McGavic drew the money as ordered, but not wishing to take the responsibility of investing it upon himself, he asked the court to appoint a trustee. Then followed an unpleasant history. Col. Dudley was appointed trustee, and loaned the money in several sums on slender security. The borrowers afterward fell into bankruptcy, and the money was declared lost. But even worse was behind. In the year 1875 the whole matter was brought out, and proceedings set on foot to punish the great apparent wrong and compel restitution. Col. Dudley and his associates were quite hardly pressed, when their attorney, W. A. Cord, opportunely discovered some persons who claimed to be heirs of Andrews. They did not know the money was all lost and were easily induced to bring suit for it. Esquire Cord himself testified that Andrews had told him that he came from Holliday's Cove, in Hancock Co., West Virginia, the former residence of these parties.   There a series of affidavits were taken in due form showing the parentage and early life of Andrews and his relationship to these claimants.   So far as the evidence on the books went, all seemed clear, though the relationship was distant; and in some way, the suit against Col. Dudley for fraudulent use of trust funds was discontinued; then the so-called heirs were informed that no money remained unexpended !   I read the evidence on file at Flemingsburg, and the whole story was far from creditable.

For me, the greatest interest of these proceedings was in the light they seemed to throw on the early life of our leader.   The picture was not flattering; the family connection, as a citizen of Holliday's Cove afterward expressively said, "was the lowest of the low," their story including squalid poverty, vagabondage and drunkenness.   But sometimes persons rise above their surroundings; and at any rate, truth, not romance, was the object of search.   I will anticipate the order of my journey to say that a week after I visited Holliday's Cove, in company with my old friend, Captain Sarratt, and speedily found that the whole story was a sham.   Our James J. Andrews was totally unknown; and as the place was only a country neighborhood where everybody had known everybody else for a lifetime, the presumption was strong that he had never been there at all. But a James Tolman Andrews was known, and to him the affidavits truthfully referred.   The only points of similarity were in the first and last names—both common—and in his having left that part of the country near the opening of the war.   But James T. Andrews was short and light, with no point of physical resemblance to James J. Andrews; he was ten years younger, and, most decisive of all, was living in Kansas long after the close of the war !   The pretended identification was absurd, but it had probably served its purpose !

While in Flemingsburg I was shown the house where Andrews first boarded, and many specimens of his work in painting houses, boxes, etc. One person after another came to me claiming to have been on terms of friendly intimacy with him.   It struck me as a little remarkable that very many of them, notwithstanding they knew him to be a spy, referred to his *truthfulness* as one of his distinguishing traits !   "You could believe every word he said;" was repeated many times; and when I said that the Confederates found differently, they would admit the contradiction, but repeat the former statement.   I made sure from their testimony that the man was not a natural deceiver, but on occasion of becoming a spy had forced himself for a great purpose to act a part not belonging to his character.

So vividly did these repeated conversations with the old acquaintances of Andrews recall my own sorrow for my friend and leader, together with all the dreadful scenes of the past, that when the evening came I was completely unnerved; a raging headache with a sense of utter prostration forcibly reminded me—not for the first time on this journey so fraught with tragical memories—that feeling is more exhausting than any form of physical exertion. I did not sleep much that Saturday night, for I was living over again the whole story, from Andrews' point of view, as I had so often gone over it from my own. I had some other investigations to make, probably of still more sad and pitiful character, connected with the betrothal of Andrews, but was glad of the Sabbath's rest before beginning them. I attended the little village church where Andrews had often attended, and heard a plain and sensible sermon; but I was listening to an inward one far more powerful, of stern and sad application, preached on the vanity of human endeavor. How Andrews had striven! with what activity, resolution, and daring courage! and with what result! His enterprises had miscarried, he had spent his last days in chains, his body was flung coffinless into the ground, and his money, the only tangible result of such sacrifices that many of his townsmen could appreciate, left as a perpetual benediction to the poor, was ruthlessly squandered; and as the last memorial of all, an empty trunk returned to his betrothed! My heart ached in view of such sorrow! But a gleam of light shone on the cloud; as I recalled the words of his wonderful letter written at the foot of the scaffold, I felt that all his sufferings had not been in vain: " I have now calmly submitted to my fate, and have been earnestly engaged in preparing to meet my God in peace. And I have found that peace of mind and tranquillity of soul which even astonishes myself. I never supposed it possible that a man could feel so complete a change under similar circumstances. . . . . . . We may meet in heaven where the troubles and trials of this life never enter." There is something better than worldly success, when it bears this fruit!

On Monday, I first procured a photograph of the village of Flemingsburg, and copied the evidence, since found to be worthless, of the pretended heirs of Andrews and their friends. Advisers differed as to whether it was better to go to Mill Creek to seek the story of the betrothal or in an opposite direction. However, the indication given in the letter regarding the empty trunk seemed to me more likely to be right than any other: so I drove in a drizzling rain out the Maysville turnpike for some eight miles, and came to the old Lindsey residence. I had previously learned of the owner's death, but the widow, now very old, still lived in the same house. When she entered the large, quaint front room, and I saw how old she was, I almost feared that she would be unable to fix her mind upon the facts I wished; but the first mention of Andrews' name aroused her deepest interest. She remembered all about the betrothal, and it was indeed through her husband and herself that the young people had been introduced. As we talked, a middle-aged lady came into the room, and for a moment my pulse quickened with the thought, "Perhaps this is she of whom I am in search, and I will hear her sad romance from her own lips?" But she was introduced as Mrs. Lindsey's own daughter. Elizabeth J. Layton was the one of whom I was inquiring. She had lived with Mrs. Lindsey some six years, as after her father's second marriage it was less pleasant at her own home. Mrs. Lindsey described her as "a member of the 'Christian Church,' a good religious girl, of medium height, fine form, dark eyes and dark, wavy hair. It was in this room that Andrews courted her. They were very affectionate and happy until the terrible war came on; she was not rich, but had a little property in her own right, and was economical and a good worker. She would have made him a good wife. And now can you tell," she said, "if it was really true that he was a spy? We would not have thought anything so bad as that of him, for we loved him as our own son. He painted all the new part of our house, and was here a good deal of his time. He and Elizabeth were strong Union people, while Mr. Lindsey and myself thought the South was right, and were very sorry to have him go into the war."

Mrs. Lindsey searched her album to find me a photograph of Miss Layton; but not succeeding, she said that if I would go to Mrs. William Rawlins, a half sister of Miss Layton's, she was sure that I could get a photograph of the latter, and probably of Andrews also. As these pictures, which I did not know to be in existence, would be very valuable, I took leave with many thanks and followed her directions. I found Mrs. Rawlins in the village of Helena, two miles distant. Here I gathered additional facts, and procured the lady's photograph, but learned that for the other I would have to go to East Maysville on the Ohio River. She was sure that her sister who lived there, Miss Elvira Layton, a full sister of Elizabeth, had Andrews's photograph. I determined to go at once.

Reaching Maysville the same evening, after a little inquiry, I boarded the right street car to take me up the river to the suburb known as Newtown or East Maysville. The street-car conductor kindly set me off at the residence of Mr. Benjamin Williams, with whom Miss Layton lived. The latter was much older than Elizabeth, indeed now far advanced in life, and when she came into the room where Mrs. Williams and some other members of the family were, it seemed as if I could not make her understand clearly what I wanted. I spoke a little louder; but she said she could hear me well enough. At length she said a little abruptly, "I guess you had better come upstairs to my room." On doing so I found an immediate change of manner; all constraint was gone in a moment. She brought the album, saying as she opened it, "I could not talk about him before them; they were rebels." Her fingers trembled with eagerness as she turned the page, and at length, holding the book open before me, she said, "Do you know that?" Lo there—the first time I had looked upon it for twenty-four years—was the face of our lost leader! "It killed my sister," she continued. "From the time she saw the newspaper account of the execution she was never the same." From Mrs. Lindsey and Mrs. Rawlins I had learned some of the particulars, and now heard the remainder of the simple, pitiful story of Elizabeth Layton's bereaved love, sickness, and death. At first when there were only scattering newspaper intimations of the capture and danger of Andrews, they had kept it from her; but when a paper was received with the full account of the execution, they allowed her to see it. While she read, her face became deadly pale. She did not faint, nor cry out, nor utter a single word; but when she had looked a good while at the article she went to her own room and remained for hours. After this they said she seemed to have little interest in anything, and that her mind was obviously shaken. Her health, which had been robust, began at once to decline. There is a little discrepancy as to the account of her death, which I had not the time to clear up; Mrs. Lindsey and Mrs. Rawlins said that she died the following February; but Miss Layton, while agreeing as to the month, was positive that she lived a year longer; but they all agreed in tracing her death directly to this awful bereavement.

In the corner of the room stood the large black trunk which Andrews had mentioned in his last letter. "A most singular bequest from a lover," Miss Layton said, "That was enough in itself to kill her." This trunk was procured for her by Parrot and Hawkins, who met her in Louisville; they had to bring a good deal of pressure to bear on the hotel-keeper before he would give it up. My own theory is that Andrews had intended to bring it back filled with wedding-presents.

Leaving Miss Layton's late in the evening, I went down to the steamboat landing—the same place where Andrews and myself had landed before the war, on our intended Southern teaching expedition, at nearly the same time, though without the slightest knowledge of each other. I had little expectation of finding a steamboat to go down the river at that hour, as all the regular boats had long since left. But the *Bonanza* happened to be at the wharf, and I was soon on my way to Cincinnati. This assured me of being able to reach the place appointed for the reunion of the survivors of our prison party at McComb, Hancock Co., Ohio, the next day.

While waiting at Deshler, Ohio (the junction with the branch road leading to McComb) I saw a fine-looking man who had a familiar and preacher-like look, and who seemed to be watching me. I thought he was probably one of the members of the East Ohio Conference, of which I had been a member before being transferred to the N. J. Conference. At length he came up to me and said, "Is not your name Pittenger?" I assented, and learned that he was J. A. Wilson, my old prison comrade—

the first time I had seen him since we had shaken hands, not without moist eyes, just before we attacked the jailor and the guard at Atlanta twenty-four years earlier. We were both boys then, and in the wonderful rush of memory and emotion which swept over me I began to realize for the first time all that this reunion—the first that our party had ever held—was likely to prove. We talked rapidly, the minutes fled like magic, and it seemed almost too soon that we were at McComb. I learned that Dorsey had arrived from Nebraska, that Knight also had come, and that more were expected. This was the home of two others—Porter and Bensinger—and we were sure of a great gathering. Though already late, we had very much to talk over before we could sleep.

The next morning the lake winds blew cold, and the falling snow was quite a contrast to the warm days I had left behind in the South. But this did not diminish our enjoyment; perhaps, as the next day was Thanksgiving and we were to feast together at the house of Comrade Porter, the cold imparted additional zest. Of the whole number the only one I could have recognized—excepting Dorsey, who had visited me in New Jersey—was Bensinger. The grave, bearded men I saw were not much like the boys who had played games and sung in sight of the scaffold! but the very first tones of the voice of each one had a familiar thrill that flashed like lightning backward over all the intervening years. The citizens of McComb felt great interest in our meeting. The Methodist Church was crowded in the evening to hear the telling of experiences by the different members of the party. This entertainment was repeated the next evening and also for one evening at Findlay, an admission fee being charged, which went far toward defraying the whole expenses of the reunion.

The Thanksgiving dinner provided by the hospitality of Porter was excellent, and the contrast between that, and our boarding in Chattanooga and Atlanta, was duly dwelt upon. In the revival of old incidents I was much amused in noticing what all persons who have had much to do with sifting testimony have been forced to recognize—the uncertainty, in minute points, of the best witnesses. None of us had any motive to recall any other than the exact facts; but we differed in a thousand little matters, though none of the discrepancies were of practical importance. With this narrative in view, I questioned the others carefully about those things which I had not jotted down at the time, or which took place beyond the bounds of my own experience. Substantial agreement, with numberless minute divergencies, was the outcome. For example (and this was the more striking from its non-essential character) we began to talk of the number of windows in the upper cell of Swims' jail in Chattanooga. There was no difference as to the lower room, for every detail of that horrible place was burned into our memories forever. But some said there were three windows in the upper room, and others that there were but two.

WILLIAM KNIGHT.   WILLIAM BENSINGER.   WILLIAM PITTENGER.   D. A. DORSEY.   J. A. WILSON.
                                                         JACOB PARROTT.   J. R. PORTER.

Seven Survivors of the Andrews Raid. Group photographed at Reunion, Findlay, O., November, 1886.

After the dispute had lasted for some time, another came in and was at once appealed to as a fresh witness, and unhesitatingly declared that there was *but one*. In the two matters following I was forced to admit myself in error, which I did the more readily as in both these I had depended on other eyes.

The two engineers, Brown and Knight, and the fireman Wilson united in saying that the "Yonah" engine was in sight when we passed. A more serious difficulty arose in apportioning the comparative labors of Brown and Knight. When the box-car contingent, with which I was, broke out the end of their car and crawled on the engine and tender, Brown was acting as engineer. This mainly had led me to write and others to speak of him as "*the* engineer," sometimes without mention of Knight at all. But it appeared that while Brown was examined as to his qualifications as engineer by Mitchel himself, and approved; yet Knight had been spoken to by Andrews and the Colonel of the Twenty-first Ohio, and was also well qualified, had gotten ahead of Brown at the start and had taken the first turn at the throttle. Knight was also a mechanician, and probably better able than any man of the party to have repaired damages to the engine. These little discussions did not impair, but rather added spice, to an occasion which on the whole was such a time of tender, sad, and triumphant memories as seldom comes in this world—perhaps not altogether unlike some of the reviews of earthly experiences that may be had in the world beyond death !

The photographs of the party were taken as a group in Findlay. Brown was not then with us, but he was seen afterward. Then, with tender and hearty farewells, we scattered on our several ways in life.

At Steubenville, Ohio, I saw my old Captain—now Major—Sarratt, who had made the detail of four men for the first expedition and so earnestly opposed my going on the second. From him I obtained the pictures of those men, and many particulars of this primary enterprise'—a gallant and daring affair which, however, would not have been heard of, if it had not been for the greater celebrity of the one that followed. He then accompanied me over into Virginia to Holliday's Cove, and helped me to hunt down that gross personation of Andrews, which, if I had not thus learned its true character, might have marred this volume. Nothing more then remained but to return home, and give six months of hard labor to weaving into the slight old volume the abundance of material accumulated during this trip, and the years that had passed since the first publication.

CHAPTER XXXVII.

A DETAILED ACCOUNT OF THE FIRST OR BUELL
RAILROAD RAID.

SOON after the events narrated in Chapter Second of this work—at the beginning of April, 1862—one of the men from Co. C, 2nd Ohio Regiment, returned to Gen. Mitchel's camp at Murfreesboro. He reported having gone as far as Chattanooga where he was recognized by a rebel acquaintance. For the sake of former friendship he hesitated to denounce him, but insisted that he should at once return northward, which he made haste to do. He reported the others still pressing southward.

About a week afterward all returned. First was my relative Mills; and on two consecutive days the remainder came also. From them I learned the particulars of their romantic journey. Of the four from Sarratt's company, one only, Frank J. Hawkins, of Columbus, O., still survives; and he has kindly furnished me an account of their adventures taken from letters written near the time, which, with the statements of Captain Sarratt, and the account given by Frank Mills, is the ground for the following narrative. It throws valuable light upon the Second expedition.

James J. Andrews had become quite well acquainted with Captain Sarratt, and depended upon his Company mainly for the volunteers for this expedition. At first Sarratt regarded Andrews with some distrust; but his plausible representations overcame this feeling, and inspired such confidence that Sarratt, if not held to his place by the duty owed his Company, would himself have gladly gone with him. Accordingly, when Andrews obtained the order from Buell for Mitchel to furnish him volunteers, he told the latter that he could easily find them in the 2nd Ohio. He went at once to Sarratt's quarters, who agreed to furnish half of the eight men needed, and sent for Hawkins, Holliday, Surles, and Durbin. These were among his best men. Andrews told them that he wanted eight men to go South with him as far as Atlanta, Ga., from which place he had just returned, and where he had made the acquaintance of a railroad engineer who ran on a wood-train, and who was also a good Union man. This engineer had agreed to run off with his train, if Andrews would furnish hands to act as brakemen, tear up track, and burn bridges. When the four soldiers had heard his statement they all volunteered, and thus half the number was made up. Captain Mitchel's Company was next in the regimental line, and was ready to second every thing that Sarratt did. Frank Mills from that Company offered his service, as did a private named Horr from Company A. The names of the other two are not remembered.

The four from Sarratt's Company made up one band who travelled by themselves, and we will mainly follow their fortunes. They were furnished with suits of citizens' clothing in place of their uniforms, and forty

FRANK B. MILLS.
FRANK HAWKINS.
B. F. DURBIN.
ALEXANDER H. SURLES.
J. W. HOLLIDAY.
Members of the first expedition; from war-time photographs.

dollars each, in gold. The same evening they went out to the reserve picket post, where they slept for the night, and started into the enemy's lines at four o'clock the next morning. They bent their steps toward Tullahoma, forty miles distant, which was the furthest point northward to which the rebels at that time ran their trains. After a walk of seven miles, they were ready for breakfast, and stopped with a strong rebel. Hawkins paid for the breakfast with a twenty-dollar gold piece, and received *thirty-eight* dollars change—in Confederate money.

Five miles further on they met three rebel citizens in a spring wagon, who inquired where they were from, and were informed that they came from Nashville, and were on their road to Atlanta to enlist. The citizens were curious to know how the four got through the Yankee pickets, but were told that they went round them. They next stated their own business, which was to find out the number of Yankees in Murfreesboro'. In other words, they were spies, and Hawkins and his friends misled them as much as they could by stating that 25,000 troops had left Nashville. This surprised them. The information of the Southern officers in regard to our numbers, in the first stages of the war, was far more accurate than our own. But they accepted this statement, and said they would go no further. They were very sorry their wagon was too small to invite their new acquaintances to a seat on the way back to Tullahoma. The four now walked on the railroad for some fifteen miles further, where they stopped for supper and lodging. The latter was given on condition that they slept on the floor, which they were tired enough to do willingly. By one o'clock of the next day they had reached Tullahoma, the place from which they expected railroad passage. Here they met the same party they had seen in the spring wagon the day before, and were informed, after friendly greeting, that the train for Chattanooga would be made up in about an hour. As soon as it was ready, they bade our friends good bye, wished them a safe journey, and recommended them to go into a Tennessee regiment. From this point the adventurers proceeded by rail to Chattanooga, but as their train was late, they had to wait in that town till morning. They spent the night at the Crutchfield House, the principal hotel, and at seven were off for Atlanta. At Big Shanty, they stopped for supper, and reached their destination at nine in the evening, finding lodging at the Trout House.

Andrews and others of the party arrived the next morning. He visited the four, and said that he had not yet found his engineer, but would go out and look for him. He did not succeed in finding him that afternoon. There were plenty of officers of the Confederate army at this hotel —Gen. Johnston, among others, eating at the same table with them. They did not go around the city much during the day, thinking it safer to keep close to their hotel; but at night they went to the Court House, and heard a fiery Secession speech from Robert Toombs; he said, among other things, that the "Yankees" were a distinct people from the Southerners—so distinct that he could tell one wherever he saw him. The four "Yankees" were looking right at him, cheering for the Confederacy as heartily as anybody! Going back to their hotel, they met some Union prisoners.

On this same evening Surles and Durbin had a very narrow escape. Before the unexplained absence of the Union engineer had deranged their plans, and while it was expected that the train would be captured the next day without fail, these two went out by Andrews's previous direction to cut the telegraph wire, that no word of their operations should be flashed ahead until they had time to take effectual precautions. They wished to remove a considerable portion, so that it could not be readily repaired. They found a solitary place, and Surles was up the pole in a moment and with the tools brought for the purpose soon severed a wire. Durbin had just taken hold of it and was coiling it up for removal, when two rebels belonging to an irregular cavalry company rode up. Our men were caught in the very act! For a moment they gave themselves up for lost, as they were not armed, while their opponents carried shotguns,—a very effective weapon at short range. But Surles was a man of the greatest coolness and fertility of resources. One of the enemy demanded very roughly, " What are you at up there?" This gave the man up the pole time to think, and he answered by a burst of passionate imprecations on the Confederacy, declaring that it was high time the Yankees came in and took charge of the whole thing, if it could not be managed better. This was so different from what the cavalryman expected that he stopped, lowered the gun he had drawn up and said, " Why what on earth is the matter with you?"

Surles said, " If I have to tend office all day, and go around mending wires all night, I don't care how soon the whole thing goes to destruction!"

The cavalryman, convinced that he had to do only with an overworked operator, encouraged him to bear up under his trouble, assuring him that affairs would soon be better! He then rode away, and they " bore up " far enough to finish " the repair," which consisted in getting down a large roll and throwing it into a neighboring corn-field. Then they returned to the hotel, feeling sure that if the train was captured in the morning, there could be no prompt notification to or from Atlanta. The next morning the whole party met with a terrible and most unexpected disappointment. After breakfast, Andrews came to their room and reported that he had at length been able to hear of his engineer, for whom he had been searching ever since they came to Atlanta. But the news was no comfort.

Although Andrews had been very prompt in his return to Atlanta after his last interview with this engineer, it was now too late. The man had been drafted off to the East Tennessee road to assist in the work of transporting troops toward Corinth, where the rebel forces were concentrating in anticipation of Pittsburg Landing battle. Andrews went from one to another to see if any of them, by a happy chance, was able to run an engine. Not one of them had the experience even of a fireman. Andrews himself knew nothing of a train except what he had learned by casual observation. It would be far too great a risk to attempt to force an unwilling engineer to do their work, for he would have it in his power, in spite of all their precautions, to ruin them. It was to Andrews a most bitter disappointment. If the men had been with him on his former trip all would have been well; now they were here, through great perils—as good and true men as a leader could have—and they were powerless ! But Andrews wasted no time in vain regrets; he told them that the fault was his own in not having taken an engineer along; but that now the only thing they could do was to start back, in small parties, and try to return to the Union lines. Part began the difficult journey the same day; but the four lingered till the next morning.

When they left Atlanta it was on a mixed freight and passenger train, which stopped for an hour at Big Shanty—a place afterward to become historical. It was then a Confederate camp of instruction. Having some time to spare, our soldiers went over to look with a veteran's interest upon the evolutions of the raw recruits. These were supplied only with pikes about four feet long, for arms were very scarce in the South at that period; and after witnessing their performances for an hour, they continued their own journey. Durbin and Surles went into the forward end of the train while the others took seats in the rear car. Andrews was with them. A brakeman came into the car saying, in an excited tone, " We've got some Yankees out here." Their natural presumption was that their comrades had been arrested, and they felt a strong desire to be elsewhere ! Hawkins and Holliday stepped out to investigate, and found that the prisoners were a number of Union soldiers on another train they had just met; and as the misfortune of these was not personal to our adventurers, they returned to the car feeling much better. As they looked around, they could see nothing of their leader. But a man was there, with his face toward the window, whom they had not seen before. When he turned and asked them what prisoners it was they were talking about outside, they started, for it was Andrews ! He had simply changed a hat for a cap which he carried, and thrown his hair in another way, so that it would have required a very close observer, indeed, to identify him.

In the evening the train arrived at Chattanooga, where the four separated for their still more perilous attempt to cross the mountains back to the Union army. Two of them took the route toward Tracy City and Manchester. Holliday and Hawkins went by the train as far as Stevenson, to which point the road had been shortened since they had last passed over it. From this place they started on foot, and getting their supper on the way, they pressed on for about sixteen miles, where, seeing some smouldering logs over in a field, they crossed the fence, and raking them together, slept by the hot embers through the night. They took breakfast with a rebel family, and then pressed on till just beyond the large tunnel at Cowan station. Here they found a rebel soldier, and on asking him for directions to Manchester, he offered to accompany them part of the way, and then to put them on the right track. They ascended the mountain, passed the military school at Sewanee, and lodged for the night in a secluded cove. He left them there, but told them where to strike the Manchester road. The man with whom they lodged that night was a strong friend to the Union cause, who urgently advised them not to return to the South any more. He gave them breakfast, dinner to carry with them, and as much tobacco as they could take. The next day they made a good journey, and that night slept on the floor by the fire of a negro cabin.

Four miles from Manchester the following day they met a rebel citizen on the road, who told them that they had better go no further, as a Yankee " critter company " had just come into the town that afternoon. They asked if there was a road by which they could go round the town so they would not be seen. He said there was none, but they could go through the fields " by looking sharp that they might not get picked up." After getting out of his sight they made rapid time with beating hearts, and found to their unutterable disappointment that the Federals had gone but a short time before ! Scarcely any body was left in the town. They did not care to stop there, though very tired. For ten miles further they journeyed, and after getting supper at a farm house, slept on the hay in an old barn. Without breakfast they started very early, hoping that day to be back in their own tents. A horseman whom they asked for a road that would lead them around Murfreesboro', where General Mitchel still lay, gave them the directions needed, and they travelled some eighteen miles more, when, being very weary, they stopped for the night with a bitter rebel. He informed them that it was only about twelve miles more to the Union lines.

The next day no hindrance was encountered. About four miles from our pickets they met a man on horseback, and told him that they were going to Nashville and wanted to get around the Yankees. He laughed at them, saying that they might be like a party that went through the pickets the preceding day. They also wanted to slip by toward Nashville, but when they reached the pickets they had a great time shaking hands ! Our comrades would not own the implication, but were very glad to hear that some of their band had reached the camp already. At six o'clock in the evening they found the Union pickets. Here they were promptly arrested, put under close guard, and conducted to General Turchin's quarters. He recognized them at once, treated them royally, and sent them with his best wishes to their regiments, where they were very warmly welcomed.

The above account shows how easy it was to penetrate in disguise either into or out of the South *before* the shock of the greater raid had filled the Confederate mind with distrust of all travelers, and led to the establishment of a most exacting and vexatious passport system.

CHAPTER XXXVIII.

AUTOBIOGRAPHY.

A FEW words of the writer's previous history may not be uninteresting:

I was born January 31, 1840, in the northern part of Jefferson Co., Ohio. My father, Thomas Pittenger, rented a small farm from my mother's family, the Mills, on the southern skirt of the small village of Knoxville, and afterward purchased a larger tract of land two miles north of the same town (still possessed by the family) to which he removed when I was twelve years of age.

I was the oldest of seven children, and while we all worked hard as soon as we were able, and economized closely with the purpose, on father's part, of completing the payment for his land, we never felt real poverty. As a child, I was strong, healthful, and fond of all rough and boisterous games, but was so near-sighted that my parents were advised not to send me to school for fear of ruining the little vision I had. When I began to read—for mother was wiser than her advisers—I could only see the letters at a distance of three inches, and the concave glasses which I long afterward procured for supplying the defect were of only two inches negative focus. I learned easily all that was taught in the village common school. Fortunately blackboard instruction was not then in vogue. I never attempted to recognize persons by their faces, but depended on their voices alone. I read all the books that could be borrowed for miles around, and in this way managed to secure considerable general information.

When twelve, father moved to his farm, and struggled for ten years to clear it of incumbrance. My annual schooling was reduced to the three winter months. In four years more it ceased altogether. I was then dreamy and visionary, reading history with delight, and hoping, like most boys, to share in great adventures. The work of education went on about as fast after school-days as before. The mysteries of shorthand were mastered daring the noon spells and rainy days of a busy harvest, and astronomy was studied with especial devotion. The constellations were learned, though I could not, when I first began, see a star of less than the third magnitude. When glasses were procured, I reveled in the new beauty of the sky, though it required a tedious education of several months before I could read with the new aids to vision. No words can describe the glory of the heavens that burst on my view with all the charm of novelty. I soon longed for the means to see still more of the wonders of the stars. Having no money, and telescopes being costly, it was necessary to secure additional means, for my father could spare nothing from the heavy annual payments on his land to gratify what he considered a whim. The readiest way of earning a little money was by teaching a district school; but the first effort in this direction met with a decided repulse. The board of examiners of Jefferson County declared that while not disqualified from a literary standpoint, I was too near-sighted to control a band of unruly children. They were more than half right. I was not then sixteen. A few months later I made another attempt, and as I had then become more accustomed to my glasses, which had been purchased just before the first examination, the certificate was granted. As a teacher I found the teaching easy; but the maintaining of quiet and order among sixty or eighty young people of all ages from twenty-one down to four or five, and of all degrees of culture, was no light task. I never liked the profession as it was practised in the large county districts of Ohio, and though I improved with each winter's teaching, yet the beginnings were unpromising, and the work always distasteful.

But the first money obtained after paying for clothing was devoted to securing a telescope. I could not afford to purchase a large one, and, not satisfied with a small one, I bought the mirror and eye-piece for a ten-feet reflector from Amasa Holcomb of Mass., and made all the rest of the instrument. Everything was rough but efficient. The telescope was constructed without a tube, in a method devised by Dr. Thomas Dick, of Scotland. It possessed great power, and showed clearly all objects commonly described in astronomical works. A post, planted in the ground, on the top of which the telescope was pivoted, and a step-ladder, constituted the observatory. Here on a level spot above the farm-house, many happy nights were spent, and the neighbors not unfrequently joined in gazing on the wonders of the sky. Probably nothing but the fact that while I could see nearly all that the books described, I required twice as much instrumental power for that purpose as my visitors, prevented me from becoming a professional astronomer. When the war broke out the instrument was sold, and the pursuit of astronomy suspended.

At eighteen, I left my native county and taught a select school of higher grade near Ravenna, Ohio. This was decidedly more pleasing than teaching miscellaneous public schools, and indicated that the work

of education might be made enjoyable.   But I was invited by Alexander
Clark, another Jefferson County teacher, to join him in editing and publishing the School-day Visitor, in Cleveland, Ohio.   This periodical
grew rapidly in circulation, but there were many severe financial struggles before it was fairly successful.   It was absorbed long after in
Scribner's *St. Nicholas.*

The next year, being nineteen, I visited a relative in Illinois.   My
first intention was to seek a school in the interior of Kentucky, and with
that end in view, I landed at Maysville.   The darkness of the night, the
rain, and the slenderness of my purse, but most of all, the intense pro-
slavery sentiment of the country, made the stage journey from that
point look very forbidding; and I decided to continue down the river.
I have since learned that J. J. Andrews, the hero of this story, landed
at the same place, near the same time, and with the same purpose—that
of teaching.   But he went one stage further, that is, to Flemingsburg,
the next county seat, while I returned to the river highway.   The circle
of our lives approached, but did not actually meet until three years
after, amid daring enterprises and tragic sufferings.

The journey from Maysville was made cheaply on the deck of a
steamboat.   While on the way we landed at Louisville, and, strolling up
the street for an hour's recreation, I saw at a distance a large child
beating another child much smaller over the shoulders with a heavy
club.   The little one was crying pitifully, and trying to shield its head
with its hands, but the blows, which I could hear as well as see, continued to fall with angry persistence.   I was surprised and shocked that
no one of the passers-by seemed to pay the smallest attention.   When I
came still closer the mystery was solved.   The large child was white,
the small one black!   I was in a slave state!   I afterward witnessed
many terrible features of the institution which the great civil war happily
buried forever, but few that impressed me more deeply.

I taught a successful school in Southern Illinois, and during the
winter indulged freely in the common pastime of debating.   Political
subjects were often introduced and discussed with no small degree of
heat, for the raid of John Brown at Harper's Ferry had wrought popular
passion almost to boiling heat.   The question debated was always some
phase of the slavery controversy, and I soon became known as an
"abolitionist"—far from a complimentary title in those days.   Many
Southerners were in that part of the State, and I had an opportunity to
learn their real sentiments, becoming convinced that there was great
danger of war.

In the spring I returned to Ohio, and made some essays toward the
business of photography.   But I lacked capital, and resolved to teach

till the lack was supplied. My parents were very kind in giving permission to push my fortunes in any way that I preferred, though I was not yet of age. They always gave me a hearty welcome, and plenty of work when I returned to the farm. In September, 1860—I was then twenty years old—I took a long and delightful tour of three weeks on foot through Northern Ohio and Western Pennsylvania, engaging a school for the winter in Beaver County, of the latter State. The teacher's desk was more agreeable than formerly, but the whole country was rocking in the waves of the agitation preceding the war. I did not vote for Lincoln for President only because I lacked a few months of twenty-one; but I exerted all my influence in that direction; and when predictions of war were made if he should be elected, I did not deny the probability, but maintained that the war would be short, leading directly to the abolition of slavery, and professed a willingness to enlist for the purpose of putting down that institution with the rebellion. During the winter, while one after another of the Southern States were seceding, the excitement grew more intense, yet many persons at the North refused to believe in the possibility of war; and when I wrote home giving my reasons for not sharing in the hopes of a peaceful settlement, and stating that in the event of war, I had made up my mind to become a soldier, father gave his consent very readily; his reasons, as he afterward explained, being that he did not believe there would be any war, and that if there was he felt sure that I would not, because of bad sight, be accepted as a soldier.

Yet to me, as to nearly all others, the bombardment of Fort Sumter and the President's call for troops that immediately followed, came as harshly as the jar of an earthquake. At the close of the school term, I returned to Jefferson county and purchased a photographic establishment from Charles Williams, but had not yet taken possession of it. For nearly a year I had been reading law in my leisure moments under the direction of Miller and Sherrard, of Steubenville. My first step, when the explosion occurred, was to go to these gentlemen and to Mr. Williams, and ask release from the engagements I had made. As the whole country was boiling with excitement, they readily cancelled our contracts, and I was free to enter upon another contract of a still more serious nature.

That very evening, a war mass meeting was held in the court-house at Steubenville and I attended, not feeling by any means sure that I would enlist with a company from the city, preferring to wait until one should be organized from my own end of the county, which might contain some acquaintances. But only seventy-five thousand troops were called for, and this made the proportion from our county very small. Possibly no more than one company might be needed, and at any rate it seemed desirable to be among the first to rally in support of the Government, especially as the need for some troops to protect Washington was urgent. I listened to the speaking, saw man after man go foward and put his name down, and heard the cheers that greeted each recruit. I was more and more inclined to go. The term was only three months, and I could easily endure that period among strangers; and then, if the war continued, I could re-enlist, choosing my company. But I did not want to decide finally under the excitement of a public meeting; so I took a long walk alone, going beyond the outskirts of the city, and reasoning with myself. I thought of all the possibilities of war, the hard marching, the facing the foe, being shot at, perhaps wounded or killed. But worst of all seemed a bayonet charge; and I remember pausing as the thought came like a cold chill to me, "Suppose the enemy was just ahead, and the command given, 'Charge bayonets,'—could I obey?" Having a strong imagination, and being aided by the darkness and silence, the whole scene rose before me—the waiting enemy, and the awful moment when steel touches steel! Never in actual conflict, or in worse dangers inside the enemy's line, have I felt more of dread and horror than at that moment,—for there was none of the supporting excitement that nearly always accompanies real danger. But I was able to assure myself that if duty called I would obey the order, though death followed. Then I went back to the court-room, where the crowd and excitement were greater than ever, and went forward asking the privilege of saying a few words. The address expressed gratification that the long suspense of the winter, when traitors were destroying the Government, and no hand was raised for its defense, was now over; and that at last the controversy between those who loved the old flag and those who would rend it was transferred to the battle-field. The conclusion, in which the speaker added his own name to the list of recruits, was especially applauded.

The company formed that night was one of twenty that were forwarded to Washington, stopping at Harrisburg on the way, where we were organized into the 1st and 2nd Ohio Regiments. We took our first and not very difficult lesson in the hardships of a soldier's life by sleeping on the marble floors of the Pennsylvania State House. Anson G. McCook was elected Captain, for the first volunteers elected nearly all their officers, and we proceeded on our way. At every railroad station along the route we were greeted by enthusiastic crowds, cheering, bringing coffee and other refreshments, the ladies waving their handkerchiefs and showering flowers upon us. We could hardly fail to feel that our cause was a noble one, and that we had the hearty support of the whole people. We had made no inquiries as to the wages we were to receive, and when, some weeks after, the question was one day raised in our mess, no one was able to answer it. We would have enlisted for the war as readily as for three months—at least the great majority of us. On the day of starting, a young man who came after the full number of 101 was enrolled, actually gave his gold watch to a fortunate volunteer for his place! My own fear was that I might be rejected on the surgical examination; but there was none worthy of the name. When we were mustered into service, the surgeon simply came into the room and looked around for a few moments, examined hastily any case that was reported to him by the officers, and all others were passed. I put my spectacles, which I always wore, in my pocket, and as there was no officer who did not wish me to go, I was not challenged. Before the close of the war, when men were seeking exemption by every possible means, examinations became very different!

In Baltimore we passed through rough and scowling crowds; but as our guns had been received and were loaded, we were not molested.

When we arrived in Washington, daily drill, tent life, guard and picket duty, began, and as we were soon organized into Brigades and Divisions, we began to feel something like soldiers. Reviews were frequent; and on one occasion President Lincoln passed along the line. The memory of the tall, awkward, but noble man, who could not keep step with the apparently little men who bobbed along on each side of him, is a vivid and pleasing memory. At length we entered Virginia, crossing Long Bridge by night. The hour was very impressive when the long column thus crept silently into a hostile State. It is not needful to linger upon the details of the march that followed, further than to say that the wide condemnation of the "On to Richmond" policy made after the event, was probably not so wise as is generally assumed. Aggressive action cost the country far less during the war than unreasonable delay. The exact time selected for the advance was unfortunate, for the term of enlistment of a great part of the three months' men had nearly expired; and while they would at first have enlisted for the war without the least question, and the great majority did re-enlist for three years afterward, yet the American love of fairness rebelled against being kept some days longer than the contract called for. This, and not the panic of which so much has been made, was the real reason for the rapid disintegration of the Union army after the battle of Bull Run. Though our army was not perfect in discipline, and had not yet fully imbibed the military spirit, it was now equal to the enemy in these things, and, as Grant declared, could acquire them in the field more rapidly than in the camp.

On the morning of July 21st we were roused at 2 o'clock, and made a long and tedious march upon the right wing of the enemy, all the time fearing that he might retreat without fighting, as at Fairfax, a few days before. At length, while we were halted on the side of a wooded hill, the report of a single cannon was heard. The signal gun for battle has a solemn sound, and we soon learned its meaning. I have no wish to repeat the story of the day, to tell how the incipient Union victory was turned into defeat by the arrival of a Confederate brigade from Johnson's army which had eluded Gen. Patterson, or to conjecture what might have been the consequences if the whole of the forces on Centerville Hill had been promptly brought into action. It was a defeat which came very near victory, and which had few evil consequences in the North save the seeming support it gave to those who advocated the policy of delay. The panic which closed the day was far more intense among the teamsters, sutlers and newspaper correspondents than the troops; and the dispersion of a large part of the army which had not been seriously engaged arose from the fear of officers and men that if they maintained a compact organization they would be held for the defense of Washington beyond their terms of enlistment, and thus miss the chances of promotion which they counted on in new organizations. This was not the noblest principle, but it did actuate many of the three months' men. The 2nd Ohio was no worse than others. It had seen no severe fighting, and was in fine order when the battle was entirely over. Save for the feeling that it had already served beyond its time, it would have gone back to Washington intact. But this consideration removed all restraint, and during the night the men straggled at their will, and when morning came the regiment had melted away. This was not creditable but it was natural, and is a type of what took place throughout a large portion of the army.

It is more pleasant to mention other things. When the day turned against us the writer decided to re-enlist. It would never do to quit defeated; and besides, it was now sure that the country would need every man. Many others expressed the same feeling, and nearly every man of the three months' volunteers, who were so determined to be let off on the exact day, were soon back under the flag again for three years.

When the 2nd Ohio re-enlisted, I became a member of Company G, Captain Sarratt. We were first ordered to Camp Dennison, and from that into Kentucky by the way of Cincinnati, and from that time our service was with the Western army. After camping for a time near Covington, we proceeded by rail to Paris, and thence on foot into Eastern Kentucky—a wild mountainous region where the bad roads and frequent bridgeless streams were the principal obstacles encountered. By a night march of thirty-seven miles we surprised and easily captured the village of West Liberty. At Prestonburg, we met the 33d and 59th first met J. J. Andrews. He was standing near the public square with a beautiful little repeating rifle—Winchester, I think—on his arm, which the soldiers were examining with much interest. He had a far-away look, and took little interest in the questions he answered about the gun. Some of us, noticing his striking appearance and the manner in which he watched the marching regiments, were disposed to think that he was probably a spy of the enemy. We little dreamed how closely his name would be associated with that of our regiment!

Under the command of General Nelson, the three Ohio and one Kentucky regiment marched up the Big Sandy until we reached Ivy Mountain, where we suddenly found ourselves in an ambuscade which had considerable resemblance to that into which Gen. Braddock fell before the revolutionary war. But we were not British regulars! The 2nd was in advance when there was heard the sharp ring of rifles from the steep and almost inaccessible mountain side. To retreat would have been to lose half our men; to stand still in ranks in the road would be worse. Somebody, whether officer or not, gave the word, "Climb the hills after them." It was a terrible scramble, but far more dangerous in appearance than in reality. To climb a steep hill in the face of an enemy is less perilous than to charge up a moderate and uniform slope, for the head is brought close to the ground in climbing, and the enemy nearly always overshoots. So we lost scarcely a man after we left the road, and had soon dispersed the enemy. It was only a little skirmish, but it was more satisfactory than Bull Run!

We advanced but a short distance beyond Piketon when we were recalled for the grand movement through Central Kentucky. As there was no foe near us we took the easiest means in our power of getting down the river. Many without orders constructed small rafts and floated nearly to the Ohio. This was the very romance of soldiering, though the cold nights and the frequent rains made the small uncovered raft the reverse of comfortable. Steamboats were in waiting for us at Louisa, and we were conveyed down the Ohio to Louisville. Here we became a part of the vast army that was assembling under General Buell, and endured the hardships of fall and winter camp life. Under the severe but intelligent discipline of Gen. O. M. Mitchel we rapidly became soldiers. There was no reason for such long delay save what was to be found in the character of our commander-in-chief. Mitchel would have willingly marched forward against the slender force of the enemy with his division alone, but was held resolutely back by the superior authority of Buell. At length, Grant at Donelson broke through the spell of inertia, and we moved almost without opposition through Kentucky, soon reaching and capturing Nashville. From this point the story has already been told.

CHAPTER XXXIX.

EDITORIAL ACCOUNT FROM THE *SOUTHERN CONFED-ERACY.*

Atlanta, Ga., April 15th, 1862.

THE GREAT RAILROAD CHASE—THE MOST EXTRAORDINARY AND AS-TOUNDING ADVENTURE OF THE WAR—THE MOST DARING UNDERTAK-ING THAT YANKEES EVER PLANNED OR ATTEMPTED TO EXECUTE—STEALING AN ENGINE—TEARING UP THE TRACK—PURSUED ON FOOT, ON HAND-CARS AND ENGINES—OVERTAKEN—A SCATTERING—THE CAPTURE—THE WONDERFUL ENERGY OF MESSRS. FULLER, MURPHY AND CAIN—SOME REFLECTIONS, ETC., ETC.

SINCE our last issue we have obtained full particulars of the most thrilling railroad adventure that ever occurred on the American Continent, as well as the mightiest and most important in its re-sults, if successful, that has been conceived by the Lincoln Government since the commencement of this war. Nothing on so grand a scale has been attempted, and nothing within the range of possibility could be conceived that would fall with such a tremendous crushing force upon us, as the accomplishment of the plans which were concocted and de-pendent on the execution of the one whose history we now proceed to narrate.

Its reality—what was actually done—excels all the extravagant *con-ceptions* of the Arrow-Smith hoax, which fiction created such a profound sensation in Europe.

To make the matter more complete and intelligible, we will take our readers over the same history of the case which we related in our last, the main features of which are correct, but are lacking in details, which have since come to hand.

We will begin at the breakfast table of the Big Shanty Hotel, at Camp McDonald, on the W. & A. R. R., where several regiments of soldiers are now encamped. The morning mail and passenger train had left here at 4 A. M. on last Saturday morning as usual, and had stopped there for breakfast. The conductor, William A. Fuller, the engineer, I. Cain—both of this city—and the passengers were at the table, when some eight men, having uncoupled the engine and three empty box-cars next to it from the passenger and baggage cars, mounted the engine, pulled open the valve, put on all steam, and left conductor, engineer, passengers, spectators, and the soldiers in the camp hard by, all lost in amazement and dumfounded at the strange, startling and daring act.

This unheard-of act was doubtless undertaken at that place and time upon the presumption that pursuit could not be made by an engine short of Kingston, some thirty miles above, or from this place; and that by cutting down the telegraph wires as they proceeded, the adventurers could calculate on at least three or four hours the start of any pursuit it was reasonable to expect. This was a legitimate conclusion, and but for the will, energy and quick good judgment of Mr. Fuller and Mr. Cain, and Mr. Anthony Murphy, the intelligent and practical foreman of the wood department of the State road shop, who accidentally went on the train from this place that morning, their calculations would have worked out as originally contemplated, and the results would have been obtained long ere this reaches the eyes of our readers—the most terrible to us of any that we can conceive as possible, and unequaled by anything attempted or conceived since this war commenced.

Now for the chase!

These three determined men, without a moment's delay, put out after the flying train, *on foot*, amidst shouts of laughter by the crowd, who, though lost in amazement at the unexpected and daring act, could not repress their risibility at seeing three men start after a train on foot, which they had just witnessed depart at lightning speed. They put on all their speed, and ran along the track for three miles, when they came across some track-raisers, who had a small truck-car, which is shoved along by men so employed on railroads, on which to carry their tools. This truck and men were at once "impressed." They took it by turns of two at a time to run behind this truck and push it along all up grades and level portions of the road, and let it drive at will on all the down grades. A little way further up the fugitive adventurers had stopped, cut the telegraph wires and torn up the track. Here the pursuers were thrown off pell-mell, truck and men, upon the side of the road. Fortunately, "nobody was hurt on our side." The truck was soon placed on the road again; enough hands were left to repair the track and with all the power of determined will and muscle, they pushed on to Etowah Station, some twenty miles above.

Here, most fortunately, Major Cooper's old coal engine, the "Yonah" —one of the first engines on the State road—was standing out, fired up. This venerable locomotive was immediately turned upon her old track and like an old racer at the tap of the drum, pricked up her ears and made fine time to Kingston.

The fugitives, not expecting such early pursuit, quietly took in wood and water at Cass Station, and borrowed a schedule from the tank-tender upon the plausible plea that they were running a pressed train, loaded with powder for Beauregard. The attentive and patriotic tank-tender, Mr. William Russell, said he gave them his schedule, and would have sent the shirt off his back to Beauregard, if it had been asked for. Here the adventurous fugitives inquired which end of the switch they should go in on at Kingston. When they arrived at Kingston, they stopped, went to the agent there, told the powder story, readily got the switch-key, went on the upper turn-out, and waited for the down *way freight train to pass.* To all inquiries they replied with the same powder story. When the freight train had passed, they immediately proceeded on to the next station—Adairsville—where they were to meet the *regular down freight train.* At some point on the way they had taken on some fifty cross-ties, and before reaching Adairsville, they stopped on a curve, tore up the rails, and put several cross-ties on the track—no doubt intending to wreck this down freight train, which would be along in a few minutes. They had out upon the engine a red handkerchief, as a kind of flag or signal, which, in railroading, means another train is behind, thereby indicating to all that the regular passenger train would be along presently. They stopped a moment at Adairsville, and said Fuller, with the regular passenger train, was behind, and would wait at Kingston for the freight train, and told the conductor thereon to push ahead and meet him at that point. They passed on to Calhoun, where they met the down passenger train, due here at 4:20 P. M., and without making any stop, they proceeded on, on, and on.

But we must return to Fuller and his party whom we have unconsciously left on the old "Yonah" making their way to Kingston.

Arriving there and learning the adventurers were but twenty minutes ahead, they left the "Yonah" to blow off, while they mounted the engine of the Rome Branch Road, which was ready fired up and waiting for the arrival of the pasenger train nearly due, when it would have proceeded to Rome. A large party of gentlemen volunteered for the chase, some at Acworth, Allatoona, Kingston and other points, taking such arms as they could lay their hands on at the moment; and with this fresh engine they set out with all speed but with great "care and caution," as they had scarcely time to make Adairsville before the down freight train would leave that point. Sure enough, they discovered this side of Adairsville three rails torn up and other impediments in the way. They "took up" in time to prevent an accident, but could proceed with the train no further. This was most vexatious, and it may have been in some degree disheartening, but it did not cause the slightest relaxation of efforts, and as the result proved was but little in the way of the *dead game*, pluck and resolution of Fuller and Murphy, who left the engine and again *put out on foot alone!* After running two miles they met the down freight train, one mile out from Adairsville. They immediately reversed the train and ran backwards to Adairsville, put the cars on the siding and pressed forward, making fine time to Calhoun, where they met the regular down passenger train. Here they halted a moment, took on board a telegraph operator, and a number of men who again volunteered, taking their guns along, and continued the chase. Mr. Fuller also took on here a company of track hands to repair the track as they went along. A short distance above Calhoun they *flushed their game* on a curve, where they doubtless supposed themselves out of danger, and were quietly oiling the engine, taking up the track, etc. Discovering that they were pursued, they mounted and sped away, throwing out upon the track as they went along the heavy cross-ties they had prepared themselves with. This was done by breaking out the end of the hindmost box-car, and pitching them out. Thus, "nip and tuck," they passed with fearful speed Resaca, Tilton, and on through Dalton.

The rails which they had taken up last they took off with them, besides throwing out cross-ties upon the track occasionally, hoping thereby the more surely to impede the pursuit; but all this was like tow to the touch of fire, to the now thoroughly aroused, excited and eager pursuers. These men, though so much excited and influenced by so much determination, still retained their well-known caution, were looking out for this danger and discovered it, and though it was seemingly an insuperable obstacle to their making any headway in pursuit, was quickly overcome by the genius of Fuller and Murphy. Coming to where the rails were torn up, they stopped, tore up rails behind them, and laid them down before, till they passed over that obstacle.[1] When the cross-ties were reached, they hauled to and threw them off, and thus proceeded, and under these difficulties gained on the fugitives. At Dalton they halted a moment. Fuller put off the telegraph operator, with instructions to telegraph to Chattanooga to have them stopped, in case he should fail to overhaul them.

Fuller pressed on in hot chase—sometimes in sight—as much to prevent their cutting the wires before the message could be sent as to catch them. The daring adventurers stopped just opposite and very near to where Colonel Glenn's regiment is encamped, and cut the wires, but the operator at Dalton *had put the message through about two minutes before.* They also again tore up the track, cut down a telegraph-pole,

---

[1] This, like many other statements in the above article, is an errror.—W. P.

and placed the two ends under the cross-ties, and the middle over the rail on the track. The pursuers stopped again and got over this impediment in the same manner as they did before—taking up rails behind and laying them down before. Once over this, they shot on, and passed through the great tunnel, at Tunnel Hill, being then only five minutes behind. The fugitives thus finding themselves closely pursued, uncoupled two of the box-cars from the engine, to impede the progress of the pursuers. Fuller hastily coupled them to the front of his engine, and pushed them ahead of him to the first turn-out or siding, where they were left—thus preventing the collision the adventurers intended.

Thus the engine-thieves passed Ringgold, where they began to fag. They were out of wood, water and oil. Their rapid running and inattention to the engine had melted all the brass from the journals. They had no time to repair or refit, for an iron horse of more bottom was close behind. Fuller and Murphy and their men soon came within four hundred yards of them, when the fugitives jumped from the engine and left it—three on the north side and five on the south side—all fleeing precipitately and scattering through the thicket. Fuller and his party also took to the woods after them.

Some gentlemen, also well armed, took the engine and some cars of the down passenger train at Calhoun, and followed up Fuller and Murphy and their party in the chase, but a short distance behind, and reached the place of the stampede but a very few moments after the first pursuers did. A large number of men were soon mounted, armed, and scouring the country in search of them. Fortunately there was a militia muster at Ringgold. A great many countrymen were in town. Hearing of the chase, they put out on foot and on horseback, in every direction, in search of the daring, but now thoroughly frightened and fugitive men.

We learn that Fuller, soon after leaving his engine, in passing a cabin in the country, found a mule having on a bridle but no saddle, and tied to a fence. "*Here's your mule,*" he shouted, as he leaped upon his back and put out as fast as a good switch, well applied, could impart vigor to the muscles and accelerate the speed of the patient donkey. The cry of "Here's your mule!" and "Where's my mule?" have become national, and are generally heard when, on the one hand, no mule is about, and on the other, when no one is hunting a mule. It seems not to be understood by any one, though it is a peculiar Confederate phrase, and is as popular as "Dixie" from the Potomac to the Rio Grande. It remained for Fuller, in the midst of this exciting chase, to solve the mysterious meaning of this national by-word or phrase, and give it a practical application.

All of the eight men were captured, and are now safely lodged in jail.

The particulars of their capture we have not received. This we hope to obtain in time for a postscript to this, or for our second edition. ˙˙˙ confessed that they belonged to Lincoln's army, and had been sent dow⁻ from Shelbyville to burn the bridges between here and Chattano᎐᎐ and that the whole party consisted of nineteen men, eleven of whom ᵥere dropped at several points on the road as they came down, to assis᷃ in the burning of the bridges as they went back.

When the morning freight train which left this city reacheᵈ Shanty, Lieutenant Colonels R. F. Maddox and C. P. Phillips took the engine and a few cars, with fifty picked men, well armed, and followed on as rapidly as possible. They passed over all difficulties, and got as far as Calhoun, where they learned the fugitives had taken to the wood᎐, and were pursued by plenty of men, with the means to catch them if it were possible.

One gentleman, who went up on the train from Calhoun, who h᷄s furnished us with many of these particulars, and who, by the way, is one of the most experienced railroad men in Georgia, says too much prais᷄ cannot be bestowed on Fuller and Murphy, who showed a cool judgment and forethought in this extraordinary affair, unsurpassed by anything he ever knew in a railroad emergency. This gentleman, we learn from another, offered, on his own account, $100 reward on each man, for the apprehension of the villains.

We do not know what Governor Brown will do in this case, or what is his custom in such matters, but if such a thing is admissable, we insist on Fuller and Murphy being promoted to the highest honors on the road —if not by actually giving them the highest positions, at least let them be promoted by *brevet*. Certainly their indomitable energy and quick, correct judgment and decision in the many difficult contingencies connected with this unheard-of emergency, has saved all the railroad bridges above Ringgold from being burned; the most daring scheme that this revolution has developed has been thwarted, and the tremendous results which, if successful, can scarcely be imagined, much less described, have been averted. Had they succeeded in burning the bridges, the enemy at Huntsville would have occupied Chattanooga before Sunday night. Yesterday they would have been in Knoxville, and thus had possession of all East Tennessee. Our forces at Knoxville, Greenville and Cumberland Gap would, ere this, have been in the hands of the enemy. Lynchburg, Va., would have been moved upon at once. This would have given them possession of the valley of Virginia, and Stonewall Jackson could have been attacked in the rear. They wᵢ have possession of the railroad leading to Charlottesville and Oran᷄ ˙ Court House, as well as the South Side Railroad, leading to Petersburg᷄ and Richmond. They might have been able to unite with McClellan's forces and attack Jo. Johnston's army, front and flank. It is not by any means improbable that our army in Virginia would have been defeated, captured or driven out of the state this week.

Then re-enforcements from all the eastern and southeastern portion of the country would have been cut off from Beauregard. The enemy have Huntsville now, and with all these designs accomplished his army would have been effectually flanked. The mind and heart shrink appalled at the awful consequences that would have followed the success of this one act. When Fuller, Murphy and Cain started from Big Shanty *on foot to capture that fugitive engine*, they were involuntarily laughed at by the crowd, serious as the matter was—and to most observers it was indeed most ludicrous; but *that foot-race saved us*, and prevented the consummation of all these tremendous consequences.

One fact we must not omit to mention is the valuable assistance rendered by Peter Bracken, the engineer on the down freight train which Fuller and Murphy turned back. He ran his engine fifty and a half miles—two of them backing the whole freight train up to Adairsville—made twelve stops, coupled to the two cars which the fugitives had dropped, and switched them off on sidings—all this, *in one hour and five minutes*.

We doubt if the victory of Manassas or Corinth were worth as much to us as the frustration of this grand *coup d'état*. It is not by any means certain that the annihilation of Beauregard's whole army at Corinth would be so fatal a blow to us as would have been the burning of the bridges at that time and by these men.

When we learned by a private telegraph despatch a few days ago, that the Yankees had taken Huntsville, we attached no great importance to it. We regarded it merely as a dashing foray of a small party to destroy property, tear up the road, etc., *à la* Morgan. When an additional telegram announced the Federal force there to be from seventeen thousand to twenty thousand, we were inclined to doubt—though coming from a perfectly honorable and upright gentleman, who would not be apt to seize upon a wild report to send here to his friends. The coming to that point with a large force, where they would be flanked on either side by our army, we regarded as a most stupid and unmilitary act. We now understand it all. They were to move upon Chattanooga and Knoxville as soon as the bridges were burnt, and press on into Virginia as far as possible, and take all our forces in that state in the rear. It was all the deepest laid scheme and on the grandest scale that ever emanated from the brains of any number of Yankees combined. It was one that was also entirely practicable on almost any day for the last year. There were but two miscalculations in the whole programme; they did not expect men to start on foot to pursue them, and they did not expect these pursuers on foot to find Major Cooper's old "Yonah" standing there already fired up. Their calculations on every other point were dead certainties, and would have succeeded perfectly.

This would have eclipsed anything Captain Morgan ever attempted. To think of a parcel of Federal soldiers, officers and privates, coming down into the heart of the Confederate States—for they were here in Atlanta and at Marietta—(some of them got on the train at Marietta that morning, and others were at Big Shanty) ; of playing such a serious game on the State Road, which is under the control of our prompt, energetic and sagacious Governor, known as such all over America ; to seize the passenger train on his road, right at Camp McDonald, where he has a number of Georgia regiments encamped, and run off with it; to burn the bridges on the same road, and go safely through to the Federal lines, all this would have been a feather in the cap of the man or men who executed it.

Let this be a warning to the railroad men and everybody else in the Confederate States. Let an engine never be left alone a moment. Let additional guards be placed at our bridges. This is a matter we specially urged in *The Confederacy* long ago. We hope it will now be heeded. Further : let a sufficient guard be placed to watch the government stores in this city ; and let increased vigilance and watchfulness be put forth by the watchmen. We know one solitary man who is guarding a house in this city, which contains a lot of bacon. Two or three men could throttle and gag him, and set fire to the house at any time; and worse, he conceives that there is no necessity for a guard, as he is sometimes seen off duty, for a few moments—fully long enough for an incendiary to burn the house he watches. Let Mr. Shakelford, whom we know to be watchful and attentive to his duties, take the responsibility at once of placing a well-armed guard of sufficient force around every house containing government stores. Let this be done without waiting for instructions from Richmond.

One other thought. The press is required by the Government to keep silent about the movements of the army, and a great many things of the greatest interest to our people. It has, in the main, patriotically complied. We have complied in most cases, but our judgment was against it all the while. The plea is that the enemy will get the news, if it is published in our papers. Now, we again ask, what's the use? The enemy get what information they want. They are with us and pass among us almost daily. They find out from us what they want to know, by passing through our country unimpeded. It is nonsense, it is folly, to deprive our own people of knowledge they are entitled to and ought to know, for fear the enemy will find it out. We ought to have a regular system of passports over all our roads, and refuse to let any man pass who could not give a good account of himself, come well vouched for, and make it fully appear that he is not an enemy, and that he is on legitimate business. This would keep information from the enemy far more effectually than any reticence of the press, which ought to lay before our people the full facts in everything of a public nature.

CHAPTER XL.

MILLEDGEVILLE, GA., Oct., 1862.

"THE people of the State have been informed through the medium of the public press of the facts connected with the daring attempt made by a band of spies, sent by the authority of the enemy, to burn the bridges on the W. & A. Railroad. The conduct of Mr. Fuller, the conductor, and of some others in the hazardous pursuit, while the spies were in possession of the train, deserves the highest commendation, and entitles them to the consideration of the General Assembly. I therefore recommend the appointment of a committee of the two houses to enquire into the facts and report upon them, and that such medals or other public acknowledgment be awarded to the parties whose conduct was most meritorious, as will do justice to their services, and stimulate others to like deeds of daring when necessary for the public security.'

"Soon after this bold attempt to burn all the bridges of the Road (two of which had been burned and replaced but a short time previous), I felt it my duty to organize a military Company to guard this valuable property. Some time after the Company had been raised, the committee of the House of Representatives visited the Road and joined in a unanimous recommendation of the committee that I add another company to the guard. In deference to the recommendation of the committee, which my own judgment approved, I directed the organization of a second company. These companies now consist of about 150 men each. There are sixteen valuable bridges, besides smaller ones, upon the Road, which is a great thoroughfare, and will be during the war a great military necessity. The destruction of two or three of these bridges over the larger streams might not only cause great derangement of the business of the Road and great inconvenience to the travelling public, but might so delay military movements as to cause the loss of an important victory. The only question with me is whether the two companies should not be increased to two regiments, and thoroughly armed, equipped and trained, and kept constantly in the service of the state until the close of the war."

---

[1] Gold medals were voted by the Legislature in accordance with this recommendation.

CHAPTER XLI.

REPORT OF JUDGE-ADVOCATE-GENERAL HOLT TO THE
SECRETARY OF WAR.

"JUDGE-ADVOCATE-GENERAL'S OFFICE, March 27, 1863.

"SIR:—I have the honor to transmit for your consideration the accompanying depositions of Corp. William Pittenger, Company G, Second Regiment, Ohio Volunteers; Private Jacob Parrot, Company K, Thirty-third Regiment, Ohio Volunteers; Private Robert Buffum, Company H, Twenty-first Regiment, Ohio Volunteers; Corporal William Reddick, Company B, Thirty-third Regiment, Ohio Volunteers; and Private William Bensinger, Company G, Twenty-first Regiment, Ohio Volunteers, taken at this office on the 25th inst., in accordance with your written instructions, from which the following facts will appear:

"These non-commissioned officers and privates belonged to an expedition set on foot in April, 1862, at the suggestion of Mr. J. J. Andrews, a citizen of Kentucky, who led it, under the authority of Gen. O. M. Mitchel, the object of which was to destroy the communications on the Georgia State Railroad between Atlanta and Chattanooga.

"The mode of operation proposed was to reach a point on the road where they could seize a locomotive and a train of cars, and then dash back in the direction of Chattanooga, cutting the telegraph wires and burning the bridges behind them as they advanced, until they reached their own lines. The expedition consisted of twenty-four men, who, with the exception of its leader, Mr. Andrews, and another citizen of Kentucky, who acted on the occasion as the substitute for a soldier, had been selected from the different companies for their known courage and discretion. They were informed that the movement was to be a secret one, and they doubtless comprehended something of its perils, but Mr. Andrews and Mr. Reddick alone seem to have known anything of its precise direction or object. They, however, voluntarily engaged in it, and made their way, in parties of two or three, in citizens' dress, and carrying only their side arms, to Chattanooga, the point of rendezvous agreed upon, where twenty-two out of the twenty-four arrived in safety. Here they took passage, without attracting observation, for Marietta, which they reached at 12 o'clock on the night of April 11. The following morning they took cars back again toward Chattanooga, and at a place called Big Shanty, while the engineers and passengers were breakfasting, they detached the locomotive and three box cars from the train, and started at full speed for Chattanooga. They were now upon the field of the perilous operations proposed by the expedition, but suddenly encountered unforeseen obstacles. According to the schedule of the road, of which Mr. Andrews had possessed himself, they should have met but a single train on that day, whereas they met three, two of them being engaged in extraordinary service. About an hour was lost in waiting for these trains to pass, which enabled their pursuers to press closely upon them. They removed rails, threw out obstructions on the road, and cut the wires from time to time, and attained, when in motion, a speed of 60 miles an hour, but the time lost could not be regained.

"After having run about 100 miles, they found their supply of wood, water and oil exhausted, while the rebel locomotive which had been chasing them was in sight. Under these circumstances they had no other alternative but to abandon their locomotive and flee to the woods, which they did under the orders of Mr. Andrews, each one endeavoring to save himself as best he might.

"The expedition thus failed from causes which reflected neither upon the genius by which it was planned nor upon the intrepidity and discretion of those engaged in conducting it. But for the accident of meeting the extra trains, which could not have been anticipated, the movement would have been a complete success, and the whole aspect of the war in the South and Southwest would have been at once changed.

"The expedition itself, in the daring of its conception, had the wildness of a romance, while in the gigantic and overwhelming results which it sought, and was likely to accomplish, it was absolutely sublime. The estimate of its character entertained in the South will be found fully expressed in an editorial from the *Southern Confederacy*, a prominent rebel journal, under date of April 15, and which is appended to and adopted as a part of Mr. Pittenger's deposition. The editor says:

"'The mind and heart sink back appalled at the bare contemplation of the consequences which would have followed the success of this one act. We doubt if the victory of Manassas or Corinth were worth as much to us as the frustration of this grand *coup d'état*. It is not by any means certain that the annihilation of Beauregard's whole army at Corinth would have been so fatal a blow to us as would have been the burning of the bridges at that time by these men.'

"So soon as those composing the expedition had left the cars and dispersed themselves in the woods, the population in the country around turned out in their pursuit, employing for this purpose the dogs which are trained to hunt down the fugitive slaves in the South. The whole twenty-two were captured. Among them was private Jacob Parrot, of Company K, Thirty-third Regiment, Ohio Volunteers. When arrested, he was, without any form of trial, taken possession of by a military officer and four soldiers, who stripped him, bent him over a stone, and while two pistols were held over his head, a lieutenant in rebel uniform inflicted with a rawhide upwards of 100 lashes on his bare back. This was done in the presence of an infuriated crowd, who clamored for his blood and actually brought a rope with which to hang him. The object of this prolonged scourging was to enforce this young man to confess to them the objects of the expedition and the names of his comrades, especially that of the engineer who had run the train. Their purpose was, no doubt, not only to take the life of the latter if identified, but to do so with every circumstance of humiliation and torture which they could devise. Three times in the process of this horrible flogging it was suspended, and Mr. Parrot was asked if he would confess, but steadily and firmly to the last he refused all disclosures, and it was not until his tormenters were weary of their brutal work that their task was abandoned as useless.

"This youth was an orphan, without father or mother, and without any of the advantages of education. Soon after the rebellion broke out, though but eighteen years of age, he left his trade and threw himself into the ranks of our armies as a volunteer, and now, though still suffering from the outrages committed on his person in the South, he is on his way to rejoin his regiment, seeming to love his country but the more for all that he has endured in its defense. His subdued and modest manner while narrating the part he had borne in this expedition showed him to be wholly unconscious of having done anything more than performing his simple duty as a soldier. Such Spartan fortitude and such fidelity to the trusts of friendship and the inspirations of patriotism deserve an enduring record in the archives of the Government, and will find one, I am sure, in the hearts of a loyal people.

"The twenty-two captives, when secured, were thrust into the negro jail at Chattanooga. They occupied a single room, half under-ground and but 13 feet square, so that there was not space enough for them all to lie down together, and a part of them were, in consequence, obliged to sleep sitting and leaning against the walls. The only entrance was through a trap door in the ceiling, that was raised twice a day to let down their scanty meals, which were lowered in a bucket. They had no other light or ventilation than that which came through two triple-grated windows. They were covered with swarming vermin, and the heat was so oppressive that they were often obliged to strip themselves entirely of their clothes to bear it. Add to this that they were all handcuffed, and with trace chains, secured by padlocks around their necks, were fastened to each other, in companies of twos and threes. Their food, which was doled out to them twice a day, consisted of a little flour wet with water and baked in the form of bread, and spoiled pickled beef. They had no opportunity of securing supplies from the outside, nor had they any means of doing so, their pockets having been rifled to their last cent by the Confederate authorities, prominent among whom was an officer wearing the rebel uniform of a major. No part of the money thus basely taken was ever returned.

"During this imprisonment at Chattanooga, their leader, Mr. Andrews, was tried and condemned as a spy, and subsequently executed in Atlanta, June 7.

"They were strong and in perfect health when they entered this negro jail, but at the end of something more than three weeks, when they were required to leave it, they were so exhausted from the treatment to which they had been subjected that they were scarcely able to walk, and several staggered from weakness as they passed through the streets to the cars.

"Finally, twelve of the number, including the five who have deposed, and Mr. Mason, of Company K, Twenty-first Regiment, Ohio Volunteers, who was prevented by illness from giving his evidence, were transferred to the prison at Knoxville, Tenn. On arriving there, seven of them were arraigned before a court martial, charged with being spies. Their trial, of course, was summary. They were permitted to be present, but not to hear the argument of their own counsel or that of the judge advocate. Their counsel, however, afterward visited the prison and read to them the written defense which he made before the court in their behalf. The substance of that paper is thus stated by one of the witnesses, Corporal Pittenger:

"'He (the counsel) contended that our being in citizens' clothes was nothing more than what the Confederate Government itself had authorized, and was only what all guerillas in the service of the Confederacy did on all occasions when it would be an advantage to them to do so, and he recited the instance of General Morgan having dressed his men in the uniform of our soldiers and passed them off as being from the Eighth Pennsylvania Cavalry Regiment, and by that means succeeded in reaching a railroad and destroying it. This instance was mentioned to show that our being in citizens' clothes did not take from us the protection awarded to prisoners of war. The plea went on further to state that we had told the object of our expedition; that it was a purely military one, for the destruction of communications, and as such, lawful, and according to the rules of war.'

"This just and unanswerable presentation of the case appears to have produced its appropriate impression. Several members of the court martial afterward called on the prisoners and assured them that from the evidence against them they could not be condemned as spies; that they had come for a certain known object, and not having lingered about or visited any of their camps, obtaining or seeking information, they could not be convicted. Soon after all the prisoners were removed to Atlanta, Ga., and they left Knoxville under the belief that their comrades, who had been tried, either had been or would be acquitted.

"In the meantime, however, the views entertained and expressed to them by the members of the court were overcome, it may be safely assumed, under the prompting of the remorseless despotism at Richmond.

"On June 18, after their arrival at Atlanta, where they rejoined the comrades from whom they had been separated at Chattanooga, their prison door was opened and the death sentence of the seven who had been tried at Knoxville was read to them. No time for preparation was allowed them. They were told to bid their friends farewell and be quick about it. They were at once tied and carried out to execution. Among the seven was Private Samuel Robinson, Company G, Thirty-third Ohio Volunteers, who was too ill to walk. He was, however, pinioned like the rest, and in this condition was dragged from the floor where he was lying to the scaffold. In an hour or more, the cavalry escort which had accompanied them was seen returning with the cart, but the cart was empty; the tragedy had been consummated! On that evening and the following morning the prisoners learned from the provost-marshal and guard that their comrades had died as all true soldiers of the Republic should die in the presence of its enemies.

"Among the revolting incidents which they mentioned in connection with this cowardly butchery was the fall of two of the victims from the breaking of the ropes after they had been for some time suspended. On their being restored to consciousness they begged for an hour in which to pray and to prepare for death, but this was refused them. The rope was readjusted, and the execution at once proceeded.

"Among those who thus perished was Geo. D. Wilson, Company B, Twenty-first Ohio Volunteers. He was a mechanic, from Cincinnati, who, in the exercise of his trade, had travelled much through the States, North and South, and who had a greatness of soul which sympathized intensely with our struggle for national life, and was in that dark hour filled with joyous convictions of our final triumph. Though surrounded by a scowling crowd, impatient for his sacrifice, he did not hesitate, while standing under the gallows, to make them a brief address. He told them that though they were all wrong he had no hostile feelings toward the Southern people, believing that not they, but their leaders, were responsible for the rebellion; that he was no spy, as charged, but a soldier regularly detailed for military duty; that he did not regret to die for his country, but only regretted the manner of his death; and he added, for their admonition, that they would yet see the time when the old Union would be restored and when its flag would wave over them again; and with these words the brave man died. He, like his comrades, calmly met the ignominious doom of a felon; but, happily, ignominious for him and them only so far as the martyrdom of the patriot and hero can be degraded by the hands of ruffians and traitors.

"The remaining prisoners, now reduced to fourteen, were kept closely confined under special guard in the jail at Atlanta until October, when, overhearing a conversation between the jailer and another officer, they became satisfied that it was the purpose of the authorities to hang them as they had done their companions. This led them to form a plan for their escape, which they carried into execution on the evening of the next day, by seizing the jailer when he opened the door to carry away the bucket in which their supper had been brought. This was followed also by the seizure of the seven guards on duty, and before the alarm was given, eight of the fugitives were beyond reach of pursuit. It has been since ascertained that six of these, after long and painful wanderings, succeeded in reaching our lines. Of the fate of the other two, nothing is known. The remaining six of the fourteen, consisting of the five witnesses who have deposed and Mr. Mason, were recaptured and confined in the barracks until December, when they were removed to Richmond. There they were shut up in a room in Castle Thunder, where they shivered through the winter, without fire, thinly clad, and with but two small blankets, which they had saved with their clothes, to cover the whole party. So they remained until a few days since, when they were exchanged; and thus at the end of eleven months, terminated their pitiless persecutions in the prisons of the South—persecutions begun and continued amid indignities and sufferings on their part, and atrocities on the part of their traitorous foes, which illustrate far more faithfully than any human language could express it the demoniac spirit of a revolt, every throb of whose life is a crime against the very race to which we belong.

"Very respectfully, your obedient servant,

"J. HOLT, Judge-Advocate-General.

"HON. E. M. STANTON, Secretary of War."

CHAPTER XLII.

B UT two disparaging criticisms have ever been offered, so far as the writer is aware, of the Mitchel railroad raid, and these emanate from substantially the one source, Gen. D. C. Buell and his Chief of Staff, Gen. J. B. Fry. That some degree of irritation existed between Gen. Buell and his adherents on the one hand and Gen. Mitchel on the other, was understood during war times, and is made still more evident by these criticisms. They would claim no notice here if the following letter of Gen. Buell had not been printed in the War Records, and the review of Gen. Fry quoted in a pension debate in the Senate of the United States by some Southern Senators as the final military verdict on the nature and objects of this railroad expedition. Gen. Buell writes:[1]

SARATOGA, August 5, 1863.

To GENERAL L. THOMAS,
 Adjutant General U. S. A., Washington City, D. C.

SIR,—In the "Official Gazette" of the 21st ultimo, I see a report of Judge-Advocate-General Holt, dated the 27th of March, relative to "an expedition set on foot in April, 1862, under the authority and "direction" as report says, "of General O. M. Mitchel, the object of which was to destroy the communication on the Georgia State Railroad between Atlanta and Chattanooga." The expedition was "set on foot" under my authority; the plan was arranged between Mr. Andrews, whom I had had in employment from shortly after assuming command in Kentucky, and my Chief of Staff, Colonel James B. Fry; and General Mitchel had nothing to do either with its conception or execution, except to furnish from his command the soldiers who took part in it. He was directed to furnish six; instead of that he sent twenty-two. Had he conformed to the instructions given him it would have been better; the chances of success would have been greater, and in any event several lives would have been saved. The report speaks of the plan as an emanation of genius; and of the results which it promised as "absolutely sublime." It may be proper, therefore, to say, that this statement is made for the

---

[1] War Records, Series 1. Vol. X. (1) page 634.

sake of truth, and not to call attention to the extravagant colors in which it has been presented.

Very respectfully, your obedient servant,

[Signed] D. C. BUELL, Major-General.

Even Gen. Fry finds it necessary to explain that Buell wrote the above in ignorance of the facts, inexcusable ignorance, as they had been widely published. Fry says, "General Buell knew only of the first expedition, the one he authorized." Of course, this deprives his statement of all value, as far as the second expedition is concerned.

The remarks of Gen. Fry are contained in the Journal of the Military Service Institution for 1882, in a very eulogistic editorial notice of the book "Capturing a Locomotive." Fry declares the book to be "the most thrilling story of the rebellion;" and adds "No romance contains more of danger, pluck, resolution, endurance, suffering, gloom, and hope than this truthful account of an actual occurrence in our War of Rebellion."

After such high praise it may seem invidious to answer the criticisms made by this writer, on the military character of the enterprise, and to show where he was mistaken as to the actual military situation. But the truth of history seems to require that confident and unwarranted statements should not be allowed to pass unchallenged. General Fry declares:

"The destruction of bridges between Marietta and Chattanooga would not have enabled General Mitchel to take the latter place (1). If his instructions (2) or the military conditions (3) had justified him in an attempt to capture Chattanooga—which they did not—the preservation of the bridge over the Tennessee would have been essential to his success (4). The enemy had only to burn that structure, as they did when Mitchel's troops approached it, April 29th (5) in order to check an advance on Chattanooga. Furthermore, if Mitchel's party had succeeded in burning bridges between Marietta and Chattanooga, that would not have prevented the reinforcement of the latter place, as the regular railroad route through East Tennessee was open, and in the enemy's possession (6), and it was from the east, and not from the south, where there were but few if any available troops (7) until Corinth was evacuated, that the place was most likely to be reinforced. Mitchel's bridge-burners, therefore, took desperate chances to accomplish objects of no substantial advantage (8)."

It would be difficult to condense more errors and baseless assertions into a single paragraph. I have marked them by figures, and the reader who will attentively consider the evidence, will perhaps conclude that the authority of General Fry is not sufficient to end all controversy.

(1) Is a mere matter of opinion.

(2) Is specious, for General Buell in a letter of instruction to General Mitchel of March 27th 1862,[1] enumerates a number of things which General Mitchel might do in certain contingencies, among which the capture of Chattanooga is not included. But he calls them mere suggestions. It was never claimed that Buell *ordered* the capture of Chattanooga; but he did not give such clear and rigid orders as would have prevented Mitchel from moving in that direction had the latter judged it wise. Buell says: "It is not necessary to point out to you how this force (that placed under Mitchel's orders) can be concentrated either for advance or defense." "I do not think it necessary to do more than suggest these general features to you." "You will understand well how to guard against them (possible advances of the enemy) or take advantage of them according to circumstances." Such language leaves a detached commander great discretion, and if Mitchel had been able to go to Chattanooga, he would not have been insubordinate in doing so, if he had also provided properly for his special trust of guarding Nashville.

(3) Is purely a matter of opinion. What military conditions will justify depends largely on the officer judging.

(4) Is a strange assertion. Mitchel *did carefully preserve this bridge,* and states that he did it with the hope of an advance to Chattanooga![2]

(5) Is also erroneous. Mitchel saved the main bridge at that time, and declares that the short span which the enemy burned is of small value, and that he can cross to the other side whenever he desires.[3]

(6) Gen. Fry seems to be ignorant of the fact that the East Tennessee R. R. crossed Chickamauga Creek very close to its junction with the Western and Atlantic R. R., and that its bridge over that stream would also have been involved in the proposed bridge-burning, thus isolating Chattanooga from the East as well as from the South.

(7) This is sufficiently answered by Gen. E. Kirby Smith, who reports *six complete regiments, already on the way from the South,* while the utmost he can spare from the East is *four* regiments![4]

---

[1] War Records, Series I., Vol. X., Part 2, page 71.

[2] Mitchel to Sec. Chase, April 19th, 1863, "I spared the Tennessee bridges near Stevenson in the hope that I might be permitted to march on Chattanooga." War Records, Series I., Vol. X., Part 1, page 113.

[3] Mitchel to Buell, April 29, 1862. War Records, Series I., Vol. X., Part 1, page 655.

[4] Official Report of E. Kirby Smith, April 13th. War Records, Series I., Vol. X., Part 1, page 643.

(8) Is the conclusion reached from the extraordinary series of statements in this paragraph. General Fry may have some information which proves the object of the Mitchel bridge-burners to have been " of no substantial advantage," but it does not appear in this paragraph, which seems to have been hastily written in entire misapprehension of the facts. Possibly the present writer is partly to blame for these misunderstandings, because " Capturing a Locomotive" was written without access to *War Records*, and did not, therefore, present the whole subject with the fullness of evidence that is now possible.

The only other point of adverse criticism made by Gen. Fry is much stronger. He claims that the treatment of the prisoners by the Confederates was not blamable, arguing that our being in disguise deprived us of the privileges of prisoners of war. The reader of the foregoing pages will see that he probably misunderstands the writer's position. He says:

"Only eight were executed. Instead of blaming the winner for taking one-third the stakes, the author should have thanked him for not enforcing his right to the other *two*-thirds." In a private letter written subsequently, Gen. Fry explains that he did not mean to say that we were under any special obligation of gratitude to the enemy for the lives that were spared, as it probably was not intentional mercy, but an oversight, adding that if Davis had known all the facts, he would have probably ordered our immediate trial and execution. This is very probable, indeed.

General Fry twice refers to the omission of Mitchel to report this expedition, and assumes that this was a tacit confession of the insufficiency of its object. There is no evidence that he did not write or forward a report. Mitchel complains that many of his reports are not received, and the reader of the War Records will not fail to notice how often documents referred to are declared to be "not found." The fact that no report by Gen. Mitchel has been discovered, where so many have been lost, is by no means conclusive proof that he made none. Such a report may have been made and lost in transmission or afterwards, as vast numbers have been; or it may yet be found; or with his communications so insecure, he may well have feared to write the facts lest they should fall into the enemy's hands, as some of his important personal papers did.

CHAPTER XLIII.

THE RE-BURIAL OF ANDREWS.

O N page 428 a brief notice is given of the finding of the body of Andrews which had long been supposed to be lost. It is fitting that an incident of such pathetic interest should be fully recorded. Fred. J. Cooke of the American Press Association accidentally heard of a family who professed to know where Andrews had been buried. This was in 1887 while I was engaged in preparing a new edition of the history of the raid; and knowing my anxiety on this subject he at once sought them out. The household of John H. Mashburne had continued in the same spot during the twenty-five intervening years, and had never lost sight of the scene of a tragedy which had deeply interested them. The father-in-law of Mr. Mashburne had witnessed the execution. The place of the scaffold was one square from Peachtree Street on the right of Ponce De Leon Avenue, on a hill-top. This father saw the body cut down and followed it down to the adjoining ravine, watching the hasty burial. For some reason he kept watch over the grave as long as he lived, often going out on Sunday afternoons and pointing out to Mr. Mashburne and others "the grave of the leader of the engine thieves." Yet the number of those who were thus shown the spot was comparatively small, and until the matter was taken up by Mr. Cooke it did not reach any one who had interest or authority to act in the matter.

After the visit described on page 427, I wrote the facts of the discovery to the United States Secretary of War, Wm. C. Endicott, who directed Major E. B. Kirke of the U. S. Army, stationed at Atlanta, to excavate for the remains with a view of removing them to Chattanooga, all expenses being borne by the government.

On the 11th of April, 1887, just 25 years after the raid, the melancholy work was undertaken. Frank M. Gregg in a pamphlet published in Chattanooga in 1891 thus describes it:

"Major Kirke, U. S. A., Dr. C. L. Wilson, President of the National Surgical Institute, Mr. Fred. J. Cooke, Mr. John H. Mashburne as guide, and a negro laborer, with pick and shovel, started out on a journey which proved to be the sequel of a journey commenced twenty-five years before by the man whose ashes were now sought for. Turn-

ing to the right from Peachtree Street, they went their way down Ponc
De Leon circle, two squares to Juniper Street; turning into this stree
they continue one square to the first cross street; into this unused high.
way about twenty steps they stopped near a large rock, beneath a pine
tree, at a depression in the ground, from which the blackberry bushes
grew a tangled mass. This depression, nearly filled with leaves caught
by the bushes, was the spot which Mr. Mashburne had been told f
years was the one where Andrews was buried. Under his directions tl
laborer began his work, handling his shovel with care, lest it shou
shatter the bones; each shovel full, as it was thrown out, was carefully
examined by all. At the depth of three feet their research was rewarded, and Mr. Mashburne's story verified by the discovery of a portion of a skeleton. One by one the bones were laid aside by Dr. Wilson, who identified them as being human, and the remains of a large
man. The skeleton exhumed was placed carefully in a box and removed, under the supervision of Major Kirke to Dr. Wilson's Surgical
Institute. There was no doubt about it now, the real facts corroborated the resident's story, and the identification of the physician present,
that the bones found were those of a man of the size of Andrews, was
ample proof that this was the body of the leader of this perilous expedition, which for twenty-five years had lain in an unknown grave, lost
from the care of admiring comrades, hid away from the decorations
which yearly crown the soldiers' graves. The remains of Andrews were
thus recovered from the earth which had been his winding-sheet and
funeral mantle, and on the record, which classed his remains as those of
a brave unknown, was inscribed the name, 'James J. Andrews.'

"The absence of the right forearm bone was accounted for by Mr.
Mashburne as follows. He was at the grave one Sunday afternoon with
a friend from East Tennessee, a Mr. McKamie; he was telling him that
it was that of the leader of the 'engine thieves.' This friend pushed
his cane down into the soft earth and forced up a bone which he carried
off with him as a relic. Maj. Kirke after having satisfed himself by
further research and inquiry that the skeleton exhumed was that of Andrews, commenced preparations for their removal to Chattanooga, there
to be reinterred in the National cemetery. Not a piece of the manacles
in which Andrews was hung was found with his body, although it is an
actual fact, so witnesses say who saw the execution and burial, that his
shackles were never removed. It is hardly probable that they would
rust away in this time. It is a rumor, only, that Andrews' remains wei
at one time dug up. Of this Mr. Averille, in the Atlanta Journal, on
April 14th, 1887, says as follows: 'Andrews' remains are said to have
been exhumed a day or two after they were first buried, for the purpose

of securing his clothing, and immediately reinterred. From the fact that several bones are missing, it is supposed that they must have afterwards been disturbed.' It is believed by many that such is the fact, but the writer could find nothing to substantiate such supposition.

"When Post 45, of Chattanooga, first heard of the recovery of Andrews' remains, they at once offered their services to take charge of the body; and with due ceremony, furnish it an escort from their ranks, and give them a soldier's burial. Their offer was at once accepted and Maj. Kirke with an escort sent the remains from Atlanta to Chattanooga, leaving the G. A. R. Post at that city to prepare the programme and arrange the ceremonies for the last interment.

"On Sunday afternoon the 16th of October, 1887, one of those perfect days, a gem of Indian summer, when the verdure begins to don her variegated autumnal robes, and the elements are at peace, along the graveled drives of the National Cemetery, moved a cortege bearing Andrews' remains, over whom the last sad rites were soon to be pronounced. At an open grave the last of a semi-circle of eight, the procession stops, surrounded by a concourse of three thousand people, the remains are lowered to rest by the side of his comrades with whom he gave his life. The roll was now complete. Ross, Wilson, Shadrack, Scott, Slavens, Robertson, Andrews, not one was missing, together in the 'narrow house of dreamless darkness' they await, to answer the bugle call of eternity.

"The tribute to Andrews, pronounced by Hon. A. H. Pettibone, Department Commander of Tennessee G. A. R., on this occasion was an eloquent eulogy, chaste in language, fertile in thought, sublime in subject. The following is the opening paragraph:

"'Comrades of the Grand Army: We are now assembled in this silent city of the dead to here finally inurn the bones and pay a tribute of respect and grateful remembrance to the memory of a patriot who had all the fervor and earnestness of Nathan Hale, and a courage equal to any of the three hundred Spartans who fell with Leonidas at the pass of Thermopylæ! It is needless to say that we have come to make a lasting grave in this beautiful National cemetery for one who was a leader in an exploit of romantic daring which equals any of the tales of mediæval chivalry! The story of his exploit will ever remain as one of the most thrilling and picturesque in the thousands of noted events which marked the progress of our great civil war.'"

CHAPTER XLIV.

AN OHIO MONUMENT TO THE RAIDERS PROPOSED.

" On fame's eternal camping ground
Their silent tents are spread
And glory guards with solemn round
The bivouac of the dead."

THE subject of a fitting monument to the Andrews Raiders was widely discussed and many propositions were made looking to that end. There was, however, a practical difficulty about the site. Whether the seizing of the train at Big Shanty, the places of execution in Atlanta, or some spot in a National Cemetery should be selected, was a matter affording room for difference of opinion, and voices were given in favor of each of these. When the bodies of the seven soldiers executed in Atlanta were interred by government authority on a beautiful hill-slope of the National Cemetery at Chattanooga, this was considered the most eligible site, for here, as nowhere else, the column and the graves would be cared for by the Nation and by the local inhabitants. In 1887 Major Smith of the American Press Association proposed to purchase the two sites where the scaffolds had stood for Memorial purposes, but they could not be secured. When the body of Andrews had been recovered and removed to Chattanooga as narrated in the preceding chapter, all doubt as to place was ended, and the agitation for building became more earnest. A proposition was made by G. A. R. Post 45, of Chattanooga, to the Posts of the United States for a general contribution by the order for the erection of the monument. The response was very favorable, but before it had gone very far the Ohio State Legislature took up the same work on the ground that the Raiders were all from three Ohio regiments. Post 45 then sent a strongly written petition to the legislature giving reasons for desiring the memorial to be at Chattanooga and offering to be responsible for its fitting care and decoration.

One of the earliest workers in the same field was Hon. Thomas Cowgill, who represented in the Ohio Senate the district in which the relatives of Marion Ross resided. But to Stephen B. Porter, editor of the Columbus Evening Dispatch, a former member of Co. G, Second Ohio Regiment, who had a personal knowledge of the events of the

The Captured "General" decorated at the G. A. R. National Encampment at Columbus, Ohio, August, 1888. From a Photograph.

Raid from the first, and was a warm personal friend of the writer, the monument scheme was more indebted than to any other person. By his exertions, all the survivors of the Raid were specially invited to attend the Grand Army Encampment at Columbus, Ohio, and one only was absent. The Western and Atlantic Railroad kindly sent the old captured engine, the "General," to the same gathering, and what was still better, put it in the charge of Conductor Fuller; so that the captors and the captured, the pursuers and the pursued were all together for the first time since the war! Public curiosity to see the men and the engine was extreme. A track was laid out to a square a short distance from the R.R. depot, and a guard placed around the locomotive to prevent it from being carried away piecemeal by the relic hunters; and during the whole encampment it was surrounded by a great crowd of eager spectators. The Raiders were accorded a place in the centre of the great procession and excited universal interest. A reunion or camp-fire was held by the Raiders in the open air in front of the state house, and was attended by many thousands of the old soldiers—at least as many as could get within hearing distance. Different members of the band made short addresses giving recollections of their strange experiences. The writer gave a complete summary of the story and asked that a monument be erected for the purpose of showing in this tangible form that the Raiders were recognized, not as men who had justly died for offences against the laws of war, but as martyrs for our country. Let such a monument be Ohio's answer to the Atlanta scaffold! Conductor Fuller was then introduced by me and delivered a speech of great power and beauty. He began:

"I shall not attempt to entertain you, with that trained rhetoric and eloquence possessed and displayed on this occasion by the learned gentleman who has preceded me. But my friends, I assure you that I regard it as one of the greatest pleasures of my life to have the privilege of *following the distinguished gentleman once more!*"

This sally was received with great laughter and applause. In a well-worded speech, Fuller set forth very clearly the Confederate side of the chase. He was listened to with almost breathless attention when he told the story of heroism on the Atlanta scaffold which he had witnessed. After this the strong appeal with which he closed had great weight, especially as coming from such a source. "Now before I close my remarks, I desire to say to the Grand Army of the Republic and especially to the people of Ohio that, though you have ample opportunity and abundance of wealth, you are unable to do too much for the surviving members of the expedition, nor can you do too much in memory of the dead."

The following incident is taken from the " Andrews Raiders" by F. M. Gregg:

" While Captain Fuller was at Columbus, a very pathetic scene transpired, by which a widow's sorrows of years' standing were alleviated, and her unkind feelings toward those whom she thought had wronged her were forgotten. Mrs. Samuel Slavens, wife of one of the raiders executed at Atlanta, was left with a family of three small children. The struggles of life did not subdue her feelings of hatred toward those who had robbed her of a protector and husband. The meeting took place at a reception at which the victims of one man's perseverance shook him kindly by the hand, without a thought of reviling him for the misery or sufferings he had caused them to endure. An eye-witness, Stephen B. Porter, gives the meeting of the widow, Mrs. Slavens, and Capt. Fuller as follows: 'All of the surviving raiders, except two, were present, together with the family of Mrs. Feltrows. All were standing in a semicircle; I was conducting Captain Fuller around the circle, introducing him to the men whom he had never seen as free men before, but had known them as captives in a prison cell. It was a dramatic scene of the most subdued nature, especially the meeting between Mrs. Slavens and Captain Fuller. She was about in the middle of the circle. As we approached her, I saw she was very much excited, her face was flushed, and the years of sorrow lingered on her brow. No one can ever tell the thoughts of this woman as she took the hand of the man who was responsible for her husband's death. When I mentioned her name, I seemed to observe a perceptible, but momentary feeling in the nature of a slight shock come into Captain Fuller's strong frame. He spoke so gently, however, and kindly, that the lady was deeply touched. They sat down beside each other, and conversed in undertones. What they said they alone know, though the house was silent, and we were all in the room.

" 'It was a touching scene, and one which those present will never forget. When they had finished their talk, Captain Fuller was introduced to the remainder of the party. Mrs. Slavens was a changed woman after her meeting with Captain Fuller, for she said she felt all right now toward the men who captured, tried and executed her husband. I have no doubt but that it made her life happier.' "

Mr. Porter, on the assembling of the sixty-eighth Ohio State Legislature, sent each member of that body a copy of the " Evening Dispatch," with the addresses of the Raiders and of Capt. Fuller. This was the final argument in favor of the bill, as on March 20th, 1889, it became a law, granting $5,000 for a monument to be erected in the National Cemetery at Chattanooga to the Andrews Raiders. Not a

single negative vote was recorded against it in either house, and in the
debate all conceded it was but doing tardy justice to those who so
eminently deserved it. To Captain Porter, as much as to any other
person, is the credit due of obtaining this appropriation, first endorsed
by the memorial from the Chattanooga Post.

Governor Foraker appointed a monument commission to select a
proper design and to have erected a fitting structure. As such the
members were chosen from the three regiments of Sills' brigade, the
2nd, 21st and 33rd Ohio, from whose ranks members of the party were
selected. They were Judge Thaddeus Minshall, a captain of the 33rd,
now on the supreme bench of the State of Ohio; Hon. Earl W. Merry,
sergeant major of the 21st Ohio, now a banker at Bowling Green, Ohio;
and Stephen B. Porter, a sergeant in Company G, 2nd Ohio. No wiser
selection of men could have been made for this purpose, than the three
who were appointed. Judge Minshall applied his years of legal and
practical life to his new mission. Mr. Merry, as a man of business,
gave his executive ability to the work, and Mr. Porter, urged by patriotism, which had caused him to champion the cause when others were
silent. Together this monument commission have each applied their
best ability to the erection of a memorial which is destined to stand for
ages; attesting the wisdom of this commission in the selection of a design, and their labors in carrying it out.

"It was thought by the commission that the most fitting monument
would be old 'General' itself and placed as the monument above the
men who made it famous, but the Western & Atlantic road valued it too
highly as a memento to sell it. After the inspection of many designs
from different artists all agreed on the one selected as being the most
fitting and appropriate to celebrate the event. The idea of a locomotive
mounted on a marble column was given by the writer of this volume at
their request and was at once adopted, thus making it different from all
other monuments." The pattern selected was that of an engine in
bronze, a miniature facsimile of the "General," the Western & Atlantic
engine on which the raiders made their trip from Big Shanty to Ringgold. Surmounting a Vermont marble pedestal nine feet six inches
long, five feet three inches wide and seven feet six inches high, the
whole to be twelve feet above the ground. The front of the die contains the inscription, "Ohio's Tribute to the Andrews' Raiders, 1862,
Erected 1890." The unveiling was to have been last October, but was
postponed until the decoration of the Nation's dead on memorial day;
this was not decided until too late to change the date. The left of the
die contains the names:

James J. Andrews, Flemingsburg, Ky.
Marion A. Ross, Co. A. 2nd Ohio Vol. Inf.
George D. Wilson, " B " " " "
Perry G. Shadrack, " K " " " "
John M. Scott, " F 21st " " "
Samuel Slavens, " E 33rd " " "
Samuel Robertson, " G " " " "
William H. Campbell, Salineville, Ohio.

" These being the members who were executed at Atlanta, Ga. Andrews was not an enlisted soldier, was a scout, spy, and contraband merchant but of great service to the army. Campbell arrived at the camp of the 2nd Ohio on the day of the departure of the raiders and left with his friend Shadrack without enlisting; he was always recognized as a member of Co. K, 2nd Ohio, and gave himself as such on trial.

" On the right of the die are the names of the eight men who escaped from jail at Atlanta, Ga.

James A. Wilson, Co. C. 21st Ohio Vol. Inf.
Mark Wood, " " " " " "
J. R. Porter, " " " " " "
W. W. Brown, " F " " " "
William Knight, " E " " " "
D. A. Dorsey, " H 33rd " " "
Martin J. Hawkins A " " " "
John Wollam, Co. C " " " "

" To the rear of the die the names of those who were exchanged from Libby Prison.

William Pittenger, Co. G. 2nd Ohio Vol. Inf.
Jacob Parrot, " K 33rd " " "
William Reddick, " B " " " "
Robert Buffum, " H 21st " " "
William Bensinger, " G " " " "
Elisha H. Mason, " K " " " "

" This design selected, its execution was let to the Smith Granite Company of Westerly, R. I., who have moulded this fruit of imagination into material, bearing on its imperishable face of marble the story of the living and dead.

" The completion and unveiling of this tardy justice to the dead and recognition of the living is the last chapter of the daring act begun 29 years ago. History knows no parallel to it, fiction touches no domain kindred to it, tradition tells naught that compares to it; as long as the spirit of chivalry and freedom finds its abode in the heart of man the daring hardihood of these men will never be forgotten."[1]

---

[1] From "Andrews Raiders," by F. M. Gregg, Chattanooga, 1891.

EXECUTED

ANDREWS ANDRENG PENNSYLVANIA

OH A 2 ND OHIO VOL INT
MARION A ROSS
WILSON W BROWN
PERRY G SHADRACH
JOHN M SCOTT
SAMUEL SLAVENS
SAMUEL ROBERTSON
WILLIAM H CAMPBELL CALHOUN OHIO

Ohio's Tribute to Andrews Raiders.

CHAPTER XLV.

DEDICATION OF THE OHIO MONUMENT.

THE unveiling of the Andrews monument on the 30th of May, .91, was a unique and interesting celebration. Surviving raiders came from Ohio, Kansas, Oklahoma, Nebraska, Iowa and California. Excursion trains were run at reduced rates from Columbus and Cincinnati. The whole town of Chattanooga, now grown to be a great city, made holiday. Great trouble was taken to mark every spot connected with the raid, and committees of citizens dispensed a lavish hospitality. Many thousand old soldiers and not a few ex-confederates were present. Two or three days in advance visitors began to arrive, and the stir and joyous bustle afforded a most striking contrast to the first time the raiders had looked upon the village of Chattanooga. During this time the newspapers teemed with sketches, incidents and illustrations. The headquarters of the party was at the Read House—a magnificent hotel opposite the Atlanta depot and on the site of the Crutchfield House to which I had been carried in chains twenty-nine years before. Such sharp contrasts could not easily grow commonplace. Very enjoyable also, though quite fatiguing, was the evening reception in which thousands of men, women and children crowded around the raiders for a word or grasp of the hand. There we met many prison comrades, especially from East Tennessee, for the first time since we had been immured together.

The 30th of May was a perfect and sabbath-like day. The well-dressed people crowded toward the National Cemetery early in the morning, and the green slopes and the shadows of the great trees were soon gay with the great crowds. A large stand was erected beside the veiled monument, and by the hour of opening at 2 P.M. it was estimated that from 8,000 to 10,000 persons were compactly seated on the smooth sward. The great crowd was as orderly and respectful as if in a church. The usual impressive services for the decoration of the 13,000 graves in the cemetery first were performed, and at the hour fixed the dedication program was entered upon. On the stand were the notables of Chattanooga and of the G. A. R. of several States; the orator of the day, ex-Governor Foraker of Ohio, under whose administration the work had

˙begun, and other speakers; the raiders; the monument commission
ᴊse labors were now concluded; and most pathetic of all, the relatives
˙ʰe executed raiders who had so long mourned the tragedy at Atlanta,
ᴊ who were now to be comforted as they learned by word and deed
ᴢat the sacrifice of the lives of their loved ones could never be forgotten

There was prayer by Rev. T. C. Warner; a welcome by Department
Commander Gahagan, which was extended to Confederate as well as
Union veterans; and Judge Minshall of the Ohio Supreme Court, one
of the monument commission, gave the reasons tor the State of Ohio
providing such a memorial.

Then followed a touching part of the program to which the people
had been looking forward for hours. Little Marion A. Ross, nephew
and namesake of the raider Marion A. Ross of the 2nd Ohio, pulled the
cord which unveiled the beautiful and graceful monument. Among the
mass of floral tributes was one sent by President Harrison.

Hon. H. B. Case then, in a ringing speech, introduced the orator of
the day, ex-Governor Foraker.

The address of this accomplished orator was nearly two hours in
length and had been thoroughly prepared. He spoke with great intensity and with his whole heart in the effort. Sometimes the people
were hushed to silence and tears and again aroused to almost uncon-
trolable enthusiasm. The whole speech was a masterpiece of condensa-
tion. The story of the raid was never better or more compactly told,
or its unique place in history more completely set forth. No epitome
can do justice to this magnificent speech, but a few salient points will be
quoted. He said:

"We are here in the name, by the authority and on behalf of the
Commonwealth of Ohio, to dedicate the monument they have erected
to the sacred purposes it is intended to subserve.

We have come, therefore, at the bidding of a great and far distant
State, to gather about these graves, as the accredited representatives of
her four millions of people.

.   .   .   .

"If these men had been great commanders, great scholars, great
statesmen, or great citizens, in any ordinary sense of the term, our
presence here would need no explanation. But they were the very opposite. They were simply typical volunteer Ohio boys, hardly out of
their teens, without name, family, influence, or station, to cause them
to be remembered and honored, as they are remembered and honored
to-day.

"Why is it, then, that we are here? What purposes are we seeking to

promote? Why should the General Assembly of a great State turn aside
from its ordinary cares and duties to take such action as has been mentioned? Why should a Justice of the Supreme Court and the two distinguished and honored citizens who are his associates on the commission labor, as they have, with zealous pride to discharge the duties that
have been intrusted to them?

"The answer is plain and simple.

.   .   .   .

"Upon these particular men fell an uncommon misfortune. They not
only lost their lives, but they lost them in such a way as to place a
stigma upon their memory.

"Ohio is here to-day to remove that stigma. By this action she re-
claims them from all imputation of crime, and effaces forever the ignominy of a felon's death. She proclaims to the world and future generations that they were neither thieves nor marauders, but brave and honorable men and soldiers; that their punishment was unmerited, and that
their names shall shine on the roll of honor among the brightest of all
that illumine the pages of our history.

"It is but another added to the many illustrations the world has given
of the impotency of blind malice to blacken virtue, disfigure worth and
pervert the truth.

"Socrates has been all the nearer and dearer to the world's generations
because compelled to drink the fatal hemlock. Cicero has grown continually greater and grander through all the centuries that his name has
outlived the wicked madness that condemned him to death. And as
truth and justice have vindicated these and thousands, so too have they
vindicated those whom we are here to honor.

"This monument is our visible and enduring testimonial of that fact.
We erect and dedicate it in an impressive presence. Not only are we
in the midst of the dead, but we are surrounded with bloody fields and
historic heights. Every spot on which the eye rests is hallowed ground."

The two points especially insisted on by Governor Foraker were the
full military character and great value of the objects of the expedition
and the cruel and unmerited severity of the treatment of the prisoners ·
culminating in the terrible death of those for whom the monument was
erected. The impression made was profound and lasting.

Captain S. B. Porter made a short but felicitous address of thanks to
local committees and the two Chattanooga G. A. R. Posts, 45 and 2,
which was responded to by Col. Wood, Jno. A. Patton, and Mrs.
Hattie S. Stewart on behalf of the G. A. R. Sons of Veterans, and
Woman's Relief Corps, after which the writer closed with prayer.

Some remarks were also made by Capt. Fuller and by Anthony Murphy and the assembly dissolved, thus closing a memorable day.

On the succeeding day the Western & Atlantic Railroad provided a special train for the raiders and their friends to run over the historic road to Atlanta, stopping just where and remaining as long as the raiders directed. Conductor Fuller and Murphy were also in the company and a large number of Ohio visitants. Most of the raiders had not been over the road since the war and were thrillingly interested in each way-mark of former days. The only circumstance to mar the dramatic completeness of this ride was that we were running southward instead of northward and took each place in reverse order. We first stopped at the point where the train was abandoned after the chase was over, and the course of the different fugitives was discussed by pursuers and pursued. The tunnel, Dalton, Calhoun, Adairsville, were also very interesting stopping-points, but at Kingston, where we had waited an hour amid gathering trains, and at Big Shanty (Kenesaw), the interest was almost painful, and it seemed as if the twenty-nine years had rolled away and we were again in the midst of the conflict. Many persons in the various towns who heard of this strange excursion came to the various points at which stops were made, and not a few claimed to have seen either the passage of the captured locomotive or to have met the prisoners in their chains. When Atlanta was reached we once more looked upon the scenes of the daring escape from prison and the heroism unto death of our comrades.

The next day we returned to Chattanooga and then, lingeringly and reluctantly, parted to our several homes in distant States—probably to meet no more till the general roll-call above!

.   .   .   .

—And yet—as I write the country is gathering toward the great Columbian Fair at Chicago, and arrangements are already made by which the captured locomotive, "The General," will be taken also. Probably most of the raiders will see each other once more!

www.ingramcontent.com/pod-product-compliance
Lightning Source LLC
Chambersburg PA
CBHW020900210326
41598CB00018B/1735